Yeast Biotechnology

Yeast Biotechnology

Valeria Verdinelli

Kruger Brentt
Publishers
2024

Kruger Brentt Publishers UK. LTD.
Company Number 9728962

Regd. Office: 68 St Margarets Road, Edgware, Middlesex HA8 9UU

ISBN: 978-1-78715-260-1

Disclaimer:

Every possible effort has been made to ensure that the information contained in this book is accurate at the time of going to press, and the publisher and author cannot accept responsibility for any errors or omissions, however caused. No responsibility for loss or damage occasioned to any person acting, or refraining from action, as a result of the material in this publication can be accepted by the editor, the publisher or the author. The Publisher is not associated with any product or vendor mentioned in the book. The contents of this work are intended to further general scientific research, understanding and discussion only. Readers should consult with a specialist where appropriate.

Every effort has been made to trace the owners of copyright material used in this book, if any. The author and the publisher will be grateful for any omission brought to their notice for acknowledgement in the future editions of the book.

For information on all our publications visit our website at http://krugerbrentt.com/

PREFACE

Yeast, a remarkable microorganism, has been a steadfast companion to humanity for centuries, contributing not only to the art of fermentation in the creation of bread, beer, and wine but also emerging as a pivotal player in the dynamic field of biotechnology. In recent years, the scientific community has witnessed an unprecedented surge in the exploration of yeast's potential, unravelling its intricate biology and harnessing its versatile capabilities for a myriad of applications. This surge forms the essence of Yeast Biotechnology – a field where the microscopic Saccharomyces cerevisiae and its counterparts transcend their traditional roles and become bioengineered workhorses, producing everything from aromatic compounds to valuable enzymes and pharmaceuticals. This book serves as a compass navigating through the expansive landscape of Yeast Biotechnology, offering a comprehensive exploration of its nuances, engineering approaches, and practical applications. Join us as we embark on a journey to unveil the hidden potential of yeast, where biology meets technology, and innovation converges with tradition.

Yeast biotechnology holds immense significance in diverse fields, serving as a cornerstone for numerous industrial, medical, and research applications. Its unrivaled role in biofuel production, particularly in the generation of bioethanol from renewable sources, contributes to sustainable energy solutions. In the food and beverage industry, yeast plays a pivotal role in fermentation processes, influencing the quality and characteristics of products like bread, beer, and wine. The pharmaceutical sector benefits from yeast biotechnology through the production of vaccines and therapeutic proteins, leveraging the organism's amenability to genetic manipulation. Additionally, yeast serves as a fundamental model organism in genetic and molecular biology research, providing valuable insights into cellular processes applicable to higher organisms. Its versatility extends to bioremediation, where certain yeast strains aid in environmental cleanup efforts. Furthermore, yeast is integral to industrial enzyme production, synthetic biology, and metabolic engineering, offering a versatile platform for the synthesis of valuable compounds and materials. In essence, yeast biotechnology stands as a linchpin in scientific and industrial advancements, addressing challenges across various domains and fostering innovation in sustainable and biotechnological solutions.

This book encompasses a broad spectrum of topics within yeast biotechnology, offering in-depth discussions on the production of aroma compounds, long-chain dicarboxylic acids, citric acid, lipolytic enzymes, and much more. Each chapter provides a thorough examination of the respective subject, ensuring a comprehensive understanding of the diverse applications of yeast in biotechnological processes. The book showcases how genetic and metabolic engineering strategies contribute to the optimization of yeast strains for specific industrial applications. It explores how yeast organisms can be harnessed for the production of valuable compounds, enzymes, and flavors using renewable raw materials. The diverse topics covered appeal to readers with a curiosity about the role of yeast in shaping modern biotechnological practices. Embark on a journey through the captivating world of Yeast Biotechnology, where scientific exploration meets real-world applications. We hope this book serves as a source of inspiration and knowledge for all those intrigued by the remarkable potential of yeast organisms in the realm of biotechnology. This comprehensive book delves into the intricate realms of yeast research and its applications in biotechnology. The compilation explores a myriad of topics, from the fundamental aspects of yeast biology to cutting-edge engineering approaches, showcasing the versatility of yeast organisms in various biotechnological processes. The comprehensive and detailed exploration of yeast biotechnology makes this book an essential reference for researchers and academics working in the fields of microbiology, biotechnology, genetics, and bioengineering. Professionals working in industries such as food and beverage, pharmaceuticals, and bioenergy will benefit from the practical insights provided in the book. The information presented can be directly applied to optimize processes and enhance product development. Individuals with a general interest in biotechnology and its applications will find the book accessible and engaging.

We are grateful to all those persons as well as various books, manuals, periodicals, newsletters, magazines, journals etc. that helped in the preparation of this book. In spite of the best efforts, it is possible that some errors may have occurred into the compilation and editing of the book. Further queries, constructive suggestions and criticisms for the further improvement of the book are always welcome and shall be thankfully acknowledged.

Valeria Verdinelli

CONTENTS

CHAPTER-1

INTRODUCTION TO YEAST BIOTECHNOLOGY

INTRODUCTION: AN OVERVIEW

- Yeasts are regarded as the first microorganisms used by humans to process food and alcoholic beverages. The technology developed out of these ancient processes has been the basis for modern industrial biotechnology.
- For thousands of years, yeasts have been used for the making of bread and the production of fermented alcoholic drinks, such as wine and beer. *Saccharomyces cerevisiae* (bakers' and brewers' yeast) is the yeast species that is surely the most exploited by man.

Nowadays, *Saccharomyces* is a cornerstone of modern biotechnology and also a top choice organism for industrial production of fuels, chemicals, and pharmaceuticals. Today, more and more different yeast species are explored for industrial applications._

YEAST: SYNTHETIC BIOLOGY AND STRAIN ENGINEERING

Recently, important progress has been made in unlocking the key elements in the biochemical pathways involved in the synthesis of aroma compounds, as well as in methods to engineer these pathways. Recent advances in bioengineering of yeasts—including *S. cerevisiae*—to produce aroma compounds and bioflavors are reviewed in Reference [1]. This review presents yeast as a significant producer of bioflavors in a fresh context and proposes new directions for combining engineering and biology principles to improve the yield of targeted aroma compounds. In a proof-of-concept study, *Yarrowia lipolytica* was used as a whole cell factory for the *de novo* production of long chain dicarboxylic acid (LCDA-16 an -18) using glycerol as the sole carbon source [2]. The results provide basis for developing *Y. lipolytica* as a safe biorefinery platform for sustainable production of high-value LCDCAs from non-oily feedstock. It was demonstrated that a mutant strain of *Y. lipolytica* can be used to produce citric acid from renewable carbon sources such as rapeseed oil, glycerol, and glycerol-containing waste of the biodiesel industry and glucose-containing aspen waste [3]. The cost-effective production of cellulosic ethanol requires robust microorganisms for rapid co-fermentation of glucose and xylose. Therefore, a recombinant diploid xylose-fermenting *S. cerevisiae* strain was developed by integrating *Piromyces* sp. E2 xylose isomerase (*PirXylA*) and *Orpinomyces* sp. ukk1 xylose (*OrpXylA*) in the genome in multiple copies [4]. The development of a counter-selection method for phenyl auxotrophy could be a useful tool in the repertoire of yeast genetics. A fluorinated precursor, i.e., 4-fluorophenylpyruvate (FPP), was found to be toxic to several strains from *Saccharomyces* and *Candida* genera [5]. The results show that FPP could effectively be used for counter-selection, but not for enhanced phenylethanol production.

New developments in efficient biomolecule production

In recent years, interest in the industrial production of yeast β-glucan has increased since it is an immunostimulant molecule for human and animal health. The β-glucan yield was optimised during anaerobic fermentation by evaluating the effect of the carbon source (glucose) and NaCl osmotic stress [6]. A yeast isolate, selected for its lipolytic activity from a meat product, was characterized as *Pichia anomala* [7]. Submerged fermentation optimization resulted in a significantly increased production of an extracellular lipolytic enzyme.

Fermented beverages: beer, wine and honey fermentation

Nowadays, wild yeasts are explored for beer and wine making to increase the natural flavour diversity of fermented beverages. Flavour was added to beer by performing mixed fermentation using non-*Saccharomyces cerevisiae/pastorianus* yeasts [8]. For this, a total of 60 strains belonging to the genera *Candida*, *Pichia*, and *Wickerhamomyces* were evaluated. Several strains produced substantially higher amounts of aroma alcohols and esters compared to a reference lager yeast strain.

Proline is the predominant amino acid in grape juice, but it is poorly assimilated by wine yeast under the anaerobic conditions of most fermentation. A novel wine yeast mutant that was obtained through ethyl methane sulfonate (EMS) mutagenesis, showed a markedly increased proline utilization and could be used to perform fermentations in nitrogen-limited conditions [9]. Icewine is a sweet dessert wine produced from grapes naturally frozen on the vine. Since acetic acid is undesired in Icewine, the yeast cytosolic redox status and its correlation to acetic acid production was investigated [10]. Yeasts involved in veil formation during the biological aging of Sherry wines are mainly *S. cerevisiae*, which are traditionally been divided in the varieties *beticus*, *cheresiensis*, *montuliensis* and *rouxii*. A microtiter plate assay method was developed to assure the identification and classification of veil-forming yeasts during Sherry wine aging [11].

Honey fermentations are usually performed using almost exclusively yeasts in the genus *Saccharomyces*. To increase the yeast biodiversity, two strains of *Torulaspora delbrueckii* were isolated from the gut of a locally collected honey bee [12]. These wild yeast fermentations displayed better sensory characteristics than mead fermentations by a champagne yeast, and mixed fermentations of the wild and the champagne yeast resulted in a rapid industrial fermentation process.

YEAST NANOBIOTECHNOLOGY

Clinical needs for novel antifungal agents have increased due to the increase of people with a compromised immune system, and the appearance of resistant fungi and infections by unusual yeasts. In recent years, several micro- and nanoscale approaches have been introduced for antifungal drug discovery. These are reviewed in the last contribution to this special issue [13].

In summary, this book compiles the current state-of-the-art of research and technology in the area of "yeast biotechnology" and highlights prominent current research directions in the fields of yeast synthetic biology and strain engineering, new developments in efficient biomolecule production, fermented beverages (beer, wine, and honey fermentation), and yeast nanobiotechnology.

REFERENCES

1. Van Wyk, N.; Kroukamp, H.; Pretorius, I. The Smell of Synthetic Biology: Engineering Strategies for Aroma Compound Production in Yeast. *Fermentation* **2018**, 4, 54. [CrossRef]
2. Abghari, A.; Madzak, C.; Chen, S. Combinatorial Engineering of *Yarrowia lipolytica* as a Promising Cell Biorefinery Platform for the de novo Production of Multi-Purpose Long Chain Dicarboxylic Acids. *Fermentation* **2017**, 3, 40. [CrossRef]
3. Morgunov, I.; Kamzolova, S.; Lunina, J. Citric Acid Production by *Yarrowia lipolytica* Yeast on Different Renewable Raw Materials. *Fermentation* **2018**, 4, 36. [CrossRef]
4. Liu, T.; Huang, S.; Geng, A. Recombinant Diploid *Saccharomyces cerevisiae* Strain Development for Rapid Glucose and Xylose Co-Fermentation. *Fermentation* **2018**, 4, 59. [CrossRef]
5. Murdoch, I.; Powers, S.; Welch, A. Fluorinated Phenylalanine Precursor Resistance in Yeast. *Fermentation***2018**, 4, 41. [CrossRef]
6. Varelas, V.; Sotiropoulou, E.; Karambini, X.; Liouni, M.; Nerantzis, E. Impact of Glucose Concentration and NaCl Osmotic Stress on Yeast Cell Wall β-D-Glucan Formation during Anaerobic Fermentation Process. *Fermentation* **2017**, 3, 44. [CrossRef]

7. Papagianni, M.; Papamichael, E. A *Pichia anomala* Strain (*P. anomala* M1) Isolated from Traditional Greek Sausage is an Effective Producer of Extracellular Lipolytic Enzyme in Submerged Fermentation. *Fermentation* **2017**, *3*, 43. [CrossRef]
8. Ravasio, D.; Carlin, S.; Boekhout, T.; Groenewald, M.; Vrhovsek, U.; Walther, A.; Wendland, J. Adding Flavor to Beverages with Non-Conventional Yeasts. *Fermentation* **2018**, *4*, 15. [CrossRef]
9. Long, D.; Wilkinson, K.; Taylor, D.; Jiranek, V. Novel Wine Yeast for Improved Utilisation of Proline during Fermentation. *Fermentation* **2018**, *4*, 10. [CrossRef]
10. Yang, F.; Heit, C.; Inglis, D. Cytosolic Redox Status of Wine Yeast (*Saccharomyces cerevisiae*) under Hyperosmotic Stress during Icewine Fermentation. *Fermentation* **2017**, *3*, 61. [CrossRef]
11. Ruíz-Muñoz, M.; Bernal-Grande, M.; Cordero-Bueso, G.; González, M.; Hughes-Herrera, D.; Cantoral, J. A Microtiter Plate Assay as a Reliable Method to Assure the Identification and Classification of the Veil-Forming Yeasts during Sherry Wines Ageing. *Fermentation* **2017**, *3*, 58. [CrossRef]
12. Barry, J.; Metz, M.; Hughey, J.; Quirk, A.; Bochman, M. Two Novel Strains of Torulaspora delbrueckii Isolated from the Honey Bee Microbiome and Their Use in Honey Fermentation. *Fermentation* **2018**, *4*, 22. [CrossRef]
13. Willaert, R. Micro- and Nanoscale Approaches in Antifungal Drug Discovery. *Fermentation* **2018**, *4*, 43. [CrossRef]

CHAPTER-2

UNVEILING THE AROMATIC ESSENCE: EMPLOYING ENGINEERING APPROACHES TO GENERATE AROMA COMPOUNDS IN YEAST

INTRODUCTION

In the captivating realm of synthetic biology, a fragrant transformation unfolds as we delve into the intricacies of yeast engineering. This journey embarks on the aromatic essence, where the manipulation of genetic elements and innovative engineering strategies converge to unlock the potential of yeast in the production of captivating aroma compounds. In this exploration, we navigate the intricate pathways of yeast synthetic biology, unraveling the secrets that empower us to orchestrate a symphony of scents through precise and deliberate genetic design.

Yeast—especially *Saccharomyces cerevisiae*—have long been a preferred workhorse for the production of numerous recombinant proteins and other metabolites. *S. cerevisiae* is a noteworthy aroma compound producer and has also been exploited to produce foreign bioflavour compounds. In the past few years, important strides have been made in unlocking the key elements in the biochemical pathways involved in the production of many aroma compounds. The expression of these biochemical pathways in yeast often involves the manipulation of the host strain to direct the flux towards certain precursors needed for the production of the given aroma compound. This review highlights recent advances in the bioengineering of yeast—including *S. cerevisiae*—to produce aroma compounds and bioflavours. To capitalise on recent advances in synthetic yeast genomics, this review presents yeast as a significant producer of bioflavours in a fresh context and proposes new directions for combining engineering and biology principles to improve the yield of targeted aroma compounds.

An overarching definition for the term "aroma compound" is one that provides a sensorial stimulus to the olfactory senses and, in certain cases, also the gustatory senses. In literature, it shares overlapping designations with words like "flavours", "scents", "odorants" and "fragrances" and these terms are often used interchangeably. Aroma compounds have various applications in the food, feed, cosmetic and pharmaceutical industries [1]. Some compounds have applications beyond their sense-activating properties, including potential as a biofuel [2], the improvement of the shelf-life of certain fruit varieties [3] and antimicrobial activities [4]. They can either be desirable or unwanted in a given product and significant efforts can be made to either eliminate or increase levels depending on the application. Aroma compounds are rarely perceived in isolation (especially in fermented foodstuffs) and thus its interaction with other compounds can greatly affect how they are identified.

Although not discussed in this review, a crucial component in the perception of aroma compounds is the olfactory receptors that recognize odorous ligands. Seminal work done by Nobel laureates Richard Axel and Linda Buck show the large and diverse nature of these membrane-bound receptors present in our olfactory neurons which are responsible for the detection of odorants and give rise to the sense of smell [5]. These receptors can be variably expressed among individuals resulting in the different perceptions of the same compound by individuals—a key consideration of consumer preference of foodstuffs [6]. Often neglected and poorly understood are the psychological aspects of odour perception as it can relate to the associative memory of the individual [7].

Aroma compounds are structurally remarkably heterogeneous. They can have cyclic or non-cyclic, saturated or unsaturated, straight-chain or branched-chain structures bearing all kinds of functional groups (e.g., alcohols, aldehydes, ketones, esters and ethers) and, in some cases, have nitrogen and sulphur within the structure. Certain aroma compounds are even inorganic in nature. If made enzymatically, aroma compounds are derived from the pool of precursor molecules from the core metabolism of the cell (i.e., the carbohydrates, fatty acid, nucleotides and amino acids). Odour thresholds (i.e., the concentration ranges at which a given aroma compound is detected or sensed) are key parameters in aroma compound studies.

Most aroma compounds on the market are produced by isolating natural compounds from plant or animal or by chemical synthesis. However, there is a clear swing away from chemically-produced aroma compounds and aroma compounds that require extensive extraction from plants or animals towards the production and use of aroma compounds of (micro) biological origin—also called bioflavours. This is despite the fact that the chemically produced compounds are identical to their natural counterparts. Reasons for such a change in market preferences include the fact that chemical synthesis can often result in environmentally detrimental production processes and in undesired racemic mixtures. Also, extraction of aroma compounds from plants or animal sources can be resource-intensive and cost-inefficient because of low yields. In addition, multiple purification steps often lead to product loss and degradation. Consumer aversion toward chemical compounds relates especially to food and home-care products. Despite changing preferences in consumer markets, the financial implication of aroma compound generation remains a strong consideration as those derived from chemical synthesis are, in general, markedly less expensive than those derived from natural sources.

In this context, researchers are directing their research efforts toward producing aroma compounds from microbial sources. This usually involves *Escherichia coli* or *S. cerevisiae* as cell factories by incorporating genes that code for enzymes that are relevant to the production of the given compound in a recombinant host [8]. Despite the campaigns against genetically-modified organisms (GMOs) in some sections of global consumer markets, there are numerous food ingredients derived from GMOs that are commercially-available the world over. However, in the case of such GM food-ingredients that comply with regulatory safeguards, high yields using cost-effective substrates have not yet been achieved in many instances.

This chapter primarily focuses on recent advances in research aimed at the production of aroma compounds in yeast. This paper is distinct from other published reviews, including those that extensively covered the use of flavour-active brewing and wine yeasts for the enhancement of the aroma of beer and wine [9–11]. Here, we focus on the exploitation of two types of yeast precursors which are responsible for a variety of aroma compounds, namely the aromatic amino acids L-tyrosine and L-phenylalanine, which are derived from the shikimate pathway and the mevalonate pathway-derived isoprenoid precursors dimethylallyl pyrophosphate and isopentenyl pyrophosphate.

YEAST AS A GENETICALLY MODIFIED HOST FOR THE PRODUCTION OF BIO-FLAVOR

Various yeasts—with *S. cerevisiae* being the model organism—have long been harnessed for the expression of recombinant genes to enhance endogenous aroma-active metabolites of the host cells or to produce novel recombinant compounds. The initial reasons why researchers opted for *S. cerevisiae* remains true, that is, this yeast species is by far the best-studied unicellular eukaryote with the genomes of several of its strains fully sequenced [12]; it is a non-pathogen that enjoys GRAS (generally recognised as safe) status; and it is amenable to genetic manipulation with a wide range of genetic tools available to alter the genetic make-up of the yeast. *S. cerevisiae* also possesses an efficient homologous recombination machinery, which greatly assists stable integration of genetic elements. This yeast is also the most robust fermented and laboratory-scale processes can be scaled up to industrial-level set-ups with relative ease. Some of the abovementioned attributes also hold true for *E. coli*. However, as a prokaryote, this bacterium lacks a sophisticated protein-folding mechanism.

This often leads to the recombinant proteins being insoluble and most likely non-functional and that might require additional recovery steps for refolding of the protein of interest.

S. cerevisiae is, however, by no means a perfect host; for example, it is not a prolific biomass producer and the way secreted proteins are glycosylated sometimes lead to pronounced reduction in bioactivity. There are also reports of recombinant genes that cannot be successfully expressed for unknown reasons. Regardless of the whether *S. cerevisiae* turns out to be appropriate as a host to produce a particular recombinant product, it remains the best starting point to move onwards to other organisms. A prudent strategy is to examine the expression levels in multiple yeast hosts and to compare titres of a protein (or metabolite) of interest. Often the methylotrophic yeast *Pichia pastoris* (now reclassified as *Komagataella phaffii*) and *Hansenula polymorpha* (now reclassified as *Ogataea polymorpha*) have shown superior protein and/or metabolite production capabilities owing to their unusually high biomass production [13]. Many other yeast species with their own special attributes can (and have) been utilised as a recombinant host with varying outcomes. Examples of such yeasts include *Kluyveromyces lactis*, *Yarrowia lipolytica* and *Schizosaccharomyces pombe*. The usefulness of non-*Saccharomyces* yeasts in the biotransformations of certain substrates into aroma compounds with whole-cell or resting cell systems are well-documented [14]. This has been a popular way of producing aroma compounds as it can allow for the assembly of regio- and stereo selective compounds under mild and mostly solvent-free conditions.

Identification of natural variation within a yeast strains and species has undeniably created a valuable source of flavour-active strains [15]. The underlying molecular determinants for a particular phenotype has been elucidated through the advances in 'omics' capability. Effective mining of genes and alternative alleles responsible for a desired phenotype have become common practice, with access to comprehensive conventional yeast libraries based on mutagenesis, breeding [16], single gene deletions [17] and over expression [18]. Yeast libraries have become more sophisticated and, in many cases, combine the genomic variation generation with a selection for the particular characteristic of interest. This includes biosensor-enabled directed evolution (discussed in later section below), rapid genome-wide editing (YOGE) or the complete reconstruction of pathways (VEGAS) and genomes (Yeast 2.0).

Yeast Oligo-mediated Genome Engineering (YOGE) enables rapid genome engineering by introducing allele variation by sequential oligonucleotide recombination [19]. Designer synthetic DNA oligonucleotides allow the combinatorial alteration of pathway genes and, with successive rounds of transformation, gradually remodel the yeast genome toward the production of a metabolite or to embody a specific phenotype. Smaller, directed libraries, only altering the pathway(s) of interest, have been demonstrated with techniques like Versatile Genetic Assembly System (VEGAS). VEGAS uses the yeast's innate preference for homologous recombination to assemble complex pathways, allowing different combinations of the pathway genes to be assembled and subsequently screened for the best production [20].

A new generation of yeasts might allow us to greatly expand yeast strain diversity beyond what has resulted to date with directed breeding and natural selection. The revolutionary synthetic biology initiative known as the Yeast 2.0 project (also known as Sc2.0) was initiated in 2007 [21] to deepen our understanding of the molecular mechanisms that drives this versatile organism. Upon completion, the Sc2.0 strain will be world's first eukaryote with a streamlined chemically-synthesised genome. In addition to the removal of repetitive sequences, the liberation of a codon and the introduction of hundreds of watermark sequences, LoxPsym sequences were introduced at the 5^t-ends of all genes considered individually non-essential [22,23]. These sites allow for inducible homologous recombination downstream of all non-essential genes, mediated by the action of the site-specific Cre-recombinase. Upon activation of the site-specific Cre-recombinase, homologous recombination is promoted between these LoxPsym sites, resulting in rapid gene deletion, duplication or inversion. This process—known as SCRaMbLE (Synthetic Chromosome Rearrangement and Modification by LoxPsym-mediated Evolution)—allows for the rapid synthetic rearrangement and evolution of the yeast genome [24] (Figure 1). In addition to this novel way of producing large libraries of genomically-divergent yeasts, SCRaMbLE also allows us to produce and explore minimum eukaryotic genomes for the first time. These libraries will be valuable assets in the screening for interesting phenotypes, like aroma compound production and the elucidation of the underlying principles governing these production pathways.

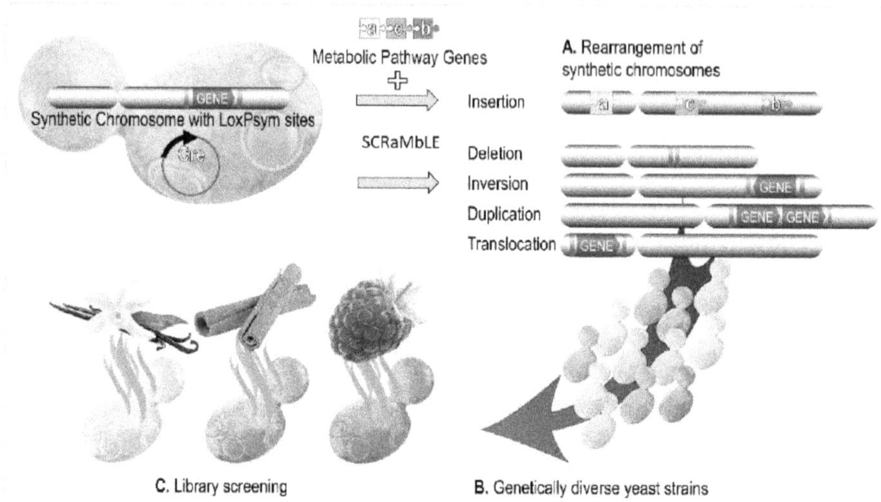

Figure 1. Depiction of aroma compound pathway optimisation through loxPsym-mediated rearrangement of synthetic chromosomes (SCRaMbLE) in yeast. (A) A yeast containing synthetic versions of their respective chromosomes with multiple loxPsym sequences would be subjected to the actions of the loxPsym-specific Cre recombinase. (B) The subsequent insertions, duplications, deletions, inversions and other genetic alterations will allow for the generation of an instantly-made library of yeast that have tremendous diversity in their respective genetic backgrounds (C) allowing for the screening of yeast with preferred phenotypes. By introducing metabolite pathway genes, flanked by loxP sequences, copy number optimised pathways can be assembled into the generated library. At the time of writing this review, 6 of the 16 chromosomes have been fully synthesized, with the rest at various stages of construction and debugging [25]. The strains harbouring these chromosomes (or combinations thereof) can currently be used for SCRaMbLE-based phenotype generation experiments.

Irrespective of the specific yeast strain used, optimisation of the recombinant production of a given protein or metabolite would require the systematic improvement of the properties of the recombinant host using analytical and computational methods to quantify fluxes and their regulation. The following guiding principle questions, regarding global and pathway-specific metabolic engineering, have been proposed previously [26]: (i) can the precursor and/or cofactor supply be increased?; (ii) can the heterologous expression of non-native genes be different or the expression thereof be improved?;

- can pathways that compete for the same precursors and co-factors be blocked or down-regulated?;
- are transcriptional regulators known and what would be the effect if they are over expressed?; and (v) can the enzyme specificity be improved? Most of these questions are directly applicable in improving a yeast's ability to

produce aroma compounds. Below we will discuss the work researchers have undertaken in addressing these questions in order to increase the levels of phenylpropanoid and terpenoid production in yeast.

YEAST PRECURSORS UTILISED

Phenylpropanoids

The aromatic amino acids L-phenylalanine, L-tyrosine and L-tryptophan serve as the precursors to many compounds of commercial interest [27]. More specifically, L-phenylalanine, L-tyrosine provide the precursors for a large group of compounds called phenylpropanoids—of which many have aroma-active properties. The biosynthesis of the aromatic amino acids proceeds via the shikimate pathway [28] (Figure 2). It is a seven-step metabolic pathway leading to the production of chorismate, the common aromatic precursor to all three amino acids. The shikimate pathway is initiated with the condensation of phosphoenolpyruvate (PEP)—an intermediate in the glycolysis pathway—and erythrose-4-phosphate (E4P)—an intermediate in the pentose phosphate pathway—to generate 3-deoxy-D-arabino-heptulosonate-7-phosphate (DAHP). Chorismate is the branching node, where L-tryptophan is separated from the other two amino acids as chorismate is converted to prephenic acid (the precursor molecule of L-phenylalanine and L-tyrosine) by a chorismate mutase. Subsequent decarboxylation and transamination events lead to the production of L-tyrosine and L-phenylalanine. In general, intracellular L-tyrosine levels in *S. cerevisiae* are about ten-fold higher than L-phenylalanine with L-tryptophan 10 times less than L-phenylalanine [29].

S. cerevisiae has a limited capacity to process aromatic amino acids beyond using them for protein synthesis. Pathways involved in using L-tyrosine and L-phenylalanine as precursors have been incorporated into yeast to produce a multitude of compounds and of these, the phenyl ring structure represents a central feature (Figure 2). A key aroma compound derived from the shikimate pathway is that of vanillin (imparting vanilla flavour) and has been the subject of many investigations in the past due to its high value and wide use. Vanillin is not synthesised from any of the aromatic amino acids, but from an intermediate in the shikimate pathway, namely dehydroshikimate. The first report of vanilla production by yeast used three recombinant genes in the fission yeast *S. pombe* to transform dehydroshikimate to vanillin [30]. In the same study, *S. cerevisiae* was also used, but an additional activation enzyme was needed. Vanillin is moderately toxic to yeast cells (it represses translational processes [31]). It was shown that adding a glycosyl moiety, by expressing a 1-UDP-glycosyltransferase, leads to the conversion of vanillin to vanillin-glucoside (VG), which markedly increased production levels. Remarkable improvements in VG titres have been achieved with rational engineering design approaches: in silico metabolic engineering algorithms have been implemented to identify yeast target genes that could enhance productivity [32]. Manipulations of two of the identified targets (*PDC1* and *GDH1*) led to a five-fold improvement of VG yields and was attributed to the recycling of the supply of cofactors. Additional modelling-based methodologies underlined the utility of in silico design for improvement in VG levels [33,34].

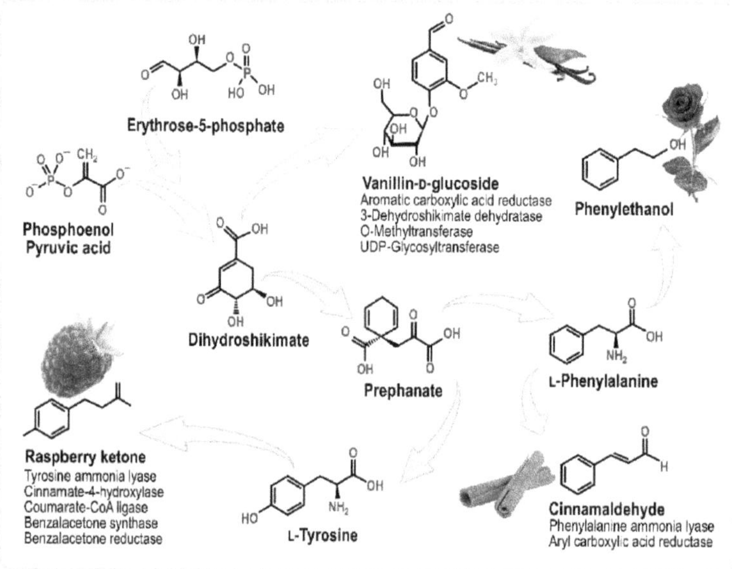

Figure 2. Biosynthetic pathway for phenylpropanoids. Yeast can synthesize all three aromatic amino acids (L-phenylalanine, L-tyrosine and L-tryptophan) via the shikimate pathway but have few processing capabilities beyond utilising them in peptide synthesis or their catabolism via the Ehrlich pathway (which can produce the aroma compound 2-phenylethanol). *S. cerevisiae* and other yeast have been exploited to convert their free aromatic amino acids to compounds with aroma properties. The recombinant enzymes that have been incorporated in yeast to convert precursors to aroma compounds of commercial value are shown below each recombinant metabolite.

The pathway for the production of *p*-hydroxycinnamic acid (also known as *p*-coumaric acid), which imparts a cinnamon aroma, has been incorporated in *S. cerevisiae* [35]. This simply involved the incorporation of various phenylalanine ammonia-lyases (PAL)/tyrosine ammonia-lyases (TAL) which deaminate L-tyrosine. Several metabolic engineering strategies have proven successful in enhancing *p*-hydroxycinnamic acid along with the levels of so-called *trans*-cinnamic derivatives (which include cinnamaldehyde, cinnamyl alcohol and hydrocinnamyl alcohol) [36]. These strategies involved removing known feedback-regulated steps of aromatic amino acid biosynthesis and directing the flux towards the production of these *trans*-cinnamic compounds by side-tracking the decarboxylation step of the competing Ehrlich pathway. A phenylacrylic acid decarboxylase (PAD1) is thought to be responsible for the decarboxylation of *trans*-cinnamic derivatives, as a *pad1* knockout strain showed no endogenous activity on trans-cinnamic acid and *p*-hydroxycinnamic acid [37]. Nevertheless, the *trans*-cinnamic derivatives are converted to less toxic compounds by the yeast via unknown mechanisms [36]. It was found, similar to vanillin, that by adding a glycosyl moiety to trans-cinnamic acid catalysed by an UDP-glucose:cinnamate glucosyltransferase reduces its toxicity and led to increased levels.

A recent addition to the phenylpropanoid aroma compounds that are recombinantly produced in yeast is that of raspberry ketone [4-(4-hydroxyphenyl) butan-2-one] [38]. This involved the incorporation of a four-gene pathway from various organisms into yeast that converted L-phenylalanine and L-tyrosine to raspberry ketone. Testing various enzyme combinations and fusions resulted in higher levels of raspberry ketone.

Improving yields of 2-phenylethanol (2-PE)—a compound with a rose-like aroma—has been investigated extensively. 2-PE is the fusel alcohol of L-phenylalanine and of the four phenylpropanoid aroma compounds discussed, 2-PE does not require the expression of recombinant genes as it arises from the catabolism of L-phenylalanine via the Ehrlich pathway. This includes its deamination, decarboxylation and reduction that are conducted by ARO9, ARO10 and various alcohol dehydrogenases (ALD1-5) in *S. cerevisiae* respectively. Metabolic engineering efforts to increase 2-PE levels included the streamlining the Ehrlich pathway which involved the overexpression of *ARO9* and *ARO10* with the concomitant removal of a competing phenylacetaldehyde oxidase (*ALD3*) [39]. A transcription factor (ARO80) is known as an activator of the *ARO9* and *ARO10* genes and its over expression, together with *ARO9* and *ARO10*, led to a four-fold increase in 2-PE levels.

Efforts have also been made to increase the intracellular levels of the precursors of the shikimate pathway PEP and E4P. Especially targeting E4P, which has lower intracellular concentrations than PEP [40] would result in an equal balance of the two precursors and could facilitate improved flux toward aromatic amino acid production. Attempts thus far to increase levels of E4P have involved alternations within the pentose phosphate pathway [41]. It was shown that the deletion of the glucose-6-phosphate dehydrogenase (*ZWF1*) gene and over expression of the transketolase (*TKL1*) gene reversed the flux from the glycolytic intermediates and led to a higher (~eight-fold) increase in E4P levels [42].

Some non-*Saccharomyces* yeasts like *Ashbya gossypii* [43], *Kluyveromyces marxianus* [44] and *Candida glycerinogenes* [45] have been investigated for 2-PE production with yields reported that were greater than for *S. cerevisiae*. Similarly, as with *S. cerevisiae*, over expressing the genes involved in the Ehrlich pathway have led to increased levels in *K. marxianus* [44], but not *A. gossypii* [43].

Media composition, especially adding L-phenylalanine, have shown in many cases to enhance the production of polypropanoids [46,47]. This implies that the yeast precursor is still a major bottleneck for phenylpropanoid production. High-throughput mass spectrometry experiments conducted on a yeast gene deletion library, which determined the intracellular concentration of each amino acid, revealed that certain gene knock-outs resulted in a two to four times higher intracellular concentration of L-phenylalanine than the wild type [29]. Many of the strains carrying these respective gene deletions also had an increased level of L-tyrosine and L-tryptophan suggesting that these gene products might have a putative role in regulating the shikimate pathway, but in most cases no obvious connection has ever been reported.

Terpenoids (also called terpenes or isoprenoids)

Terpenoids (also called terpenes or isoprenoids) are the largest and most diverse group of natural compounds. They are derived from the basic five-carbon (C5) precursor unit isopentenyl diphosphate (IPP) and its double-bond isomer dimethylallyl diphosphate (DMAPP) that can be assembled and modified in over 60,000 different types of terpene-like structures. Apart from the exceptional flavour qualities of many terpenoids, certain terpenoids have promising applications in biofuel and antimicrobial research [48].

Terpenoids are either produced via the mevalonate biosynthesis pathway (MVA) or the 2-C-methyl-D-erythritol-4-phosphate pathway (MEP) with the former being the best-studied and found in yeast. In this pathway, acetyl-CoA is condensed to produce the universal isoprene building unit (C5), isopentenyl diphosphate (IPP). Subsequent condensations conducted by prenyltransferases of IPP and DMAPP result in terpenoid precursors called polyisoprenoid diphosphates of different lengths: geranyl diphosphate (GPP) for monoterpenoids (C10), farnesyl diphosphate (FPP) for sesquiterpenoids (C15), geranylgeranyl diphosphate (GGPP) for diterpenoids (C20), 2 units of FPP for triterpenoids (C30) and 2 units of GGPP for tetraterpenoids (C40). The C30 and C40 precursors lead to the biosynthesis of sterols and carotenoids, respectively. Cyclisation of the abovementioned polyisoprenoid diphosphates are catalysed by a large group of enzymes called terpene synthases to generate terpenoids with single or multiple ring structures (with some remaining open). These enzymes often display a high level of promiscuity with regards to their substrate preference leading to the large diversity among terpenoid structures. In addition, tailoring enzymes like oxygenases, methyltransferases, acetyltransferases and glycosyltransferases can add functional groups to different positions of the terpenoid structure.

As mentioned, yeast including *S. cerevisiae*, do possess an MVA pathway but does not have terpene synthases that are able to produce monoterpenoids, sesquiterpenoids and diterpenoids. *S. cerevisiae* implements the MVA pathway to produce sterols (specifically ergosterol) that are structural elements of the cell membrane and impart modulation to the membrane fluidity. It is an essential pathway for yeast as strains with mutated genes in this pathway require exogenous sterol for survival [49].

Significant efforts have been made to create a yeast platform that would be able to produce terpenoids as discussed in multiple papers and reviews [50–54] (Figure 3). A yeast without any modification within its MVA pathway would produce negligible levels of recombinant terpenoids. A key strategy to improve levels would be to direct carbon flux away from producing sterols without the complete elimination of the pathway. On a transcriptional level, the native promoter of the squalene synthase *ERG9* gene (that encodes the enzyme that catalyses the first reaction of converting farnesyldiphosphate to ergosterol) was replaced with repressible promoters which led to subsequent lower concentrations of ERG9, facilitating the increased levels of recombinant terpenoid production [55]. Recently, a degradation tag attached to ERG9 was shown to destabilise the protein, which also led to a dramatic improvement in recombinant terpenoid production without compromising the cell viability to any significant extent [56].

The over expression of a truncated version of the hydroxymethylglutaryl-CoA reductase (tHMG1), that is devoid of its transmembrane moiety and is thus present in the cytosol, led to an increased amount of squalene. This confirmed that HMG-CoA reductase, the enzyme that produces mevalonate, is a rate-limiting step [57]. Many subsequent attempts using yeast to produce recombinant terpenoids contain this feature [58,59]. Similarly, the over expression of the *ERG20* gene encoding an enzyme with both dimethylallyltranstransferase and geranyltransferase activities and the sterol regulator *upc2-1* [60] has generally led to improved yields in recombinant terpenoid production.

Improving the intracellular levels of the precursor molecule acetyl-CoA—an intermediate central to many metabolic pathways—has also been investigated. In one study by Meadows et al. [61], a *S. cerevisiae* strain was developed by completely overhauling the native metabolic network involved in acetyl-CoA supply by incorporating several synthetic pathways which directly resulted in increased recombinant terpenoid levels.

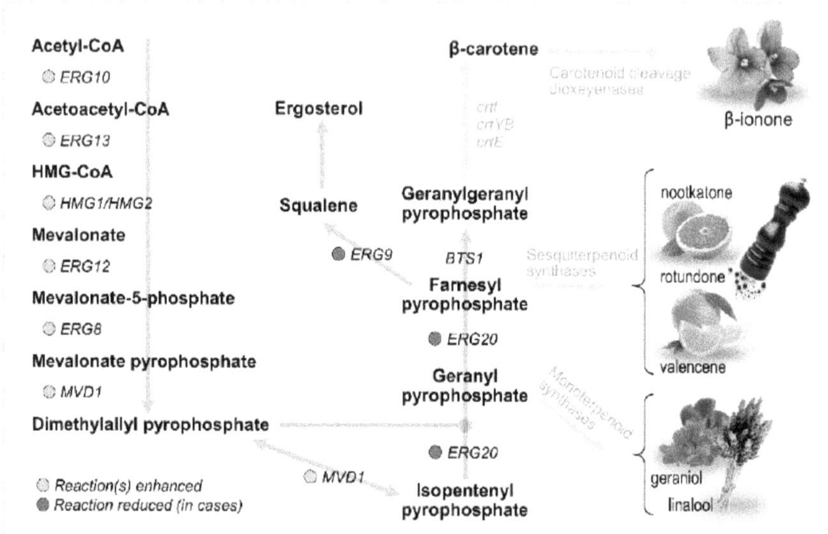

Figure 3. Manipulations in mevalonate pathway for recombinant terpenoid production. Enzymes directly involved in the MVA pathway of *S. cerevisiae* have been upregulated, downregulated or altered to increase the flux towards the production of various terpenoid-based aroma compounds. Additional recombinant enzymes needed for the catalysis of the terpenoid production are shown in orange with the respective aroma compounds being produced. Adapted from [53]. ERG10, acetyl-CoA *C*-acetyltransferase; ERG13, hydroxymethylglutaryl-CoA synthase; HMG1/HMG2, hydroxymethylglutaryl-CoA reductase 1/2; ERG12, mevalonate kinase; ERG8, Phosphomevalonate kinase; ERG19, mevalonate diphosphate decarboxylase; IPP, isopentenyl diphosphate; DMAPP, dimethylallyl diphosphate; IDI1, isopentenyl diphosphate isomerase; ERG20, geranyl/farnesyl diphosphate synthase; BTS1, GGPP synthase; ERG9, squalene synthase; crtYB, phytoene synthase and lycopene cyclase; crtI, crtE phytoene desaturase; geranylgeranyl diphosphate (GGPP) synthase.

It was shown that by fusing mitochondria signals to sesquiterpenoid synthases, along with the introduction of a recombinant farnesyl diphosphate synthase, resulted in a marked increase of the citrus aroma compound valencene [62]. This demonstrates the prowess of compartmentalisation approaches where enzymes, substrates and intermediates are close to each other and competing pathways in the cytosol are avoided.

Other successful attempts to produce terpenoids in yeast include the highly sought-after nootkatone, which imparts a strong grapefruit aroma [63]. This was achieved in *P. pastoris* with a similar engineering strategy which has been proven fruitful for *S. cerevisiae* to over express a truncated version of its *HMG1* gene to improved levels.

Carotenoids (specifically β-carotene) have also been produced in *S. cerevisiae* by expressing carotenogenic genes from strains from the ascomycete *Xanthophyllomyces dendrorhous* (previously *Phaffia rhodozyma*) [64]. Increased concentrations of β-carotene were observed when the *BTS1* gene encoding a geranylgeranyl diphosphate synthase was over expressed. These strains appear bright orange and although itself not an aroma compound, β-carotenes do serve as a substrate for carotenoid cleavage oxygenase, which releases compounds known as apocarotenoids. The expression of a carotenoid cleavage oxygenase from *Petunia hybrida* in a strain already producing β-carotene, led to the release of detectable levels of a compound known as β-ionone, which has a highly desired violet scent [65]. Interestingly, a polycistronic version (genes separated from each other by viral T2A sequences) of the abovementioned carotenogenic genes was successfully expressed in yeast [66].

Assessing the feasibility of replacing elements of the MVA with that of the MEP pathway—theorized to yield a higher stoichiometric maximum plus having a lower requirement for oxygen—has been investigated. The yeast strain was developed where its MVA pathway was replaced with an MEP pathway. The resulting strain showed a slight growth defect made less biomass compared to the wild type, implying slight incompatibility [67].

DETECTING AROMA COMPOUNDS THROUGH BIOSENSING IN YEAST

Rational engineering strategies to redistribute carbon flux towards the production of a specific metabolite or the introduction of novel synthetic biosynthesis pathways, have been successfully employed in the past to produce the desired molecule of interest [68,69]. Aroma compound production has also benefitted from these methodologies when ample precursor molecules are available [70]. However, the inherent volatile nature of some precursor molecules have excluded them from many directed evolution endeavours, as these are mostly limited to growth-selectable phenotypes, with limited high-throughput possibilities for the rapid screening of mutant libraries [71].

A promising, more recent addition to the synthetic biologist's toolbox is biosensors [72]. These genetic circuits translate a metabolic 'input' into a measurable 'output' signal like fluorescence; decoupling metabolite production from cellular growth. Biosensor designs are becoming increasingly more complex, with higher order circuits combining multiple interacting components and logic gate arrays to allow enhanced pathway regulation and output sensitivity. Table 1 shows some recent examples of biosensors developed for the direct or indirect detection of aroma compounds produced in yeast.

In one, early biosensor study [73], it was shown how biosensors can be employed to develop a high-throughput screen for yeast strains producing high concentrations of β-phenylethanol. This indirect method used the flux through the Ehrlich pathway as an indicator of high end-product concentrations. Endogenous biosensors have previously been employed to increase the flux toward the production of precursors of aroma compounds in yeast [74–77]. Synthetic biosensors have been constructed to allow feedback-regulated evolution of high IPP-producing strains [74].

There is much more scope for the development of aroma biosensors for yeast when compared to the diversity of available bacterial sensors, with *E. coli* sensors allowing for the detection of benzaldehyde, cinnamaldehyde, salicylaldehyde, syringaldehyde and vanillin [78,79]. Attempts have been made to translate some of these concepts in an endeavour to sense aroma compounds produced by bioengineered yeast, using encapsulated *p*-coumaric acid-sensing *E. coli* to screen for yeast cells producing *p*-coumaric acid [80].

Table 1. Examples of yeast biosensors to detect phenylpropanoids and terpenoids produced in *Saccharomyces cerevisiae*.

Aroma Compound	Molecule(s) Sensed	Description	Reference
Shikimate pathway			
β-phenylethanol	Aromatic amino acids	Allosteric transcription factor sensor. Transcriptional regulation *LacZ* reporter gene by the aromatic amino acid responsive *ARO9* promoter. Increased β-galactosidase activity correlated with elevated β-phenylethanol levels.	[73]
Precursor	Betaxanthin	Enzyme-coupled sensor. Highly yeast-active heterologous L-tyrosine hydroxylases were identified, based on increased betaxanthin fluorescence intensities in yeast expressing the plant DOPA dioxygenase.	[70]
p-Coumaric acid	*p*-Coumaric acid	Exogenous bacterial sensor. Droplet sorting of encapsulated *p*-coumaric acid producing yeast cells and *p*-coumaric acid sensing *E. coli* cells, to select producers based on bacterial YFP-fluorescence output.	[80]

Aroma Compound	Molecule(s) Sensed	Description	Reference
Precursor	Muconic acid	Heterologous allosteric transcription factor. Used an *Acinetobacter* sp. transcriptional regulator to drive GFP expressing in the presence of muconic acid.	[77]
Mevalonate pathway			
Precursor	Malonyl-CoA	Precursor Malonyl-CoA Recombinant allosteric transcription factor sensor. Used a bacterial FapR transcription factor and FapO operator pair to identify strains from a genome-wide overexpression library that produce high levels of malonyl-CoA.	[75,76]
Precursor	Precursor Isopentenyl diphosphate	Synthetic transcription factor to allow feedback-regulated evolution of phenotype. Higher intracellular IPP concentrations resulted in increased *GAL10* transcription, generating an evolvable growth phenotype on galactose	[74]

FUTURE VISION

In only the past decade has research been focussed on incorporating biochemical pathways in yeast to produce key aroma compounds. Beyond proving the concept—as with all biotechnological approaches—researchers will always try to find ways to improve the overall yield of a given aroma compound. Simple fermentation optimisation has proven, on many occasions, to be impactful but the most profound improvements were achieved with the metabolic engineering of the yeast. Metabolic engineering can be defined and categorised in many ways, but, it is generally regarded as any kind of genetic modification applied onto a cell that would cause a preferential change in its phenotype. It is a cornerstone characteristic of synthetic biology: a confluence of many different streams of science and engineering with the scope of building artificial biological systems. Elements that appear to be central to the field of synthetic biology have been part of science for decades: the earliest recombinant genes that were expressed in *E. coli* (that of human insulin and the human growth hormone) were indeed chemically synthesised in the 1970s (before the advent of PCR-based cloning). Even slightly more complex genetic elements like synthetic promoters have been described to function in yeast since the late 1980s [81]. Yet, it is well-recognised that the dramatic drop in the cost for synthesising DNA-sequences in the past decade was the catalyst to jump-start the current wave of synthetic biology. Large pieces of DNA can now be designed and purchased from a multitude of companies which has drastically alleviated the often-time-consuming effort of constructing genetic elements.

As has been explained in instructive reviews [82,83], central for the rapid advancement of synthetic biology is the application of the concept of design-build-test-learn (or DBTL) (Figure 4) to address biological and/or engineering questions. This idea, adapted from other engineering fields, aims to streamline and accelerate the iterative process of improving a biological system as well as to minimise human input to eliminate bias in the interpretation of the output. As mentioned by Hollywood et al. [84], the testing part is often reliant on a high-throughput screen which makes this component—if available—often the most time-consuming and costly. This is particularly true for the determination of most aroma compounds, as the screening of millions of cells and assessing individual cells for aroma compound production is not yet feasible. This is why the development of aroma biosensors with appropriate sensitivity and specificity is most desired. A desired attribute of the biosensor would be for it to have a 'dynamic range,' that is, besides detecting aroma compounds it would have the ability to provide a differing signal depending on the concentration of the aroma. This might be achieved, although not yet tested in such a capacity, by coupling either a bioluminescence resonance energy transfer (BRET)-based sensor [85] or metal oxide [86] sensor, developed to detect volatile organic compounds, within a high-throughput screening process to test a yeast library.

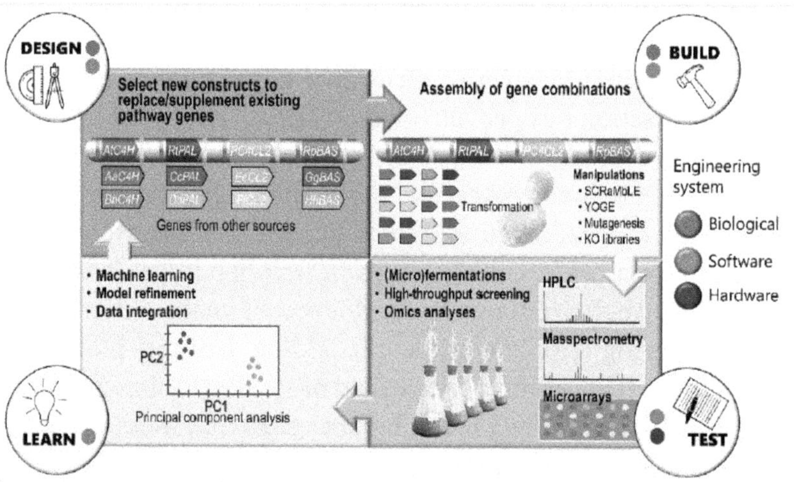

Figure 4. An example of how DBTL could be applied to build a yeast with increased production levels of raspberry ketone.

During the "design" part of the project, researchers would need to plan all the experiments with a particular focus on the types of metabolic engineering tactics that will be employed to alter the target pathway within the yeast. The raspberry ketone synthesis pathway has already been introduced in ***S. cerevisiae*** [29], thus an avenue to follow is to test several similar enzymes from different organisms to assess their compatibility within the pathway. The "build" part encompasses all the aspects of constructing the strains that need to be tested. Multiplex CRISPR-based techniques for yeast have been developed and could easily be adapted to introduce combinations of the recombinant raspberry ketone pathway genes [76,77]. CRISPR-based techniques enjoy the benefit of its precision in editing the genome that would lead to minor modifications being made to the genome. The system can also allow for the strain to be 'markerless'—free of any antibiotic resistance genes normally used in conventional genetic manipulations. Additional mutagenesis, ranging from chemical mutagenesis to using a strain in which one could induce SCRaMbLE, could be included which will add more genetic diversity to the strains that need to be tested. The 'testing' part would, in most cases, involve some level of culturing of the strains coupled with determining the levels of the metabolite of interest. This would require high-throughput screening machinery. Raspberry ketone titres will need to be measured with mass spectrometry-based techniques. Eventual sequencing of high performers would be needed to ascertain what mutagenesis allowed for the superior production levels. Once the testing is concluded, the 'learning' part—where interpretation of the data (in this case identifying strains with superior raspberry ketone titres) and assessment of how one could improve upon the titres based on which strain performed better—is executed before commencing with a new cycle of DBTL. This is a simple example of a single iteration of DBTL with each step reliant on human intervention. The eventual goal—especially with the advent of genome or biofoundries—is to curtail human input and allow for automation and machine learning to dominate proceedings. Adapted from [83,87].

As has been discussed in this review, nearly all of the reported metabolic engineering approaches for improvement in aroma compound production have been by way of rational design principles. This is because the pathways of the precursors have been largely elucidated and known bottlenecks or rate-limiting steps have been identified. Rational design will continue to be a potent pursuit as we learn more about relevant pathways and the regulation thereof. However, for fast-tracking improvements in aroma compound production in yeast, applying a DBTL-type approach holds a lot of potential. Indeed, it is perfectly suited for such an approach to target the production levels of a single metabolite. This is in contrast, however, to recent work in aroma compound development in yeast as part of an alcoholic beverage set-up, where hops flavour (a combination of geraniol and linalool) was introduced in yeast for beer production [87]. An elegant

DBTL approach was followed to build strains with ideal hops flavour (similar to commercial beer). Although the researchers were eventually successful in obtaining strains that could ferment malt and impart hops flavour, many of their initial strain building attempts failed as strains selected for ideal hops flavour could not completely consume the malt sugars. This was unintentional and not easily explained but emphasised the limitations of employing DBTL on strains with multi-parametric purposes. Nevertheless, to date, no DBTL-like approach has been performed on yeast for the sole purpose of producing enhanced levels of an aroma compound.

A key element complementary to metabolic engineering within the synthetic biology realm, is that of individual enzyme engineering in order to alter the catalytic activity and increase the flux toward the production pathway of a given aroma compound [88]. Few cases exist where mutants of the enzymes within certain pathways were examined for improved production. Site-directed mutagenesis in enzymes involved in the MVA pathway has led to the general increase in terpenoids [89] or just specific classes of terpenoids [90]. Although not studied in yeast, mutants of a sesquiterpenoid synthase (TPS24) from Syrah grape (*Vitis vinifera*) produced significantly higher levels of α-guaiene, which is a non-enzymatic precursor of rotundone, the active compound of black pepper [91].

There are also non-metabolic engineering-based challenges that exist in order to enhance aroma compound development. The overwhelming majority of interesting aroma compounds are derived from plants and it would thus make sense to explore the genes relevant to the production of the given product and assess their suitability within a recombinant yeast host. Yet, genomic data of many plants are still lacking and even though the enzymes involved in the production of a certain aroma compound are known, the genes that encode these enzymes have still not been identified. The complexities in assembling plant genomes explain the current dearth of publicly available fully-annotated whole genome sequences of plants [92,93] but more options will open up as more become available. Many recombinant pathways (like with *p*-coumaric acid and raspberry ketone) employed *Arabidopsis thaliana* genes, mainly due its sequence availability as it is a model organism and hardly because it is the 'best' enzyme candidates. Related to this is the mystery of how certain recombinant genes are expressed in high levels whereas similar ones are expressed poorly [94]. This underlines the utility of still employing a trial-and-error approach and exploring many recombinant genes to assess their ability to be expressed in yeast.

Non-*Saccharomyces* yeast have been shown, on many occasions, to be a more suitable host for the production of aroma compounds. However, the tools available to manipulate these so-called "non-conventional" yeast are not as extensive. As discussed in Wagner and Alper [95] tremendous strides have been made in this field as even CRISPR-based tools have been developed for a number of species. Further development is required, with a possible aim to incorporate large synthetically designed genetic elements in the non-*Saccharomyces* genomes similar to the SCRaMbLE-set-up developed for *S. cerevisiae.*

As already shown with the development of a hoppy yeast [87], many of the strategies employed and discussed here could easily be used to engineer aroma development in strains involved in the fermented beverage industry [96–98]. The engineering strategies should not, however, interfere with all other processes involved in fermentation. Strains tailor-made to produce certain aroma compounds could streamline the fermentation process. An ultimate goal would be to have a winemaker or brewmaster use a specific yeast strain to achieve a more predictable outcome in the aroma profile of the fermented product. An example would be to develop a yeast strain that could synthesise the oak lactones found in oak barrels. Its biosynthesis pathway is complicated and fine-tuning within the yeast strain would be necessary but such a strain would eliminate the use of expensive oak barrels in winemaking and dramatically shorten the production time.

Recombinant aroma compound development in yeast is still in its infancy especially when compared to chemical synthesis or extraction from plants—both methods in their various incarnations are centuries old. The field of synthetic biology is primed to evolve by aiming to achieve levels that can compete with the status quo. An even newer competitor is that of cell-free enzyme pathways [99] and it would be interesting to witness how each approach would evolve to compete in producing aroma compounds in a cost-effective manner. A revolutionary step, as one could imagine, would be a large yeast fermentation set-up producing a given aroma compound eventually replacing the acres and acres of flower fields and all the required inputs, dedicated for eventual aroma compound extraction.

REFERENCES

1. Schrader, J. *Flavours and Fragrances: Chemistry, Bioprocessing and Sustainability*; Berger, R.G., Ed.; Springer-Verlag: Berlin/Heidelberg, Germany, 2007; pp. 507–574. [CrossRef]

2. Kempinski, C.; Jiang, Z.; Bell, S.; Chappell, J. *Biotechnology of Isoprenoids*; Schrader, J., Bohlmann, J., Eds.; Springer-Verlag: Berlin/Heidelberg, Germany, 2015; pp. 161–199. [CrossRef]

3. Lanciotti, R.; Gianotti, A.; Patrignani, F.; Belletti, N.; Guerzoni, M.E.; Gardini, F. Use of natural aroma compounds to improve shelf-life and safety of minimally processed fruits. *Trends Food Sci. Technol.* **2004**, *15*, 201–208. [CrossRef]

4. Belletti, N.; Kamdem, S.S.; Patrignani, F.; Lanciotti, R.; Covelli, A.; Gardini, F. Antimicrobial activity of aroma compounds against *Saccharomyces cerevisiae* and improvement of microbiological stability of soft drinks as assessed by logistic regression. *Appl. Environ. Microbiol.* **2007**, *73*, 5580–5586. [CrossRef] [PubMed]

5. Buck, L.; Axel, R. A novel multigene family may encode odorant receptors: A molecular basis for odor recognition. *Cell* **1991**, *65*, 175–187. [CrossRef]

6. Swiegers, J.H.; Chambers, P.J.; Pretorius, I.S. Olfaction and taste: Human perception, physiology and genetics.*Aust. J. Grape Wine Res.* **2005**, *11*, 109–113. [CrossRef]

7. Stevenson, R.J.; Boakes, R.A. A mnemonic theory of odor perception. *Psychol. Rev.* **2003**, *110*, 340–364. [CrossRef] [PubMed]

8. Carroll, A.L.; Desai, S.H.; Atsumi, S. Microbial production of scent and flavor compounds. *Curr. Opin. Biotechnol.* **2016**, *37*, 8–15. [CrossRef] [PubMed]

9. Steensels, J.; Snoek, T.; Meersman, E.; Nicolino, M.P.; Voordeckers, K.; Verstrepen, K.J. Improving industrial yeast strains: Exploiting natural and artificial diversity. *FEMS Microbiol. Rev.* **2014**, *38*, 947–995. [CrossRef] [PubMed]

10. Dzialo, M.C.; Park, R.; Steensels, J.; Lievens, B.; Verstrepen, K.J. Physiology, ecology and industrial applications of aroma formation in yeast. *FEMS Microbiol. Rev.* **2017**, *41*, S95–S128. [CrossRef] [PubMed]

11. Hirst, M.B.; Richter, C.L. Review of aroma formation through metabolic pathways of *Saccharomyces cerevisiae*in beverage fermentations. *Am. J. Enol. Vitic.* **2016**, *67*, 361–370. [CrossRef]

12. Peter, J.; Chiara, M.D.; Friedrich, A.; Yue, J.; Pflieger, D.; Bergström, A.; Sigwalt, A.; Barre, B.; Freel, K.; Llored, A.; et al. Genome evolution across 1011 *Saccharomyces cerevisiae* isolates. *Nature* **2018**, *556*, 339–344. [CrossRef] [PubMed]

13. Gellissen, G.; Kunze, G.; Gaillardin, C.; Cregg, J.M.; Berardi, E.; Veenhuis, M.; Klei, I.V.D. New yeast expression platforms based on methylotrophic Hansenula polymorpha and Pichia pastoris and on dimorphic Arxula adeninivorans and Yarrowia lipolytica—A comparison. *FEMS Yeast Res.* **2005**, *5*, 1079–1096. [CrossRef] [PubMed]

14. Forti, L.; Mauro, S.D.; Cramarossa, M.R.; Filippucci, S.; Turchetti, B.; Buzzini, P. Non-conventional yeasts whole cells as efficient biocatalysts for the production of flavors and fragrances. *Molecules* **2015**, *20*, 10377–10398. [CrossRef] [PubMed]

15. Verstrepen, K.J.; Derdelinckx, G.; Dufour, J.P.; Winderickx, J.; Thevelein, J.M.; Pretorius, I.S.; Delvaux, F.R. Flavor-active esters: Adding fruitiness to beer. *J. Biosci. Bioeng.* **2003**, *96*, 110–118. [CrossRef]

16. Kroukamp, H.; den Haan, R.; la Grange, D.C.; Sibanda, N.; Foulquié-Moreno, M.R.; Thevelein, J.M.; van Zyl, W.H. Strain breeding enhanced heterologous cellobiohydrolase secretion by *Saccharomyces cerevisiae* in a protein specific manner. *Biotechnol. J.* **2017**, *12*, 1–10. [CrossRef] [PubMed]

17. Brachmann, C.B.; Davies, A.; Cost, G.J.; Caputo, E.; Li, J.; Hieter, P.; Boeke, J.D. Designer deletion strains derived from *Saccharomyces cerevisiae* S288C: A useful set of strains and plasmids for PCR-mediated gene disruption and other applications. *Yeast* **1998**, *14*, 115–132. [CrossRef]

18. Hvorecny, K.; Prelich, G. Characterization of chromosomal integration sites for heterologous gene expression in *Saccharomyces cerevisiae. Yeast* **2010**, *27*, 861–865. [CrossRef] [PubMed]

19. DiCarlo, J.E.; Conley, A.J.; Penttilä, M.; Jäntti, J.; Wang, H.H.; Church, G. Yeast gligo-mediated genome Engineering (YOGE). *ACS Synth. Biol.* **2013**, *2*, 742–749. [CrossRef] [PubMed]

20. Mitchell, L.A.; Chuang, J.; Agmon, N.; Khunsriraksakul, C.; Phillips, N.A.; Cai, Y.; Truong, D.M.; Veerakumar, A.; Wang, Y.; Mayorga, M.; et al. Versatile genetic assembly system (VEGAS) to assemble pathways for expression in *S. cerevisiae. Nucleic Acids Res.* **2015**, *43*, 6620–6630. [CrossRef] [PubMed]

21. Pennisi, E. Building the ultimate yeast genome. *Science* **2014**, *343*, 1426–1429. [CrossRef] [PubMed]

22. Annaluru, N.; Muller, H.; Mitchell, L.A.; Ramalingam, S.; Stracquadanio, G.; Richardson, S.M.; Dymond, J.S.; Kuang, Z.; Scheifele, L.Z.; Cooper, E.M.; et al. Total synthesis of a functional designer eukaryotic chromosome. *Science* **2014**, *344*, 55–58. [CrossRef] [PubMed]

23. Richardson, S.M.; Mitchell, L.A.; Stracquadanio, G.; Yang, K.; Dymond, J.S.; DiCarlo, J.E.; Lee, D.; Huang, C.L.V.; Chandrasegaran, S.; Cai, Y.; et al. Design of a synthetic yeast genome. *Science* **2017**, *355*, 1040–1044. [CrossRef] [PubMed]

24. Shen, Y.; Stracquadanio, G.; Wang, Y.; Yang, K.; Mitchell, L.A.; Xue, Y.; Cai, Y.; Chen, T.; Dymond, J.S.; Kang, K.; et al. SCRaMbLE generates designed combinatorial stochastic diversity in synthetic chromosomes. *Genome Res.* **2015**, *26*, 36–49. [CrossRef] [PubMed]

25. Pretorius, I.S.; Boeke, J.D. Yeast 2.0—Connecting the dots in the construction of the world's first functional synthetic eukaryotic genome. *FEMS Yeast Res.* **2018**, *18*. [CrossRef] [PubMed]

26. Pickens, L.; Tang, Y.; Chooi, Y.-H. Metabolic engineering for the production of natural products. *Annu. Rev. Chem. Biomol. Eng.* **2014**, *2*, 211–236. [CrossRef] [PubMed]

27. Lee, J.H.; Wendisch, V.F. Biotechnological production of aromatic compounds of the extended shikimate pathway from renewable biomass. *J. Biotechnol.* **2017**, *257*, 211–221. [CrossRef] [PubMed]

28. Braus, G.H. Aromatic amino acid biosynthesis in the yeast *Saccharomyces cerevisiae*: A model system for the regulation of a eukaryotic biosynthetic pathway. *Microbiol. Rev.* **1991**, *55*, 349–370. [PubMed]

29. Mulleder, M.; Calvani, E.; Alam, M.T.; Wang, R.K.; Eckerstorfer, F.; Zelezniak, A.; Ralser, M. Functional metabolomics describes the yeast biosynthetic regulome. *Cell* **2016**, *167*, 553–565. [CrossRef] [PubMed]

30. Hansen, E.H.; Møller, B.L.; Kock, G.R.; Bünner, C.M.; Kristensen, C.; Jensen, O.R.; Okkels, F.T.; Olsen, C.E.; Motawia, M.S.; Hansen, J. De novo biosynthesis of vanillin in fission yeast (*Schizosaccharomyces pombe*) and baker's yeast (*Saccharomyces cerevisiae*). *Appl. Environ. Microbiol.* **2009**, *75*, 2765–2774. [CrossRef] [PubMed]

31. Iwaki, A.; Ohnuki, S.; Suga, Y.; Izawa, S.; Ohya, Y. Vanillin inhibits translation and induces messenger ribo-nucleo-protein (mRNP) granule formation in *Saccharomyces cerevisiae*: Application and validation of high-content, image-based profiling. *PLoS ONE* **2013**, *8*, 2–11. [CrossRef] [PubMed]

32. Brochado, A.R.; Matos, C.; Møller, B.L.; Hansen, J.; Mortensen, U.H.; Patil, K.R. Improved vanillin production in baker's yeast through in silico design. *Microb. Cell Fact.* **2010**, *9*, 1–15. [CrossRef] [PubMed]

33. Brochado, A.R.; Patil, K.R. Over expression of *O*-methyltransferase leads to improved vanillin production in baker's yeast only when complemented with model-guided network engineering. *Biotechnol. Bioeng.* **2013**, *110*, 656–659. [CrossRef] [PubMed]

34. Yin, L.H.; Choon, Y.W.; Chai, L.E.; Chong, C.K.; Deris, S.; Illias, R.M.; Mohamad, M.S. *Advances in Biomedical Infrastructure*; Sidhu, A.S., Dhillon, S.K., Eds.; Springer-Verlag: Berlin/Heidelberg, Germany, 2013; Volume 477, pp. 101–116. [CrossRef]

35. Vannelli, T.; Qi, W.W.; Sweigard, J.; Gatenby, A.A.; Sariaslani, F.S. Production of p-hydroxycinnamic acid from glucose in *Saccharomyces cerevisiae* and *Escherichia coli* by expression of heterologous genes from plants and fungi. *Metab. Eng.* **2007**, *9*, 142–151. [CrossRef] [PubMed]

36. Gottardi, M.; Grün, P.; Bode, H.B.; Hoffmann, T.; Schwab, W.; Oreb, M.; Boles, E. Optimisation of trans-cinnamic acid and hydrocinnamyl alcohol production with recombinant *Saccharomyces cerevisiae* and identification of cinnamyl methyl ketone as a by-product. *FEMS Yeast Res.* **2017**, *9*, 142–151. [CrossRef] [PubMed]

37. Jiang, H.; Wood, K.V.; Morgan, J.A. Metabolic engineering of the phenylpropanoid pathway in *Saccharomyces cerevisiae*. *Appl. Environ. Microbiol.* **2005**, *71*, 2962–2969. [CrossRef] [PubMed]

38. Lee, D.; Lloyd, N.D.R.; Pretorius, I.S.; Borneman, A.R. Heterologous production of raspberry ketone in the wine yeast *Saccharomyces cerevisiae* via pathway engineering and synthetic enzyme fusion. *Microb. Cell Fact.* **2016**, *15*. [CrossRef] [PubMed]

39. Kim, B.; Cho, B.R.; Hahn, J.S. Metabolic engineering of *Saccharomyces cerevisiae* for the production of 2-phenylethanol via Ehrlich pathway. *Biotechnol. Bioeng.* **2014**, *111*, 115–124. [CrossRef] [PubMed]

40. Vaseghi, S.; Baumeister, A.; Rizzi, M.; Reuss, M. In vivo dynamics of the pentose phosphate pathway in *Saccharomyces cerevisiae*. *Metab. Eng.* **1999**, *1*, 128–140. [CrossRef] [PubMed]

41. Curran, K.A.; Leavitt, J.M.; Karim, A.S.; Alper, H.S. Metabolic engineering of muconic acid production in*Saccharomyces cerevisiae*. *Metab. Eng.* **2013**, *15*, 55–66. [CrossRef] [PubMed]

42. Deaner, M.; Alper, H.S. Systematic testing of enzyme perturbation sensitivities via graded dCas9 modulation in *Saccharomyces cerevisiae*. *Metab. Eng.* **2017**, *40*, 14–22. [CrossRef] [PubMed]

43. Ravasio, D.; Wendland, J.; Walther, A. Major contribution of the Ehrlich pathway for 2-phenylethanol/rose flavor production in Ashbya gossypii. *FEMS Yeast Res.* **2014**, *14*, 833–844. [CrossRef] [PubMed]

44. Kim, T.Y.; Lee, S.W.; Oh, M.K. Biosynthesis of 2-phenylethanol from glucose with genetically engineered Kluyveromyces marxianus. *Enzym. Microb. Technol.* **2014**, *61-62*, 44–47. [CrossRef] [PubMed]

45. Lu, X.; Wang, Y.; Zong, H.; Ji, H.; Zhuge, B.; Dong, Z. Bioconversion of L-phenylalanine to 2-phenylethanol by the novel stress-tolerant yeast Candida glycerinogenes WL2002-5. *Bioengineered* **2016**, *7*, 418–423. [CrossRef] [PubMed]

46. Mierzejewska, J.; Tymoszewska, A.; Chreptowicz, K.; Krol, K. Mating of 2 laboratory *Saccharomyces cerevisiae* strains resulted in enhanced production of 2-phenylethanol by biotransformation of L-Phenylalanine. *J. Mol. Microbiol. Biotechnol.* **2017**, *27*, 81–90. [CrossRef] [PubMed]

47. Eshkol, N.; Sendovski, M.; Bahalul, M.; Katz-Ezov, T.; Kashi, Y.; Fishman, A. Production of 2-phenylethanol from L-phenylalanine by a stress tolerant *Saccharomyces cerevisiae* strain. *J. Appl. Microbiol.* **2009**, *106*, 534–542. [CrossRef] [PubMed]

48. Zwenger, S.; Basu, C. Plant terpenoids: Applications and future potentials. *Biotechnol. Mol. Biol. Rev.* **2008**,*3*, 1–7. [CrossRef]

49. Rodriguez, R.J.; Low, C.; Bottema, C.D.K.; Parks, L.W. Multiple functions for sterols in *Saccharomyces cerevisiae*. *Biochim. Biophys. Acta* **1985**, *837*, 336–343. [CrossRef]

50. Liao, P.; Hemmerlin, A.; Bach, T.J.; Chye, M.L. The potential of the mevalonate pathway for enhanced isoprenoid production. *Biotechnol. Adv.* **2016**, *34*, 697–713. [CrossRef] [PubMed]

51. Paramasivan, K.; Mutturi, S. Progress in terpene synthesis strategies through engineering of *Saccharomyces cerevisiae*. *Crit. Rev. Biotechnol.* **2017**, *37*, 974–989. [CrossRef] [PubMed]

52. Ignea, C.; Pontini, M.; Maffei, M.E.; Makris, A.M.; Kampranis, S.C. Engineering monoterpene production in yeast using a synthetic dominant negative geranyl diphosphate synthase. *ACS Synth. Biol.* **2014**, *3*, 298–306. [CrossRef] [PubMed]

53. Vickers, C.E.; Williams, T.C.; Peng, B.; Cherry, J. Recent advances in synthetic biology for engineering isoprenoid production in yeast. *Curr. Opin. Chem. Biol.* **2017**, *40*, 47–56. [CrossRef] [PubMed]

54. Wriessnegger, T.; Pichler, H. Yeast metabolic engineering—Targeting sterol metabolism and terpenoid formation. *Prog. Lipid Res.* **2013**, *52*, 277–293. [CrossRef] [PubMed]

55. Asadollahi, M.A.; Maury, J.; Schalk, M.; Clark, A.; Nielsen, J. Enhancement of farnesyl diphosphate pool as direct precursor of sesquiterpenes through metabolic engineering of the mevalonate pathway in *Saccharomyces cerevisiae*. *Biotechnol. Bioeng.* **2010**, *106*, 86–96. [CrossRef] [PubMed]

56. Peng, B.; Plan, M.R.; Chrysanthopoulos, P.; Hodson, M.P.; Nielsen, L.K.; Vickers, C.E. A squalene synthase protein degradation method for improved sesquiterpene production in *Saccharomyces cerevisiae*. *Metab. Eng.* **2017**, *39*, 209–219. [CrossRef] [PubMed]

57. Polakowski, T.; Stahl, U. Overexpression of a cytosolic hydroxymethylglutaryl-CoA reductase leads to squalene accumulation in yeast. *Appl. Microbiol. Biotechnol.* **1998**, *8*, 66–71. [CrossRef]

58. Rico, J.; Pardo, E.; Orejas, M. Enhanced production of a plant monoterpene by overexpression of the 3-hydroxy-3-methylglutaryl coenzyme a reductase catalytic domain in *Saccharomyces cerevisiae*. *Appl. Environ. Microbiol.* **2010**, *76*, 6449–6454. [CrossRef] [PubMed]

59. Pardo, E.; Rico, J.; Gil, J.V.; Orejas, M. De novo production of six key grape aroma monoterpenes by a geraniol synthase-engineered *S. cerevisiae* wine strain. *Microb. Cell Fact.* **2015**, *14*. [CrossRef] [PubMed]

60. Vik, A.; Rine, J. Upc2p and Ecm22p, dual regulators of sterol biosynthesis in *Saccharomyces cerevisiae*. *Mol. Cell. Biol.* **2001**, *21*, 6395–6405. [CrossRef] [PubMed]

61. Meadows, A.L.; Hawkins, K.M.; Tsegaye, Y.; Antipov, E.; Kim, Y.; Raetz, L.; Dahl, R.H.; Tai, A.; Mahatdejkul-meadows, T.; Xu, L.; et al. Rewriting yeast central carbon metabolism for industrial isoprenoid production. *Nature* **2016**, *537*, 694–697. [CrossRef] [PubMed]

62. Farhi, M.; Marhevka, E.; Masci, T.; Marcos, E.; Eyal, Y.; Ovadis, M.; Abeliovich, H.; Vainstein, A. Harnessing yeast subcellular compartments for the production of plant terpenoids. *Metab. Eng.* **2011**, *13*, 474–481. [CrossRef] [PubMed]

63. Wriessnegger, T.; Augustin, P.; Engleder, M.; Leitner, E.; Müller, M.; Kaluzna, I.; Schürmann, M.; Mink, D.; Zellnig, G.; Schwab, H.; et al. Production of the sesquiterpenoid (+)-nootkatone by metabolic engineering of Pichia pastoris. *Metab. Eng.* **2014**, *24*, 18–29. [CrossRef] [PubMed]

64. Verwaal, R.; Wang, J.; Meijnen, J.P.; Visser, H.; Sandmann, G.; Van Den Berg, J.A.; Van Ooyen, A.J.J. High-level production of beta-carotene in *Saccharomyces cerevisiae* by successive transformation with carotenogenic genes from Xanthophyllomyces dendrorhous. *Appl. Environ. Microbiol.* **2007**, *73*, 4342–4350. [CrossRef] [PubMed]

65. López, J.; Essus, K.; Kim, I.-K.; Pereira, R.; Herzog, J.; Siewers, V.; Nielsen, J.; Agosin, E. Production of β-ionone by combined expression of carotenogenic and plant CCD1 genes in *Saccharomyces cerevisiae*. *Microb. Cell Fact.* **2015**, *14*, 84. [CrossRef] [PubMed]

66. Beekwilder, J.; van Rossum, H.M.; Koopman, F.; Sonntag, F.; Buchhaupt, M.; Schrader, J.; Hall, R.D.; Bosch, D.; Pronk, J.T.; van Maris, A.J.A.; et al. Polycistronic expression of a β-carotene biosynthetic pathway in *Saccharomyces cerevisiae* coupled to β-ionone production. *J. Biotechnol.* **2014**, *192*, 383–392. [CrossRef] [PubMed]

67. Kirby, J.; Dietzel, K.L.; Wichmann, G.; Chan, R.; Antipov, E.; Moss, N.; Baidoo, E.E.K.; Jackson, P.; Gaucher, S.P.; Gottlieb, S.; et al. Engineering a functional 1-deoxy-D-xylulose 5-phosphate (DXP) pathway in *Saccharomyces cerevisiae*. *Metab. Eng.* **2016**, *38*, 494–503. [CrossRef] [PubMed]

68. Siddiqui, M.S.; Thodey, K.; Trenchard, I.; Smolke, C.D. Advancing secondary metabolite biosynthesis in yeast with synthetic biology tools. *FEMS Yeast Res.* **2012**, *12*, 144–170. [CrossRef] [PubMed]

69. Jensen, M.K.; Keasling, J.D. Recent applications of synthetic biology tools for yeast metabolic engineering.*FEMS Yeast Res.* **2015**, *15*, 1–10. [CrossRef] [PubMed]

70. Deloache, W.C.; Russ, Z.N.; Narcross, L.; Gonzales, A.M.; Martin, V.J.J.; Dueber, J.E. An enzyme-coupled biosensor enables (*S*)-reticuline production in yeast from glucose. *Nat. Chem. Biol.* **2015**, *11*, 465–471. [CrossRef] [PubMed]

71. Leavitt, J.M.; Wagner, J.M.; Tu, C.C.; Tong, A.; Liu, Y.; Alper, H.S. Biosensor-enabled directed evolution to improve muconic acid production in *Saccharomyces cerevisiae*. *Biotechnol. J.* **2017**, *12*, 1–9. [CrossRef] [PubMed]

72. Adeniran, A.; Sherer, M.; Tyo, K.E.J. Yeast-based biosensors: Design and applications. *FEMS Yeast Res.* **2015**,*15*, 1–15. [CrossRef] [PubMed]

73. Ravasio, D.; Walther, A.; Trost, K.; Vrhovsek, U.; Wendland, J. An indirect assay for volatile compound production in yeast strains. *Sci. Rep.* **2014**, *4*. [CrossRef] [PubMed]

74. Chou, H.H.; Keasling, J.D. Programming adaptive control to evolve increased metabolite production.*Nat. Commun.* **2013**, *4*, 1–8. [CrossRef] [PubMed]

75. Li, S.; Si, T.; Wang, M.; Zhao, H. Development of a synthetic malonyl-CoA sensor in *Saccharomyces cerevisiae* for intracellular metabolite monitoring and genetic screening. *ACS Synth. Biol.* **2015**, *4*, 1308–1315. [CrossRef] [PubMed]

76. David, F.; Nielsen, J.; Siewers, V. Flux Control at the Malonyl-CoA node through hierarchical dynamic pathway regulation in *Saccharomyces cerevisiae*. *ACS Synth. Biol.* **2016**, *5*, 224–233. [CrossRef] [PubMed]

77. Skjoedt, M.L.; Snoek, T.; Kildegaard, K.R.; Arsovska, D.; Eichenberger, M.; Goedecke, T.J.; Rajkumar, A.S.; Zhang, J.; Kristensen, M.; Lehka, B.J.; et al. Engineering prokaryotic transcriptional activators as metabolite biosensors in yeast. *Nat. Chem. Biol.* **2016**, *12*, 951–958. [CrossRef] [PubMed]

78. Ho, J.C.H.; Pawar, S.V.; Hallam, S.J.; Yadav, V.G. An improved whole-cell biosensor for the discovery of lignin-transforming enzymes in functional metagenomic screens. *ACS Synth. Biol.* **2018**, *7*, 392–398. [CrossRef] [PubMed]

79. Fiorentino, G.; Ronca, R.; Bartolucci, S. A novel *E. coli* biosensor for detecting aromatic aldehydes based on a responsive inducible archaeal promoter fused to the green fluorescent protein. *Appl. Microbiol. Biotechnol.* **2009**, *82*, 67–77. [CrossRef] [PubMed]

80. Siedler, S.; Khatri, N.K.; Zsohár, A.; Kjærbølling, I.; Vogt, M.; Hammar, P.; Nielsen, C.F.; Marienhagen, J.; Sommer, M.O.A.; Joensson, H.N. Development of a bacterial biosensor for rapid screening of yeast *p*-coumaric Acid production. *ACS Synth. Biol.* **2017**, *6*, 1860–1869. [CrossRef] [PubMed]

81. West, R.W.; Chen, S.M.; Putz, H.; Butler, G.; Banerjee, M. GAL1-GAL10 divergent promoter region of *Saccharomyces cerevisiae* contains negative control elements in addition to functionally separate and possibly overlapping upstream activating sequences. *Genes Dev.* **1987**, *1*, 1118–1131. [CrossRef] [PubMed]

82. Chao, R.; Mishra, S.; Si, T.; Zhao, H. Engineering biological systems using automated biofoundries. *Metab. Eng.* **2017**, *42*, 98–108. [CrossRef] [PubMed]

83. Nielsen, J.; Keasling, J.D. Engineering cellular metabolism. *Cell* **2016**, *164*, 1185–1197. [CrossRef] [PubMed]

84. Hollywood, K.A.; Schmidt, K.; Takano, E.; Breitling, R. Metabolomics tools for the synthetic biology of natural products. *Curr. Opin. Biotechnol.* **2018**, *54*, 114–120. [CrossRef] [PubMed]

85. Dacres, H.; Wang, J.; Leitch, V.; Horne, I.; Anderson, A.R.; Trowell, S.C. Greatly enhanced detection of a volatile ligand at femtomolar levels using bioluminescence resonance energy transfer (BRET). *Biosens. Bioelectron.* **2011**, *29*, 119–124. [CrossRef] [PubMed]

86. Berna, A. Metal oxide sensors for electronic noses and their application to food analysis. *Sensors* **2010**,*10*, 3882–3910. [CrossRef] [PubMed]

87. Denby, C.M.; Li, R.A.; Vu, V.T.; Costello, Z.; Lin, W.; Chan, L.J.G.; Williams, J.; Donaldson, B.; Bamforth, C.W.; Petzold, C.J.; et al. Industrial brewing yeast engineered for the production of primary flavor determinants in hopped beer. *Nat. Commun.* **2018**, *9*. [CrossRef] [PubMed]

88. Erb, T.J.; Jones, P.R.; Bar-Even, A. Synthetic metabolism: metabolic engineering meets enzyme design.*Curr. Opin. Chem. Biol.* **2017**, *37*, 56–62. [CrossRef] [PubMed]

89. Mantzouridou, F.; Tsimidou, M.Z. Observations on squalene accumulation in *Saccharomyces cerevisiae* due to the manipulation of HMG2 and ERG6. *FEMS Yeast Res.* **2010**, *10*, 699–707. [CrossRef] [PubMed]

90. Fischer, M.J.C.; Meyer, S.; Claudel, P.; Bergdoll, M.; Karst, F. Metabolic engineering of monoterpene synthesis in yeast. *Biotechnol. Bioeng.* **2011**, *108*, 1883–1892. [CrossRef] [PubMed]

91. Drew, D.P.; Andersen, T.B.; Sweetman, C.; Møller, B.L.; Ford, C.; Simonsen, H.T. Two key polymorphisms in a newly discovered allele of the Vitis vinifera TPS24 gene are responsible for the production of the rotundone precursor α-guaiene. *J. Exp. Bot.* **2016**, *67*, 799–808. [CrossRef] [PubMed]

92. Schatz, M.C.; Witkowski, J.; McCombie, W.R. Current challenges in *de novo* plant genome sequencing and assembly. *Genome Biol.* **2012**, *13*. [CrossRef]

93. Claros, M.G.; Bautista, R.; Guerrero-Fernández, D.; Benzerki, H.; Seoane, P.; Fernández-Pozo, N. Why Assembling plant genome sequences is so challenging. *Biology* **2012**, *1*, 439–459. [CrossRef] [PubMed]

94. Van Wyk, N.; Trollope, K.M.; Steenkamp, E.T.; Wingfield, B.D.; Volschenk, H. Identification of the gene for β-fructofuranosidase from Ceratocystis moniliformis CMW 10134 and characterization of the enzyme expressed in *Saccharomyces cerevisiae*. *BMC Biotechnol.* **2013**, *13*. [CrossRef] [PubMed]

95. Wagner, J.M.; Alper, H.S. Synthetic biology and molecular genetics in non-conventional yeasts: Current tools and future advances. *Fungal Genet. Biol.* **2016**, *89*, 126–136. [CrossRef] [PubMed]
96. Jagtap, U.B.; Jadhav, J.P.; Bapat, V.A.; Pretorius, I.S. Synthetic biology stretching the realms of possibility in wine yeast research. *Int. J. Food Microbiol.* **2017**, *252*, 24–34. [CrossRef] [PubMed]
97. Pretorius, I.S. Solving yeast jigsaw puzzles over a glass of wine. *EMBO Rep.* **2017**, *18*, 1–10. [CrossRef] [PubMed]
98. Pretorius, I. Conducting wine symphonics with the aid of yeast genomics. *Beverages* **2016**, *2*. [CrossRef]
99. Moore, S.J.; Tosi, T.; Hleba, Y.B.; Bell, D.; Polizzi, K.; Freemont, P. A cell-free synthetic biochemistry platform for raspberry ketone production. *bioRxiv* **2017**. [CrossRef]

CHAPTER-3

COMBINATORIAL ENGINEERING OF YARROWIA LIPOLYTICA AS AN ADVANCED CELL BIOREFINERY PLATFORM FOR THE SYNTHESIS OF VERSATILE LONG CHAIN DICARBOXYLIC ACIDS

INTRODUCTION: AN OVERVIEW

Embarking on the frontier of cellular biorefinery, this study delves into the innovative landscape of combinatorial engineering applied to Yarrowia lipolytica. Positioned as a promising platform, Yarrowia lipolytica emerges as a transformative host for the de novo synthesis of versatile long-chain dicarboxylic acids, signaling a paradigm shift in bioproduction strategies. In this exploration, we navigate the intricate realm of combinatorial engineering, unraveling the potential of Yarrowia lipolytica as a cellular biorefinery powerhouse poised to yield multi-purpose long-chain dicarboxylic acids.

This proof-of-concept study establishes *Yarrowia lipolytica* (*Y. lipolytica*) as a whole cell factory for the *de novo* production of long chain dicarboxylic acid (LCDCA-16 and 18) using glycerol as the sole source of carbon. Modification of the fatty acid metabolism pathway enabled creating a pool of fatty acids in a β-oxidation deficient strain. We then selectively upregulated the native fatty acid β-oxidation pathway for the enhanced terminal oxidation of the endogenous fatty acid precursors. Nitrogen-limiting conditions and leucine supplementation were employed to induce fatty acid biosynthesis in an engineered Leu^- modified strain. Our genetic engineering strategy allowed a minimum production of 330 mg/L LCDCAs in shake flask. Scale up to a 1-L bioreactor increased the titer to

3.49 g/L. Our engineered yeast also produced citric acid as a major by-product at a titer of 39.2 g/L. These results provide basis for developing *Y. lipolytica* as a safe biorefinery platform for the sustainable production of high-value LCDCAs from non-oily feedstock.

Growing demand for petroleum dependent chemicals, in addition to surging environmental concerns, has inspired increased use of renewable resources. The sustainable production of monomers and polymers is of particular interest for reducing petroleum feedstock dependency and CO_2 emissions. The environmental cost of using petrochemical polymers can be lowered by replacing them with bio-based polymers, produced through fermentation using biomass or byproducts as feedstock [1]. Glycerol is the main byproduct of biodiesel production process, and its efficient valorization would help offset biodiesel production costs. Such bioconversion also decreases the environmental impact of waste streams. Despite various technological successes, it is still challenging to establish an integrated bioprocess, at commercial scale, for the valorization of such hydrophilic substrates into desirable compounds [2]. Nonetheless, the development of high productive strains and high value co-products advance the commercialization of such bioprocesses.

Bulk monomers of high value are commonly targeted co-products for industrial biotechnology. LCDCA monomers have a much higher value than regular free fatty acids. These building block monomers are used to make various polyamides, polyester and, polyurethanes [3] in the bio-plastic and coating industries. Their use enables the synthesis of various novel polymers with enhanced hydrolytic stability, solvent resistance, optical clarity, and flexibility properties [4]. They can also be used in the automotive, food, medical, and chemical industries as adhesives, corrosion inhibitors, lubricants, fragrances, surfactants, antiseptics, and personal care ingredients [5]. Non-renewable petroleum-derived alkanes or plant-derived fatty acids that compete with food production are currently the main hydrophobic feedstock for the production of these multi-purpose building blocks [6–8].

A chemical-based approach for the production of these monomers from hydrophobic substrates has already been established. For instance, the chemical degradation of plant oil fatty acids to LCDCAs via self-metathesis reaction is one of the major chemical-based production routes [9]. The chemical-based production of LCDCA is an energy-intensive and environmentally unfriendly process, which requires expensive catalysts and cost intensive purification steps due to the generation of various by-products. Contrary to chemical approaches, the bio-based approach offers higher selectivity and sustainability [10]. The microbial production of short to medium chain DCAs has been reviewed and partly industrialized [10–13]. In terms of LCDCAs, the major producing strain is a pathogenic yeast, *Candida tropicalis* [14].

The biosynthetic pathway to LCDCA monomers is mediated by the P450 system. In general, cytochrome P450 monooxygenases (P450 or CYP) systems are involved in the oxygenation of various hydrophobic compounds. For instance, the CYP52 family performs terminal ω-oxidation of alkanes, alcohols, and fatty acids of various chain lengths [15]. This family belongs to yeast species, including *Candida* and *Yarrowia*, and is comprised of members whose products have different specificities for alkanes versus fatty acids [16,17]. The yeast CYP52 has higher productivity than the bacterial ω-hydroxylases [18]. Terminal oxidation by P450 is carried out at high region-selectivity and stereo-selectivity. This can result in the synthesis of high value compounds whose production is nearly impossible through chemical approaches due to the instability of double bonds or the generation of unwanted byproducts [19]. Several examples of cytochrome P450 monooxygenases with terminal ω-oxidation activity toward fatty acids have been reported [20–27]. Some are also capable of fatty acid over-oxidation to their corresponding dicarboxylic acids [23,26,28]. The mechanism of a P450 mediated ω-oxidation process has been discussed [29], and is briefly presented here.

This nicotinamide adenine dinucleotide phosphate (NADPH) dependent reaction takes place under mild conditions and requires molecular oxygen. In summary, an active intermediate is formed when the heme group of P450 monoxygenase is coupled to molecular oxygen. As a result, oxygen is added to the attached substrate, the C-H bond of fatty acid. Subsequently, another atom of oxygen is released in the form of a water molecule, and a hydrogen atom is transferred from the substrate to the heme group, leading to the generation of hydroxylated substrates. During this process, the P450-diflavin reductase partner (CPR) is bound to the NADPH cofactor to derive electrons from that while positioning the redox active nicotinamide ring and flavin adenine dinucleotide (FAD) at a close distance. The FAD accepts hydride ions from the NADPH and transfers electrons to the flavin mononucleotide (FMN) to reduce the heme group of the P450 monooxygenase [30]. This complex enzymatic system has some limitations, such as low activity and stability, dependency on cofactor and redox partner proteins, and low product yields for broad industrial usage [19,31]. Moreover, eukaryotic CYPs are often associated with ER membranes [32] and have lower solubility and stability compared to bacterial ones. High-density cultivation of P450-carrying cells or modification of the P450 system for higher expression and stability can help overcome these limitations in a bioreactor application [15].

Yarrowia lipolytica is an advantageous biorefinery platform for the production of fatty acid-based bioproducts [33] and for P450 catalytic bioconversion [34]. The P450-dependent hydroxylation of alkane, fatty alcohol, and fatty acid precursors is predominant in this yeast [35,36]. This assists with detoxification and degradation of these potentially toxic molecules. The over expression of homologous and heterologous P450 has been employed in *Y. lipolytica* [37] resulting in a remarkable proliferation of ER [38]. There are at least 17 P450 genes and some electron transfer proteins in this oleaginous yeast [38]. Among them, twelve P450 Alk genes (*YlALK*) and a single NADPH P450 reductase gene,*YlCPR*, have been isolated and characterized [32,36,38,39]. They have various substrate requirements with respect to length, degree of unsaturation, and functional groups [40]. For instance, *YlALK* 3, *YlALK* 5, and *YlALK* 7 can catalyze the ω-oxidation of fatty acids [41]. The resulting ω-hydroxy fatty acids undergo further steps of oxidation that could also constitute rate limiting steps in this yeast. These steps can be carried out by fatty alcohol oxidase (*FAO1*; YALI0B14014g) and by the alcohol dehydrogenase genes (*FADHs*) in the ER and peroxisome [42,43]. The fatty acid precursors and their corresponding oxygenated fatty acid monomers are susceptible to the effective peroxisomal β-oxidation degradation in *Y. lipolytica*, resulting in the generation of energy, water, and CO_2. The *POX* genes are involved in the degradation of these monomers [44,45]. The disruption of the β-oxidation pathway via deletion of *POX* genes is commonly employed to prevent fatty acid and LCDCA degradation and unwanted chain modification [14,46,47]. In this yeast, the overproduction of fatty acids can be efficiently achieved in the form of free fatty acids rather than fatty acyl-CoAs, since the production of the latters are highly regulated by feedback inhibition [48,49]. In fact, the ω-oxidation of free fatty acids can favour their degradation and alleviate their toxicity. The ω-oxidation pathway acts as a rescue route in the peroxisome-deficient cell by generating oxygenated fatty acids of higher solubility in water [50,51].

Engineering *Y. lipolytica* for the *de novo* production of LCDCAs enables utilization of various unrelated feedstock, including sugars, glycerol, and short chain volatile fatty acids, for this process.

Y. lipolytica has a preference for using glycerol as a carbon source [32]. Glycerol and glucose have a transcription-repressive effect, with various degrees of strength, on background P450 activities in this yeast [52]. Glycerol has a strong repressor effect, so the over expression of selected P450 members in the presence of this substrate can enable higher selectivity through the minimization of background ω-oxidation activity. *Y. lipolytica* accumulates a larger proportion of endogenous unsaturated long chain fatty acids when grown on glucose [53]. The preservation of double bonds through the terminal ω-oxidation results in the generation of multi-functional unsaturated LCDCAs. These double bonds enable the attachment of additional functional groups [54] and the generation of cross-linked degradable polymer networks with adjustable properties [55].

In this study, we engineered *Y. lipolytica*, a "safe-to-use" yeast [56], for the *de novo* production of high value LCDCA monomers. This involved a one-step bioprocess, using glycerol as the sole source of carbon. This was done by inactivating the endogenous cytosolic fatty acyl-CoA synthetase, *YlFAA1* YALI0D17864g, and redirecting the resulting pool of free fatty acids toward an engineered ω-oxidation pathway, in a deficient β-oxidation background. The ω-oxidation pathway was engineered by over-expressing P450 reductase (YALI0D04422g), fatty alcohol oxidase *YlFAO1*, and a selected member of P450 monoxygenases (*YlALK5*, YALI0B13838g). After preliminary screening and shake flask cultivations, we used a 1-L bioreactor for the P450-based biocatalytic system and achieved a productivity of 0.04 g/L·h, which is more than the process productivity of 0.001 g/L·h acceptable for P450-based biocatalytic production of pharmaceuticals [19]. Our results demonstrated the promise of *Y. lipolytica* for use as an oleaginous yeast cell factory for LCDCA production. The safe status of this yeast enhances the suitability of the resulting LCDCA monomers for food and pharmaceutical uses, where high security standards should be met. To the best of our knowledge, this is the first reported study on the biosynthesis of LCDCAs from glycerol by engineered *Y. lipolytica*. The findings of this work present a unique biosynthetic route that is expected to advance *Yarrowia* platform for the sustainable production of these multi-purpose long chain building blocks.

MATERIALS AND METHODS

Strains and Culture Conditions

Escherichia coli top 10 was used for plasmid DNA construction and propagation. According to the standard protocols, Luria-Bertani (LB) broth or agar medium was made and supplemented with ampicillin at the concentration of 100 μg·mL^{-1} [57]. The recombinant *Y. lipolytica* strains of this study were all derived from *Y. lipolytica* H222 (a potential citric acid-producer wild-type German strain) [42] and are presented in Table 1.

The yeast nitrogen base (YNB) medium with following composition was used to select auxotrophic Ura$^-$ and Leu$^-$ transformants and to prepare the seed cultures: 6.7 g/L yeast nitrogen base (without amino acids w/ammonium sulfate) (Becton, Dickinson and Company, Sparks, MD, USA), 20 g/L glucose, and 1.92 g/L synthetic mix minus uracil (YNB-Ura) or 1.62 g/L synthetic mix minus leucine (YNB-Leu) (US Biological, Salem, MA, USA). YNB-Leu was also used to grow the strain carrying the plasmid with the *LEU2* marker and *CRE* gene to restore the uracil auxotrophic derivatives of the F and P strains. Rich medium (YPD) containing 20 g/L glucose, 20 g/L bacto peptone (BD), and 10 g/L bacto yeast extract (BD) was used for non-selective propagation of strains. Agar (US Biological, Salem, MA, USA) was added at a concentration of 20 g/L for preparing plates.

Table 1. ***Y. lipolytica*** **strains used in this study.**

Y. lipolytica Strains Names	Strain Genotypes	Gene Configurations	Reference
H222 (W) [wild type]	MatA		[42]
H222ΔP (HP-U) [leu$_+$, ura$_-$]	MatA ura3-302::SUC2 Δpox1 Δpox2 Δpox3 Δpox4 Δpox5 Δpox6		[42]
H222ΔPΔLΔSΔF (F) [leu$_-$, ura$_+$]	Same as HP-U, +Δleu2 Δsnf1 Δfaa1 Δsnf1::URA3	loxR-URA3-loxP flanked by SNF1 upstream and downstream	This study
H222ΔPΔLΔSΔF ΔL +ALK5 YlCPR YlFAO1 (P) [leu$_-$, ura$_+$]	Same as F, +Δleu2::URA3 YlALK5 YlCPR YFAO1	loxR-URA3-loxP flanked by SNF1 upstream and downstream	This study
H222ΔPΔLΔSΔF ΔL +ALK5 YlCPR YlFAO1 ++ALK5 (M) [leu$_-$, ura$_+$]	Same as P, +YlALK5	Multiple-copy integration of YlALK5 using zeta based integrative vector pINA1291(this strain was selected after screening of 10–20 transformants for their growth and production capacities)	This study

Production media in shake flasks and bioreactor were performed in minimal media with the following defined composition: 1.7 g/L Yeast Nitrogen Base (without amino acids and without ammonium sulphate) (BD), 1.5 g/L $MgSO_4 \cdot 7H_2O$, and1.92 g/L drop-out synthetic mix minus uracil. The $(NH_4)_2SO_4$ was used as the major source of nitrogen at a concentration of 1.9 g/L. The production media formulations contained 52 g/L glycerol, to reach the target carbon to nitrogen (C/N) ratio of 50. The $CaCO_3$ was added to the shake flasks at a concentration of 6 g/L to provide buffering conditions. This carbonate plays a role in maintaining a stable pH [58]. The shake flask cultivations were carried out in 250 mL Erlenmeyer flasks containing 50 mL of the medium at an agitation rate of 180 ± 5 rpm and a temperature of 28 ± 1°C. Fresh colonies from the YNB-Ura plates were precultured in YNB-Ura broth. Exponentially growing cells were harvested by centrifugation, washed with water, and then resuspended in water. The production media were inoculated with the resuspended cells to an initial optical density (OD_{600}) of 0.1.

Batch Fermentation

Batch cultivation was carried out in a 1-L benchtop fermenter BioFlo 110 (New Brunswick Scientific, Enfield, CT, USA). Cells from a 24 h-shake flask YNB-Ura culture were harvested, washed using water, and inoculated into 700 mL of the fermentation medium (with C/N of 50) to reach a minimum OD600 of 0.1. The temperature was kept at 28 °C. The pH was controlled and maintained at 6 during the biomass propagation (typically for 1 day) and then was increased to 8 during the production phase using 5 M NaOH. This pH adjustment was also beneficial to lessen the rate of citric acid production. A minimum dissolved oxygen level of 50% was maintained by cascading with an agitation ranging from 250 to 800 rpm, and by supplying sterile air at a flow rate of 2 vvm. Samples with a volume of 15–25 mL were taken daily during 4 days of fermentation. An antifoam Y-30 emulsion (Sigma, St. Louis, MO, USA) solution of 10% was made and added at the beginning of each batch cultivation, as well as during the fermentation, when required.

Genetic Techniques

Standard molecular manipulation techniques were used to develop the vectors [57]. All plasmids and their functions are presented in Table 2.

Table 2. Vectors used in this study.

Vector Names	Map	Features
Cre (CR)	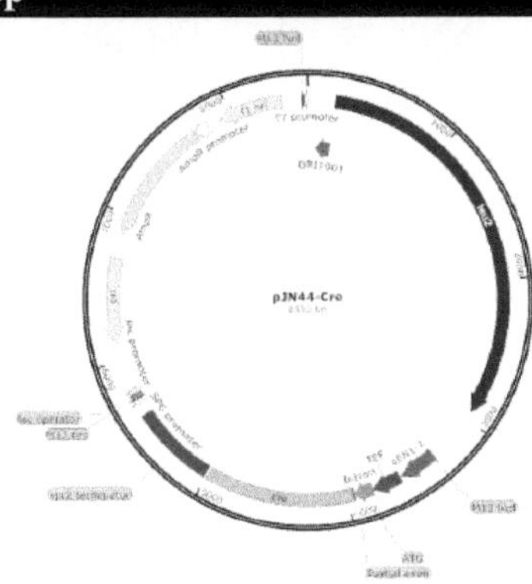	Cre recombinase flanked by TEFin promoter and Xpr2 terminator
pGR12 YlALK5 (5)	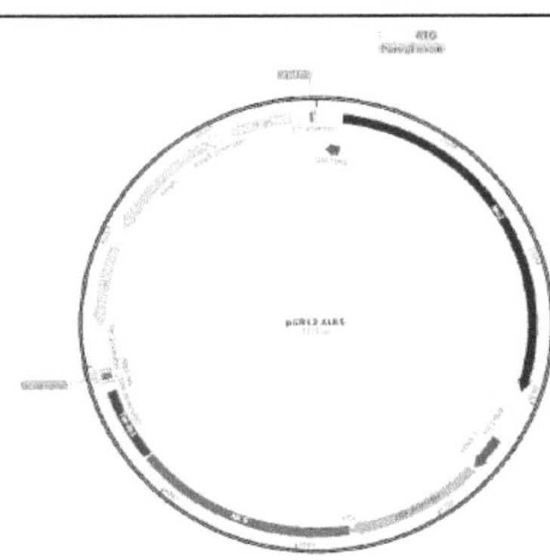	Centromeric (CEN) replicative vector, low-copy CEN plasmids (1–2 copies/cell˜1.6 plasmid copies/cell), YlALK5 controlled by PFBA-Tlip1, leucine selection marker

pGR51 YlCPR (C)	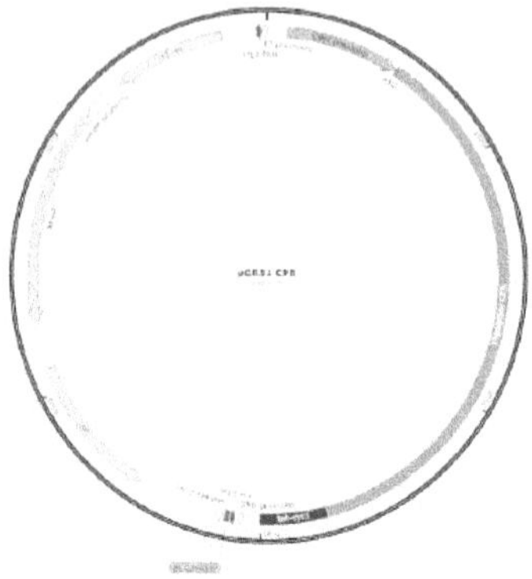	Centromeric (CEN) replicative vector, YlCPR controlled by PGPM-Tcyc1
pJN44 YlFAO (F)		Centromeric (CEN) replicative vector, YlFAO1 controlled by PTEFin-Txpr2
ALK5 CPR FAO (5CF)	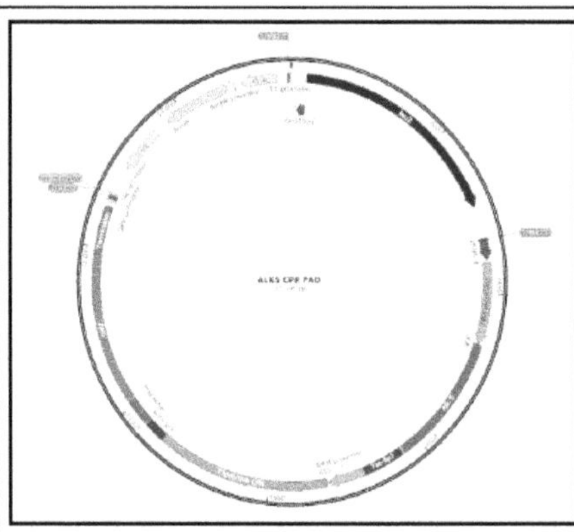	CEN replicative vector, PFBA-YlALK5 Tlip1 PGPM-YlCPR Tcyc1 PTEFin-YlFAO1 Txpr2
Leu 5cf (L5CF)	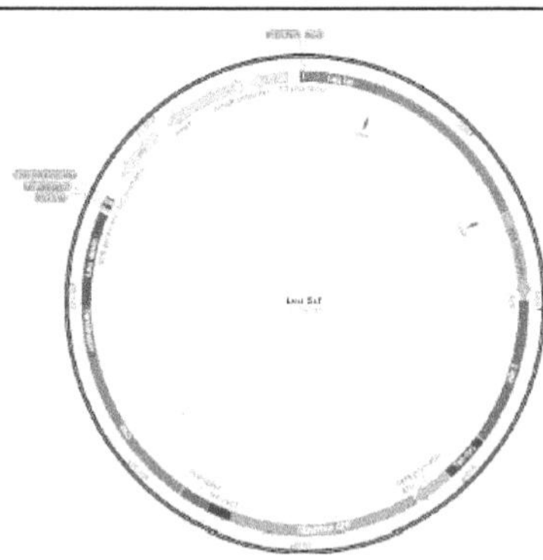	Uracil selection marker flanked by LEU2 upstream and PFBA-YlALK5 Tlip1 PGPM-YlCPR Tcyc1 PTEFin-YlFAO1 Txpr2 LEU2 downstream sequences

pINA1291 ALK5 (Z5)	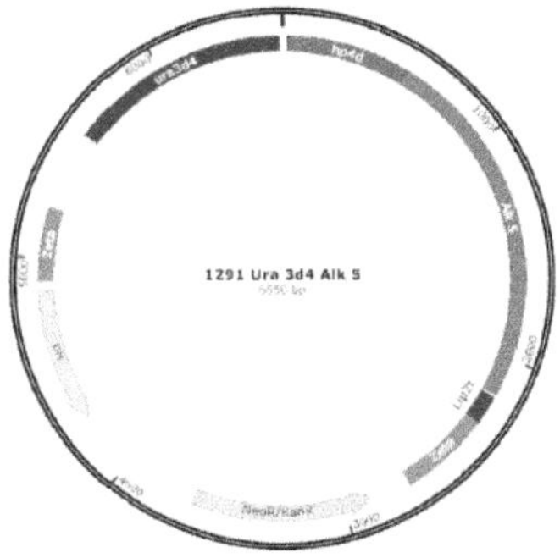	Integrative vector, hp4d-YlALK5 Tlip2, ura3d4 defective marker, zeta region for multi-copy integration
LEU (LU)	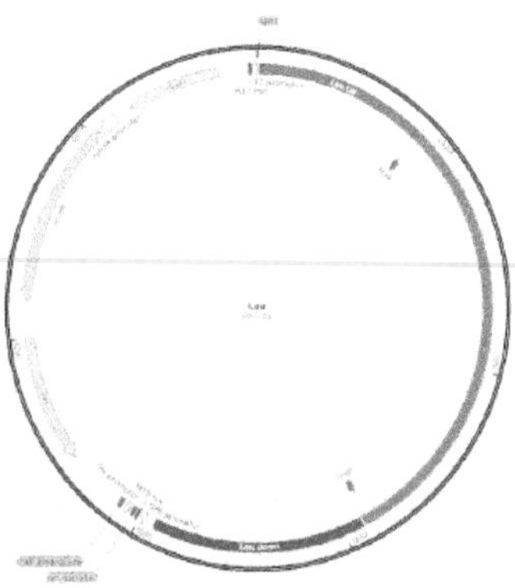	Uracil selection marker flanked by LEU2 upstream and downstream sequences
SNF (SU)	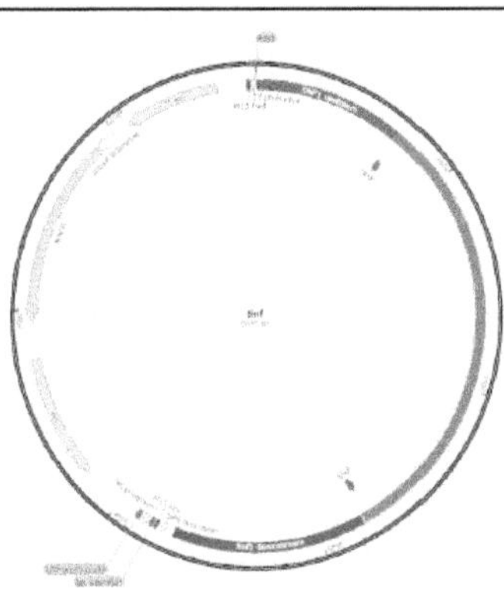	Uracil selection marker flanked by SNF1 upstream and downstream sequences
FAA (FU)	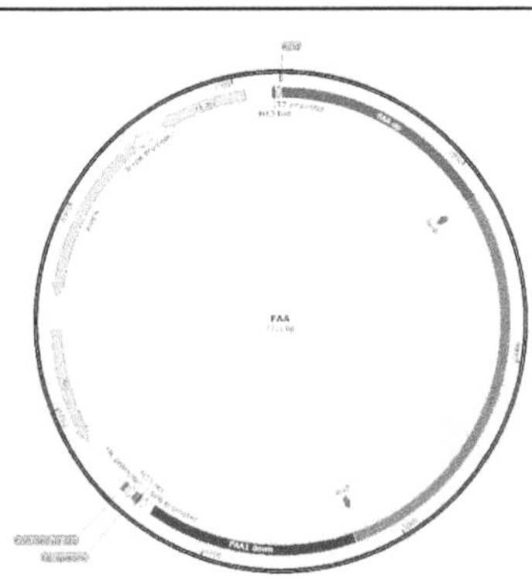	Uracil selection marker flanked by FAA1 upstream and downstream sequences

The construction of the triple gene expression cassette was carried out by the amplification of *Yarrowia lipolytica* cytochrome P450 *YlALK5*, NADPH-P450 reductase *YlCPR*, and fatty alcohol oxidase *FAO1* gene segments. Cloning inserts were obtained from polymerase chain reaction (PCR) using Q5 high fidelity DNA polymerase (New England Biolabs, Ipswich, MA, USA), the gDNA of Po1f (ATCC MYA-2613) as the template, along with the primers presented in Table 3. The amplicons were individually digested and inserted into *Y. lipolytica* plasmid pGR12 (PFBA-Tlip1), pJN44 (PTEFin-Txpr2), and pGR51 (PGPM1-Tcyc1). These are the constitutive promoters with the following orders of strength: FBA1 > GPM1~TEF [59]. The segments of PGPM-*YlCPR* Tcyc1 and PTEFin-*FAO1*-Txpr2 were separated from the non-shuttle vectors by digestion with XbaI and SpeI. For a more distinguishable separation of the plasmid digestion products on the gel, the unique sites of XmnI and ScaI were also used in a triple digestion process. The target gel recovered fragments were consecutively inserted into the SpeI-digested and Fast Alkaline Phosphatase treated *ALK5*–pGR12 plasmid. Gene knock-out was carried out using the uracil maker containing vectors. The marker was surrounded by LoxP sites. For knock-out plasmid constructions, the 0.6–1.1 kb 5'- and 3'-flanking regions of the *Y. lipolytica LEU2, SNF1*, and *FAA1* genes were amplified with the primers listed in Table 3. The amplicons were inserted into the sites upstream and downstream of uracil gene upon digestion and purification. The triple gene expression cassette segment was double digested using XbaI and SpeI and purified on gel for subsequent insertion into the SpeI-digested and Fast Alkaline Phosphatase treated *LEU2* knock-out vector. The triple gene cassette was successfully integrated into the genomic *LEU2* locus, in a one step site-specific gene knock in/out. This was carried out to reach long-term genetic stability. *YlALK5* gene segment was also inserted into a vector under the control of the growth phase-dependent hp4d promoter. This promoter shows high quasi-constitutive activity [60] at the start of the stationary phase, and is relatively independent of medium composition and pH [61]. Application of the hp4d growth dependent promoter lessens the metabolic burden of *YlALK5* over expression during the growth phase. The corresponding vector contains zeta sequences (Ylt1 retrotransposon LTRs), enabling its multiple random integration into the genome of strains devoid of Ylt1 retrotransposon, such as the P strain. This zeta-based vector bears the defective marker, ura3d4 allele, thus allowing a random ectopic integration of about ten copies into all H222 derivatives, which are devoid of Ylt1 retrotransposon [62].

Table 3. PCR primers used in this study.

No.	Name	Sequence (5ˡ →3ˡ, Underlined Restriction Site)
1	*Alk* 5 F HindIII	GAGCGAAAGCTTATGCTACAACTCTTTGGCGTCC
2	*Alk* 5 R PstI	CTTAGA CTGCAG CTACGCCTTCTCACCCTTATACA
3	*Alk* 5 Zeta F A	AATGCTACAACTCTTTGGCGTCC
4	Alk 5 Zeta KpnI R	TTGCAAGGTACCCTACGCCTTCTCACCCTTATACATCT
5	*YlCPR* F HindIII	GAGCGAAAGCTT ATGGCTCTACTCGAC TCTC
6	*YlCPR* R SmaI	GTTAT CCCGGG CTACCACACATCTTCCTGG
7	*FAO1*F NdeI	CCTCA CATAT GATGTCTGACGACAAGCACACT
8	*FAO1*R SmaI	GTTAT CCCGGG AGGATCTCCGACCTCGAATC
9	*LEU2* up F ApaI	CTATAGGGCCC ACCGGCAAGATCTCGTTAAGACAC
10	*LEU2* up R XbaI	GATCCTCTAGATGTGTGTGGTTGTATGTGTGATGTGG
11	*LEU2* down F SpeI	CTGGACTAGTCTCTATAAAAAGGGCCCAGCCCTG
12	*LEU2* down R NdeI	CCTCACATAT G GACAGCCTTGACAACTTGGTTGTTG
13	*LEU2* F Ura	TACAGTTGTAACTATGGTGCTTATCTGGG
14	*LEU2* Ura R	CCTTGGGAACCACCACCGT
15	*LEU2* Ura F	ACTTCCTGGAGGCAGAAGAACTT
16	*LEU2* R Ura	ATAGCAAATTTAGTCGTCGAGAAAGGGTC
17	*SNF1* up F ApaI	CAATTGGGCCCGTGATCAAAGCATGAGATACTGTCAAGG
18	*SNF1* up R XbaI	GATCCTCTAGAGAGGTGGTGGAAGGAGTGGTATGTAGTC
19	*SNF1* down F SpeI	CTGGACTAGT TCATTAATACGTTTCCCTGGTG
20	*SNF1* down R NdeI	CCTCACATAT GGGAATTCGTGCAGAAGAACA
21	*SNF1* F Ura	GCGGGAAATCAAGATTGAGA
22	*SNF1*Ura R	CGGTCCATTTCTCACCAACT
23	*SNF1* Ura F	CCTGGAGGCAGAAGAACTTG
24	*SNF1* R Ura	ACTACTGGCGGACTTTGTGG
25	*FAA1* up F ApaI	CAATT GGGCCC CCAGGTCTCAGTTGCACTTGC
26	*FAA1* up R XbaI	GATCC TCTAGA CAAATTATACCCCTCATCTCTCTAGGACA
27	*FAA1* down F SpeI	CT GGACTAGTTTGGTGAGCCCACCGC
28	*FAA1* down R NdeI	CCTCACATAT GAACCTCCAGCAGACTAACTAGAACA
29	*FAA1* F Ura	ACTGTAGCTAGATGGGTGCC
30	*FAA1* Ura R	CGGTCCATTTCTCACCAACT
31	*FAA1* Ura F	CCTGGAGGCAGAAGAACTTG
32	*FAA1* R Ura	ACCAGCCCAGCCGG
33	*LEU2* up F Ura	TCATGTTCGTGGAGGGGAG
34	*LEU2* up Ura R	AAAACGCAGCTGTCAGACC
35	*LEU2* down F FAO	GGAGTCCAAGCCCTTCGA
36	*LEU2* down R	CACAAGACGTCAACTAAAGCGT

Site-specific gene knock in/out was achieved by transforming the linearized vectors containing homologous upstream and downstream sequences. *Y. lipolytica* was transformed with the linearized plasmids, consisting of NdeI-digested LU, ApaI-digested SU, NdeI-digested FU, and NsiI-digested L5CF. The zeta-containing vector was linearized using FD NotI enzyme. Transformation was performed using a Zymogen Frozen EZ yeast transformation kit II (Zymo Research, Irvine, CA, USA) according to the manufacturer's protocol. We modified this protocol to enhance the efficiency of multi-copy integration transformation. The *loxR-URA3-loxP* modules were rescued for subsequent genetic modification by the LoxP-Cre system as previously reported [63]. Gene deletions and expression cassette insertion were confirmed by PCR using the primers listed in Table 3.

Construction of vectors was carried out using standard cloning methods, with FastDigest restriction enzymes (Thermo Fisher Scientific, Waltham, MA, USA). Yeast genomic DNA was isolated as described previously [64] and was used for PCR amplification and verification. A DNA Clean &Concentrator-5 Kit (Zymo Research, Irvine, CA, USA) was used for the purification of DNA products from PCR and digestion reactions. A GeneJET Gel Extraction Kit (Thermo Fisher Scientific) was used for the recovery of nucleic acid fragments from agarose gels.

ANALYTICAL METHODS

Dry Biomass

Seven milliliters of culture broth were collected daily for dry cell weight, extracellular metabolites, fatty acids, and LCDCAs measurements. Five milliliters were centrifuged for 5 min at 13,300 rpm. The cell pellet underwent two stages of washing with a saline solution (0.9% NaCl) and distilled water. The biomass yield was measured gravimetrically by drying samples at 105 C to reach a consistent weight. This was presented in grams of dry cell weight per liter (g·DCW/L).

Glycerol and Citric Acid Concentrations

High performance liquid chromatography (HPLC) with an Aminex HPX-87H column was used to measure the concentration of glycerol and citric acid. After centrifugation, supernatants were filtered using 0.22 µm pore-size membranes and eluted with 5 mM H2SO4 at a flow rate of 0.6 mL/min, at 65°C. Refractive index (RI) and UV (210 nm) detectors were employed for the detection of target signals. Standards were used for the identification and quantification of glycerol and citric acid. Our methods did not adequately separate citric acid from its isomer, iso-citric acid. Thus, the reported concentration corresponds to the sum of these isomers, expressed as citric acid.

Qualitative and Quantitative Analysis of Fatty Acids and LCDCAs

Fatty acid methyl esters (FAME) and dicarboxylic dimethyl esters were made in hexane after lipid extraction and transesterification, in compliance with the previously described method [65]. Analysis of esterified compounds was conducted using an Agilent 7890A gas chromatography instrument coupled with a flame-ionization detector (FID) and a FAMEWAX column (30 m × 320 μm × 0.25 μm) (Restek Corporation, Bellefonte, PA, USA). The injection temperature and volume were set at 250°C and 1 μL, respectively. The injection was performed with a split mode (ratio 20:1). The initial temperature of oven was 190°C and progressed to 240°C at a rate of 5°C/min. The final temperature was maintained for 20 min. The FID temperature was 250°C. FAME standards were used to identify the fatty acid peaks in the chromatograms. Quantification of fatty acids and LCDCAs were conducted by comparison with tridecanoic acid (C13:0) (0.5 mg/mL) (Sigma-Aldrich, St. Louis, MO, USA) as internal standard and by the calculation of response factor. The supernatant was filtered and used for possible extracellular metabolite extraction and quantification. Results show that the GC-FID method did not sufficiently separate C16 and C16:1 fatty acids, C18 and C18:1 fatty acids, LCDCA16 and LCDCA16:1, and LCDCA 18 and LCDCA 18:1. Therefore, the reported concentration represents the sum of these acid pairs, expressed as total fatty acid 16, total fatty acid 18, total LCDCA 16, and total LCDCA 18. LCDCA18:2 was also quantified and added to the category of LCDCA 18. An Agilent GC-MS was also used for the qualitative analysis of other dicarboxylic acids (DCAs), DCA12 and DCA14. The initial temperature of 170°C with a holding time of 1 min, was increased to 250°C at a rate of 3°C/min, and held at the final temperature for 20 min. The split ratio was 30:1.

RESULTS

Screening Process

We developed several single, double, and triple gene expression vectors with different promoters and terminators carrying homologous P450 monoxygenase (*YlAlk 2, 3, 5, 7,* and *10*), P450 reductase *YlCPR*, fatty alcohol oxidase *YlFAO1*, and fatty aldehyde dehydrogenase *YlHFD1* gene segments. We tested these vectors for their performance in the ω-oxidation of endogenous fatty acids tothe corresponding LCDCA monomers. YlAlk gene products catalyze the rate-limiting step of the ω-oxidation pathway. Therefore, it is important to use strong FBA promoter for their over expression when replicative vectors with low-copy numbers are used. Among YlALK 2, 3, 5, 7 and 10, we noted that *YlALK3*, *YlALK* 5, and *YlALK* 7 each have an effect on the production of LCDCA-16 with the following order of strength: YlALK5 > YlALK7 > YlALK3. We also observed that co-expression of *YlALK5* and *YlCPR* led to the generation of ω-hydroxy fatty acids of 16 carbons in the F strain.

We selected *YlALK5* for single as well as multiple integration because it exhibited the highest activity toward the ω-oxidation of accumulated fatty acids among the tested monoxygenases. We constructed the F strain by deleting *YlLEU2*, *YlSNF1*, and *YlFAA1* in the HP-U (*POX*-deleted) strain. Then, the P strain was constructed by performing the site specific single-copy integration of the *YlALK5*, *YlCPR*, and *YlFAO1* expression cassette into the locus of the *LEU2* gene. It is notable that this single-copy integration improved the growth behaviour of the resulting strain over the F strain in the minimal YNB-Ura media. Then, the M strain was constructed by random multiple integration of *YlALK5* into the P strain. Ten to twenty colonies were screened for their growth on selective YNB-Ura plate and LCDCA production. All of the F, P, and M strains were uracil prototrophic and leucine auxotrophic. The YNB-Ura culture of the M strain was additionally supplemented with an additional 400 mg/L leucine to study the possible stimulatory effect of this amino acid on the metabolite production by the M strain. This was labelled as the ML strain/treatment. Preliminary shake flask studies were conducted to find the best strain for a fermentation scale-up. However the production of LCDCAs in shake flask cultures is generally challenging due to insufficient aeration, agitation, and poor control over pH. The next section presents the data from the 6-day shake flask cultivations.

Preliminary Shake Flask Experiments for Fatty Acid and LCDCA Production

We tested four strains (wild type W and engineered F, P, and M) for fatty acids and monomer production in shake flask cultures. Glycerol was added to YNB-Ura at a concentration of 52 g/L to reach a C/N ratio of 50. This medium contained about 0.38 g/L of leucine and was supplemented with an additional 400 mg/L of leucine for the ML strain/treatment. Results of biomass production in the 6-day shake flask culture are summarized in Figure 1.

The data present a significant ($p < 0.05$) difference between the wild type (W) strain and all mutant strains. This can be explained in part by the final pH of media, which can affect the $CaCO_3$ solubility. Although all shake flask media contained an equal amount of calcium carbonate (6 g/L), their final pH varied depending on the strain. For example, the final pH of the W strain was above 4.1, while it was in the range of 3.2 to 3.9 for other strains because of a higher citric acid production. The pH of media can affect $CaCO_3$ solubility and this interferes with dry biomass measurements. In addition to this explanation, we also noted that glycerol was almost completely consumed by the end of the 6-day period for all mutant strains. However, we observed a residual glycerol level of 15 g/L for the W strain. The integration of the fatty acid-oxidizing genes, did not significantly ($P > 0.05$) enhance the final biomass level under nitrogen limiting conditions. However, the supplementation of the YNB-Ura medium with an additional leucine caused the M strain to produce a significantly higher biomass level than F and P strains.

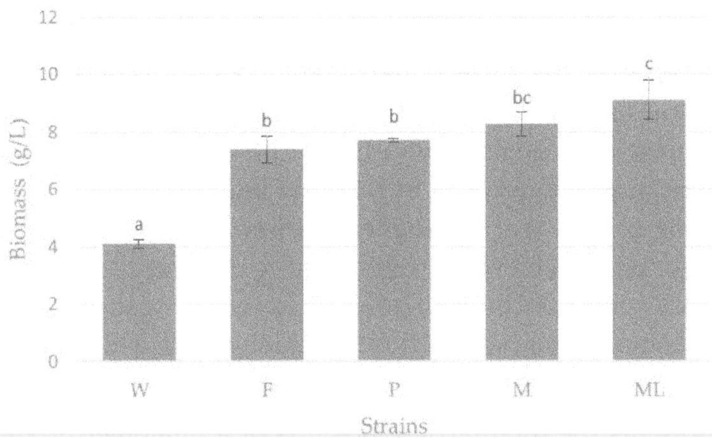

Figure 1. Study of 6-day shake flask cultures for biomass production by the W, F, P, M and ML strains/treatment. Glycerol based media under nitrogen-limiting conditions (C/N = 50). Error bars represent standard deviation for n = 3. W: H222 wild type strain, F: Δ*POX1-6* Δ*LEU2* Δ*SNF1* Δ*FAA1* Δ*SNF1::URA3*, P: F + Δ *LEU2:: URA3 YlALK5 YlCPR YlFAO1*, M: P + *YlALK5:: URA3*,ML: M + 400 mg/L leucine. abc Columns with dissimilar letters at the top are significantly different ($p < 0.05$).

We also assessed the capability of the strains for citric acid production. Our methods could not distinguish between citric acid and isocitric acid, so the reported data in Figure 2 is the sum of these two isomers.

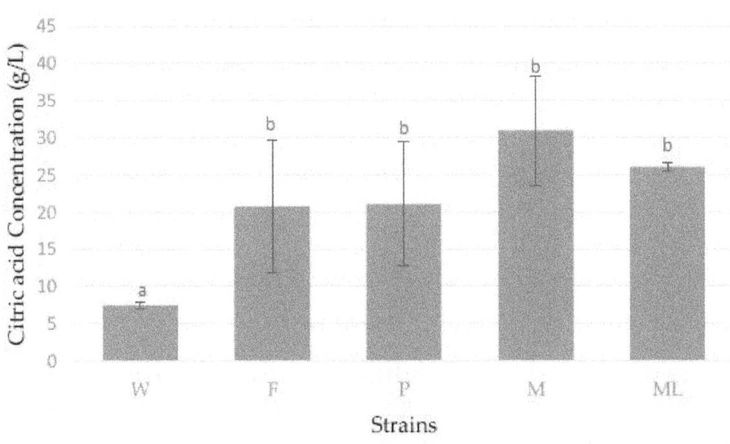

Figure 2. Study of 6-day shake flask cultures for citric acid production by the H, F, P, M and ML strains/treatment. Glycerol based media under nitrogen-limiting conditions (C/N = 50). Error bars represent a standard deviation for n = 3. W: H222 wild type strain, F: Δ*POX1-6* Δ*LEU2* Δ*SNF1* Δ*FAA1* Δ*SNF1::URA3*, P:F+ Δ *LEU2:: URA3 YlALK5 YlCPR YlFAO1*, M:P+ *YlALK5:: URA3*, ML: M + 400 mg/Lleucine. ab Columns with dissimilar letters at the top are significantly different ($p < 0.05$).

All mutant strains produced a significantly ($p < 0.05$) higher level of citric acid than the W strain, in the range of 21 to 31 g/L. A small amount of the detected acid can be in the form of its isomer, iso-citric acid because this yeast is able to produce citric and iso-citric acid and their proportion is mainly influenced by feedstocks and strains [35]. Although the M strain produced the highest level of citric acid, there was no significant ($p > 0.05$) difference among the mutant strains. Moreover, the leucine supplementation did not have a significant effect on citric acid production. Citric acid is the major by-product of the lipid biosynthesis in *Y. lipolytica* under nitrogen-limiting conditions [66]. Higher re-direction of carbon flux from citric acid to fatty acid can be achieved by upregulating the first committed step of fatty acid biosynthesis, controlled by *ACC1* [67], and by removing the feedback inhibition effect of endogenous fatty acyl-CoA through thioesterase over expression or acyl-CoA synthetase inactivation [49].

Figure 3 presents the total fatty acid and LCDCA monomer production by the strains from 6-day shake flask cultivation.

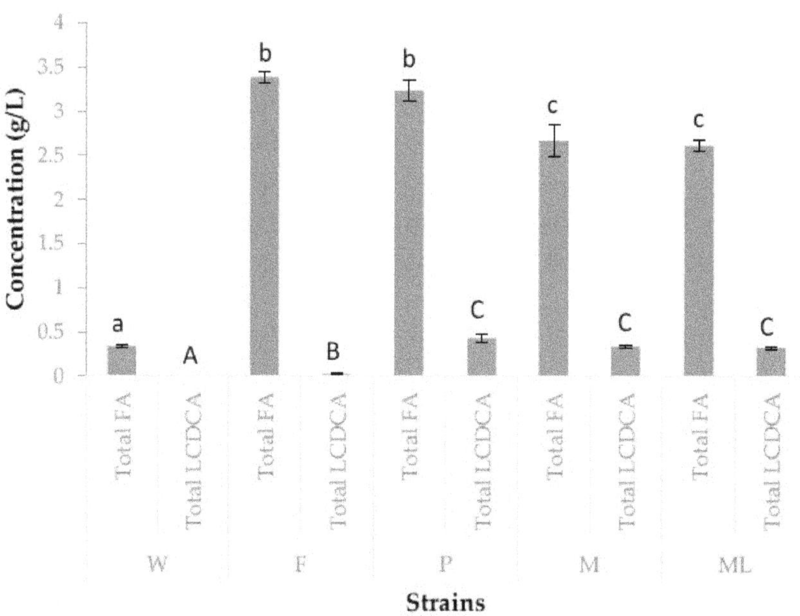

Figure 3. Study of 6-day shake flask cultures for fatty acid and LCDCA production by the H, F, P, M, and ML strains/treatment. Glycerol based media under nitrogen-limiting conditions (C/N = 50). Error bars a represent standard deviation of n = 3. W: H222 wild type strain, F: Δ*POX1-6* Δ*LEU2* Δ*SNF1* Δ*FAA1* Δ*SNF1::URA3*, P: F + Δ *LEU2:: URA3 YlALK5 YlCPR YlFAO1*, M: P + *YlALK5:: URA3*,ML: M + 400 mg/L leucine. Total FA: total fatty acids of 16 and 18 carbons, Total LCDCA: total long chain dicarboxylic acids of 16 and 18 carbons. ABC/abc Columns with dissimilar letters at the top are significantly different ($p < 0.05$).

The wild-type W strain accumulated fatty acids at a low concentration of 340 mg/L with oleic acid (FA18:1) as the main constituent. The concentration of total fatty acids in the mutant strains was significantly ($p < 0.05$) higher than in the W strain. The deletion of *YlFAA1* in the F strain significantly ($p < 0.05$) increased the TFA level to 3300 mg/L in the F strain. These endogenous fatty acids were precursors for the LCDCA production. The LCDCAs were not produced by the W strain. The production of these monomers was initiated by the deletion of *YlFAA1* in the F strain. This shows the involvement of background ω-oxidation activity in LCDCA-16 and 18 production. This contribution was higher toward the end of 6-day shake flask cultivations. Single and multiple copy integration of the fatty acid-oxidizing genes significantly ($p < 0.05$) increased LCDCA production. This LCDCA level was similar for the P, M, and ML strains/treatment. Generally, LCDCA production in shake flasks is not stable and faces several challenges, including inadequate aeration, agitation, and pH control. Oxygen limitations are typical for shake flask cultures. Moreover, addition of 6 g/L $CaCO_3$ cannot sufficiently maintain the stability of pH value in shake flask cultures. These limitations undermine the effect of genetic or cultivation modification on LCDCA production. For instance, we observed the positive effect of leucine supplementation on biomass and LCDCA production by the M strain in bioreactor fermentations. However, application of similar treatment did not significantly change the LCDCA production level of the M strain in shake flask cultivations. Additionally, an extensive post-screening for choosing the best LCDCA-producing strain is beneficial to reach the strongest effect of multiple-integration. The preliminary shake flask studies were carried out, nonetheless, to select the appropriate strain and conditions for bioreactor fermentation. We obtained a LCDCA titer of 330 to 430 mg/L from the P, M and ML strains/treatment under nitrogen-limiting conditions. Chromatograms from the F and P culture indicate that a negligible amount of FA-17 was also produced by these strains.

In general, the deletion of *POX* genes prevents the degradation of fatty acid precursors and monomers. Additionally, the inactivation of *YlFAA1* promotes the generation of free fatty acids, which do not have the feedback inhibitory effect of acyl-CoA on the fatty acid biosynthesis pathway. The combination of these genetic modifications with the deletion of *YlSNF1* can promote the re-direction of the carbon flux towards malonyl-CoA and eventually free fatty acid accumulation. In addition to background ω-oxidation activity, the upregulation of the ω-oxidation pathway can enhance the transformation of free fatty acid pool into the monomers of less toxicity. Therefore, our genetic engineering served the dual purpose of enhancing fatty acid accumulation and producing LCDCA monomers. The ML strain and treatment was selected for further bioreactor fermentation experiments.

Batch Fermentation

The batch bioreactor fermentation was carried out using YNB-Ura medium with a C/N of 50. Glycerol was added to the medium at a concentration of 52 g/L. The production medium was also supplemented with 400 mg/L leucine. The time course study of biomass and citric acid production as well as glycerol utilization during the 4-day fermentation is presented in Figure 4.

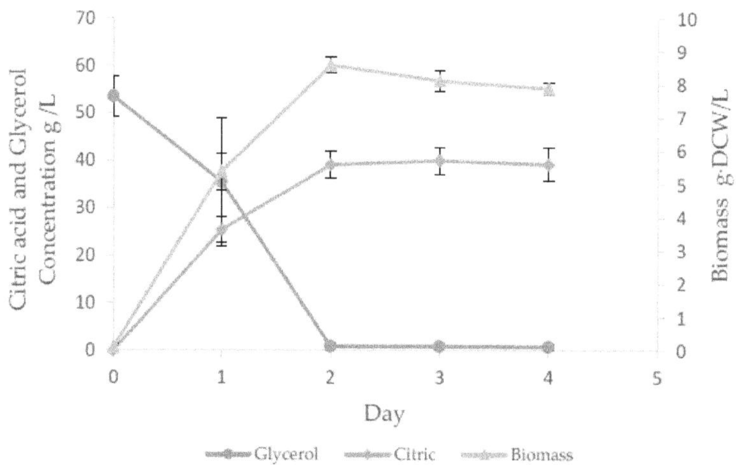

Figure 4. Time course study of glycerol utilization, biomass proliferation, and citric acid production by the ML strain/treatment. Glycerol based YNB-Ura media under nitrogen-limiting conditions (C/N = 50). Data from 2 rounds of batch fermentation. ***ML: ΔPOX1-6 ΔLEU2 ΔSNF1 ΔFAA1*** **+** ***Δ LEU2:: YlALK5 YlCPR YlFAO1*** **+** ***YlALK5:: URA3*** **+ 400 mg/L leucine.**

Glycerol was consumed rapidly within 2 days of fermentation. During this period, the citric acid and biomass reached their peak at about 39 and 8.6 g/L, respectively. Afterwards, there was a slight reduction in the concentration of biomass, possibly due to some cell attachment to the fermenter vessel. The ML strain and treatment produced citric acid at a yield of more than 0.75 g/g of consumed glycerol. During the last two days of fermentation, the citric acid concentration did not undergo a marked change. It is notable that the mutant strains of this study are derived from H222 stain, which has a strong potential for citric acid production. Additionally, nitrogen-limiting conditions reroute carbon flux to lipid accumulation for storage and to citric acid for secretion. Both citric acid and fatty acids compete for the building units of acetyl-CoA [68]. To gain insight into monomers production conditions, we analyzed the concentration of fatty acids and monomers during the 4-day fermentation. Results are shown in Figure 5.

Batch fermentation in the 1-L bioreactor significantly enhanced the titer of LCDCA from 0.33 g/L, which was obtained in the shake flask culture, to 3.49 g/L for the ML strain and treatment. This corresponds to a minimum yield of 0.067

g/g of consumed glycerol and to a productivity of 0.04 g/L·h. This significant improvement was mainly due to adequate aeration, agitation, and precise control over the pH. The increase in LCDCA level continued until the last day of the 4-day fermentation at the expense of fatty acid pool.

The LCDCAs are more soluble than their fatty acid counterparts. This feature facilitates their transportation across the cell. We also analyzed the concentration of extracellular LCDCA monomers in cell-free production media to find more about the spatial distribution of LCDCAs. Figure 6 presents the results from this analysis.

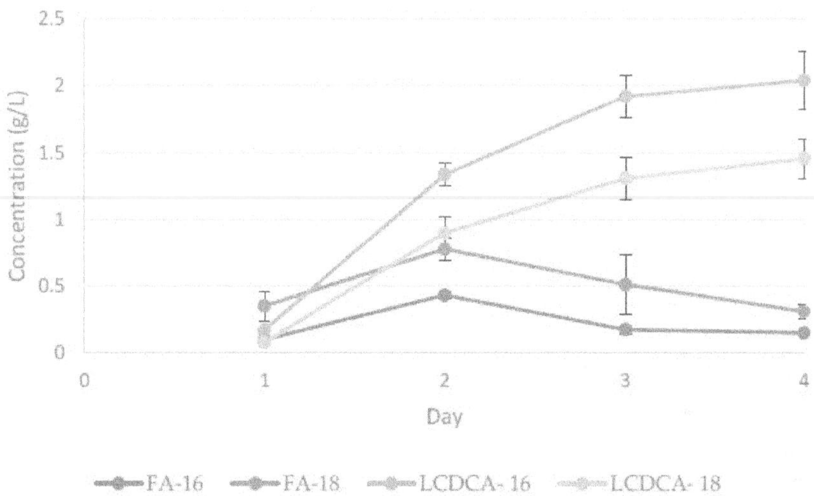

Figure 5. Time course study of fatty acids and LCDCA production by the ML strain/ treatment. Glycerol based media under nitrogen-limiting conditions (C/N = 50). ML: Δ*POX1-6* Δ*LEU2* Δ*SNF1* Δ*FAA1* Δ*LEU2:: YlALK5 YlCPR YlFAO1* + *YlALK5:: URA3* + 400 mg/L leucine. FA-16: fatty acid of 16 carbons, FA-18: fatty acid of 18 carbons, LCDCA-16: long chain dicarboxylic acids of 16 carbons, LCDCA-18: long chain dicarboxylic acids of 18 carbons. Data from two rounds of batch fermentation.

According to these results, more than half of the LCDCAs were present in the extracellular medium while the fatty acids were intra-cellularly used as precursors for the monomer production. Figure 6 also indicates that the availability of intracellular free fatty acids is the rate limiting step for *de novo* LCDCA production in the 1-L bioreactor fermentation. It should be noted that a small amount of dicarboxylic acids of 12 and 14 carbons was also found in the extracellular medium (less than 0.2 g/L). This can be due to the activity of *YlALK5* that was integrated into the genomes at multiple copies. The composition of metabolites was also determined. Results are presented in Figure 7.

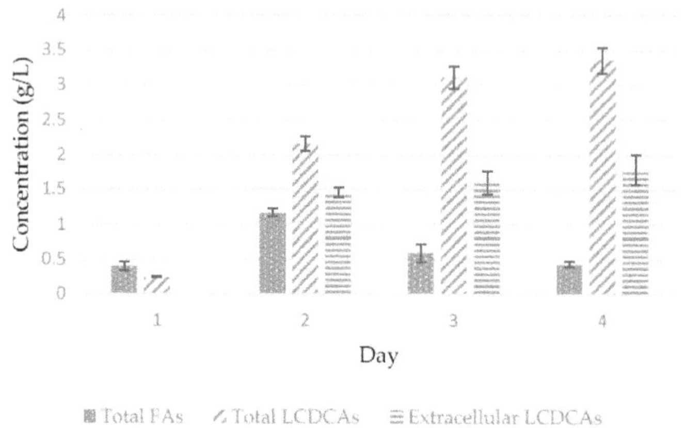

Figure 6. Time course study of total fatty acid, total LCDCA, and extracellular LCDCA production by the ML strains/treatment. Glycerol based media under nitrogen-limiting conditions (C/N = 50). Data from 2 rounds of batch fermentation. ML: *ΔPOX1-6 ΔLEU2 ΔSNF1 ΔFAA1 ΔLEU2::YlALK5 YlCPR YlFAO1 +YlALK5:: URA3* + 400 mg/L leucine. Total FAs: fatty acids of 16 and 18 carbons, Total LCDCAs: long chain dicarboxylic acids of 16 and 18 carbons. Data from 2 rounds of batch fermentation.

Figure 7 shows that LCDCAs became predominant fatty acid-based metabolites at the end of the 4-day fermentation. Also, LCDCA-16 accounted for more than 50% of LCDCAs, meaning that the ω-oxidation rate of 16 carbon-fatty acid was higher than that of 18 carbon-fatty acids. This rate is influenced by the background ω-oxidation activity, as well as the activity of those genes that have been over expressed. Table 4 presents the summary of data from biomass, citric acid, and LCDCAs production.

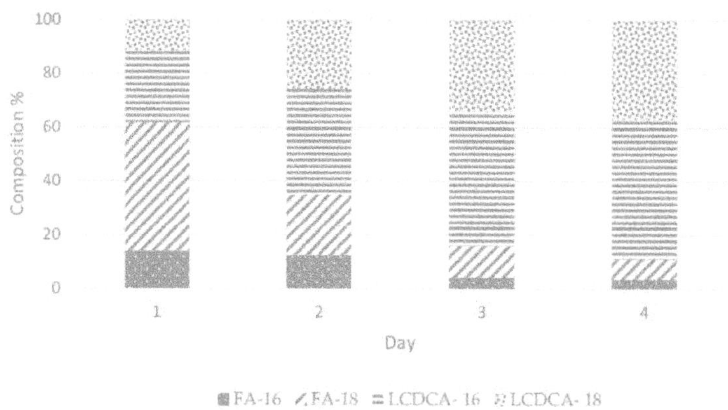

Figure 7. Percentage composition of fatty acids and LCDCAs produced by the ML strain/treatment. Glycerol based media under nitrogen-limiting conditions (C/N = 50). ML: *ΔPOX1-6 ΔLEU2 ΔSNF1 ΔFAA1 ΔLEU2::YlALK5 YlCPR YlFAO1 +YlALK5:: URA3* + 400 mg/L leucine. FA-16: fatty acid of 16 carbons, FA-18: fatty acid of 18 carbons, LCDCA-16: long chain dicarboxylic acids of 16 carbons, LCDCA-18: long chain dicarboxylic acids of 18 carbons. Data from 2 rounds of batch fermentation.

Table 4. Summary of the production level for biomass, citric acid, and LCDCAs.

DCW (g/L)	Citric Acid (g/L)	LCDCA Titer (g/L)	LCDCA Yield (g/g)	LCDCA Productivity (g/L·h)
8.58 ± 0.23	39.2 ± 3.5	3.49 ± 0.14	0.06	0.04

DISCUSSION

The *de novo* production of fatty acid-derived monomeric compounds from unrelated feedstock has been previously demonstrated. For example, the over expression of a plant thioesterase and of an engineered self-sufficient monooxygenase fusion protein in a β-oxidation deficient *E. coli* strain resulted in 234 μM ω-hydroxy octanoic acid production in a fed-batch fermentation [69]. In another study, *E. coli* was engineered for sugar-based hydroxyl fatty acid production [70]. That was achieved by the co-expression of acetyl-CoA carboxylase and acyl-CoA thioesterase in an inactivated acyl-CoA synthetase background, followed by the expression of a heterologous fatty acid hydroxylase. This resulted in the production of different hydroxyl fatty acids with chain lengths of 10 to 18 carbons at a final titer of 58.7 and 548 mg/L in culture broth and fed-batch fermentation, respectively. The engineered reversal of the ω-oxidation cycle, in combination with the expression of heterologousthioesterase and monooxygenase systems, has led to the biosynthesis of medium chain ω-hydroxy fatty acids and dicarboxylic acids at a titer of 0.8 and 0.5 g/L, respectively, from glycerol in *E. coli* [71]. This bacterial host has also been engineered for the biosynthesis of long chain hydroxylated fatty acids at a titer of 117 mg/L [72]. In that study, both thioesterase and fatty acid hydroxylase *P450BM3* were involved in the production of free fatty acids and in their conversion into oxygenated monomers. Bowen et al. (2015) also engineered *E. coli* for sugar-based C12 and C14 ω-hydroxy FAs and α,ω-DCAs production [6]. This was achieved by expressing high-specificity acyl ACP thioesterases (TEs) to detach free fatty acids from the fatty acid synthase (FAS II). The application of p450 (CYP) and alcohol/aldehyde dehydrogenases led to the generation of ω-hydroxy FAs and subsequently α,ω-DCAs at a titer and productivity of 600 mg/L and 0.026 g/L·h, respectively. Figure 8 presents a summary of aforementioned strategies.

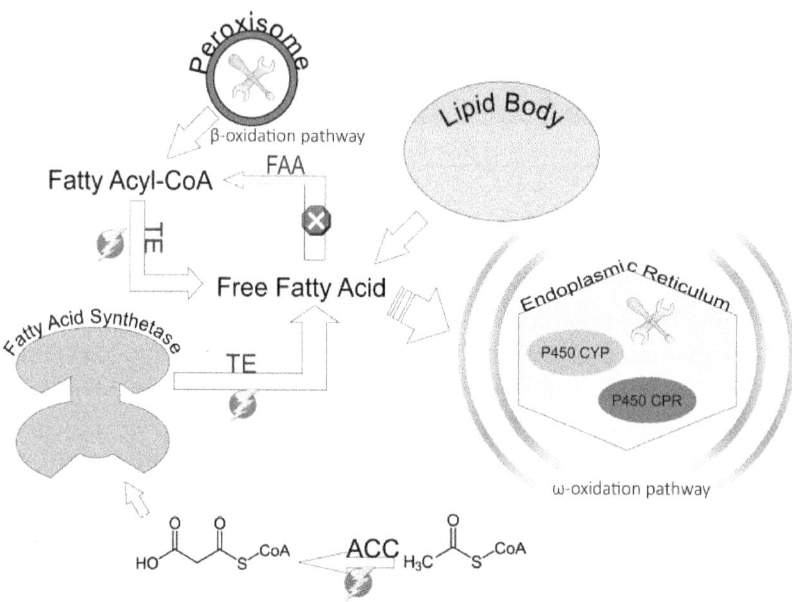

Figure 8. Genetic engineering strategies for the *de novo* production of oxygenated fatty acid monomers: overexpression of acetyl-CoA carboxylase (ACC) for higher carbon flux re-direction toward fatty acid biosynthesis pathway, overexpression of length specific thioesterases (TE) for the hydrolysis of acyl-CoA and consequently mitigation of their feedback inhibition effect, inactivation of the fatty acyl-CoA synthase (FAA) for the prevention of free fatty acid re-activation, inactivation or reversal of the β-Oxidation cycle for enhancing fatty acid pool and preventing degradation of fatty acid derivatives, and selective upregulation of the ω-oxidation pathway for the terminal oxidation of endogenous fatty acids to their oxygenated counterparts.

The above-mentioned yields and titers are notably less than those reported for whole-cell biotransformation of exogenously added long chain fatty acids to the corresponding monomers [8,73]. For instance, engineering *Y. lipolytica* for the production of LCDCA from C18 oil allowed to reach a titer of 16 g/L. This was accomplished through the blockage of β-oxidation and over expression of *ALK1,2* together with the *CPR* gene under the control of p*POX2* [46].

This study reports the construction of *Y. lipolytica* as an oleaginous yeast cell platform for the *de novo* production of LCDCA monomers from renewable hydrophilic feedstock. Our LCDCA-producing platform is independent of oily feedstock whose utilization has potential disadvantages such as toxicity, low solubility, and weaker cellular uptake [7]. Instead, this sustainability-driven biosynthetic route utilizes hydrophilic substrates for the dual purpose of energy production and LCDCA biosynthesis under nitrogen-limiting conditions.

We developed our platform through a combinatorial genetic engineering strategy in a β-oxidation deficient strain. Our strategy involved the inactivation of *SNF1* for enhancing malonyl-CoA biosynthesis, disruption of the major cytosolic acyl-CoA synthetase (*FAA1*) for generating free fatty acid precursors, and upregulation of native ω-oxidation pathway for improving oxidation of the endogenous free fatty acid precursors into the corresponding long chain monomers. This strategy was developed based on the functions of the targeted genes.

In general, it is necessary to delete the *POX* genes in order to inactivate the β-oxidation pathway as the major degradation route. This prevents the degradation of fatty acid precursors and LCDCAs [45,47]. However, complete disruption of the peroxisome creates unwanted consequences. Also, defects in peroxisome biogenesis may have an inhibitory effect on the native P450 system [23,74], and may interfere with the oxidation of various metabolites [75]. *SNF2* deletion has been found to enhance lipid accumulation [76]. Inactivation of the *SNF1* gene led to an enhanced constitutive fatty acid accumulation and citric acid production in *Y. lipolytica* [66,77]. An alternative strategy to enhance carbon flux re-direction from citric acid production to fatty acid biosynthesis is to over express *ACC1* or use hyperactive Acc1p [67,78]. This strategy can prevent the possible unintended metabolic consequences resulting from *SNF1* deletion. Inactivation of *FAA1* can interfere with the re-activation of released fatty acids that could serve in energy production, membrane maintenance, or triglyceride (TAG) homeostasis [49,79]. Deletion of the *FAA1* and *FAA4* genes in *S. cerevisiae* have been found to interfere with fatty acid incorporation in phospholipids or TAG, while generating a free fatty acid-producing phenotype [80].

In *Y. lipolytica*, the free fatty acids that are activated by YlFaa1p are partially incorporated into the lipid bodies [81]. In this yeast, several gene products that differ in their substrate specificities are involved in the activation of fatty acids in different organelles [82]. Faa1p is the major cytosolic long chain fatty acid activator [83]. This gene plays a crucial role in the utilization of both *ex novo* fatty acids and those obtained from the metabolism of *n*-alkanes [84]. *FAA1* also plays a role in the storage of fatty acids in lipid bodies [81,83]. Besides YlFat1p, which is mainly involved in fatty acid remobilization, there are at least 10 *AAL* genes involved in the activation of fatty acids in the peroxisome and lipid bodies. The inactivation of these genes can prevent the cell from storing fatty acids in lipid bodies [82]. The deficiency of fatty acid activation systems can also negatively affect yeast growth in batch cultures with glucose [85], due to either the detergent-like properties of free fatty acids or metabolic disorder [86].

We observed some residual fatty acids at the end of the 4-day batch fermentation. Since these fatty acids may occur in the esterified form in lipid bodies, they do not act as substrates for the ω-oxidation reactions. Our genetic engineering did not involve the disruption of TAG biosynthesis. This is because the generation of free fatty acids, particularly unsaturated ones, causes toxicity in the cell with deficient lipid body formation [87]. This occurs for several reasons: the destabilizing effect of fatty acid overproduction on the cellular membrane [80,88], the harmful effects of oxidative species derived from the oxidation of unsaturated fatty acids [89], and a lack of protection against excess lipid intermediates [90].

Free fatty acid accumulation or starvation can also trigger the ω-hydroxylation of fatty acids [91,92]. The resulting hydroxylated fatty acid monomers may also interfere with the TAG homeostasis [93]. We propose that ω-oxidation serves as a rescue route to detoxify free acids in the β-oxidation-disrupted strains, especially when *FAA1* is inactivated. The stress relieving effect of fatty acid oxidizing enzyme has also been observed in plants [23]. In *Y. lipolytica*, *ALK* genes are involved in the detoxification of fatty alcohol, dodecanol [36]. Thus, this background ω-oxidative activity processes free fatty acids to oxygenated monomers of higher solubility and less toxicity. This can explain the stimulatory effect of *ω*-oxidation upregulation on biomass proliferation in the F strain with deficient FAA1p activity. In fact, the natural enzymatic machinery of the host cell plays an influential role in fatty acid over-oxidation [18].

Results show that our combinatorial genetic engineering for the biosynthesis of free fatty acids and their terminal oxidation to the corresponding LCDCAs served multiple purposes. These included the reduction of free fatty acid toxicity, increase of product solubility, facilitation of product transportation across the membrane, and the creation of a pull for further fatty acid flux direction toward ω-oxidation.

We observed that free fatty acids, generated from the *FAA1* inactivation, are suitable substrates for the native *ω*-oxidation pathway. This accords with a previous study, which reported the hydroxylation of the free fatty acid pool by a fatty acid hydroxylase (CYP102A1) from *Bacillus megaterium* [70]. Free fatty acids do not have the feedback inhibition regulatory effect of fatty acyl-CoAs on some genes of lipid synthesis [48,86]. To obtain free fatty acids, the inactivation of native fatty acid activation systems or over expression of heterologous/homologous thioesterases has been suggested in the literature. For example, inactivation of *FAA1* and *FAA4* genes in *S. cerevisiae* was found to mitigate the feedback inhibition of acyl-CoA on fatty acid synthesis and undermine the β-oxidation upregulating-effect of fatty acid overproduction [86]. Over expression of thioesterases has also been employed to release the free fatty acids of specific chain lengths [6].

Studies show that *YlALK3*, *YlALK5*, and *YlALK7* exhibit ω-oxidation activity against dodecanoic acid [36,41]. However none could result in a detactable amount of products from ω-oxidation of hexadecanoic acid in a Δ*alk1-12* strain [36]. It was postulated that Alk3, 5, and 7p may prefer shorter fatty acids, while larger fatty acids are preferrably used for cell maintenance. In this study, the DCAs of 12 and 14 carbons were detected as minor products. Their production can be due to the over expression of *YlALK5*. Additionally, the over expression of this gene, using the replicative vector with the strong constitutive promoter, contributed toward LCDCA-16 and ω-hydroxy fatty acid-16 production in our preliminary shake flask cultivations. In fact, a high gene dosage is necessary to achieve sufficient P450 expression in *Y. lipolytica* [94]. This contribution was in addition to the background ω-oxidation activity in the F strain toward LCDCA-16 and 18 production. We also observed the contribution of *YlCPR* and *YlFAO1* to the multiple-step oxidation of endogenous long chain free fatty acids precursors to the monomers. In the same fashion, amplification of the P450 reductase gene was found to enhance LCDCA production from *C. tropicalis* [14]. Similarly, it was reported that the over expression of *YlFAO1* improves LCDCA production from exogenous hydrophobic substrates [42]. This accords with the contribution of *FAO1* to diacid production observed in *C. tropicalis* [95]. Various fatty aldehyde dehydrogenases have been identified in *Y. lipolytica* that are involved in the detoxification of fatty acid aldehyde intermediates by oxidizing them to dicarboxylic acids [96–98]. These native genes likely contribute to the background ω-oxidation activity.

A bioreactor is much more efficient than a shake flask for DCA production [7]. Adequate aeration, agittion, and control over pH play major roles in the P450-based LCDCA production. The inner pH of the cell, as well as aeration, both affect the induction of yeast CYP monooxygenase and CYP reductase [10]. Studies show that a higher dissolved oxygen (DO) level can remarkably improve P450 activity [99,100]. This level was kept at 80% saturation for LCDCA production from *C. ropiclais* [14]. In fact, the native P450 system utilizes molecular oxygen that is abundant in a bioreactor fermentation. In our study, fermentation scale-up to 1-L bioreactor under the DO level of at least 50% significantly enhanced the rate of free fatty acid oxidation to LCDCAs. The majority of free fatty acids were converted to LCDCA during the four days of bioreactor fermentation, reaching a titer of 2 g/L for LCDCA-16 and 1.45 g/L for LCDCA-18. Adjustment of the pH is also required for the biosynthesis of LCDCAs. During the production phase, the pH was adjusted to 8 since the LCDCAs are completely diasscoiated at this level. For instance, the diassociation constant of DCA13 is less than 5.8. A higher extracellular pH level increases the solubility of LCDCAs, improves their transportation efficiency, and consequently mitigates the corresponding product inhibition [101]. Adjustment of pH as well as metabolite production in the production media increases the demand for antifoam addition. Thus, the antfoam was also periodically added to the media.

Transport of LCDCAs out of the cell can improve P450 activity [100] and simplify recovery for an enhanced downstream process at a lower cost. Our engineered strain showed good secretion efficiency, since more than half of the synthetized monomers were found in the media [102]. In fact, DCA of 13 carbons has been found to undergo passive diffusion due to the concentration difference between the inside and outside cells [101]. At a low intracellular pH, the undissociated form of the carboxylic acid is expected, which can mitigate their transportation across the cell membrane [103]. In this study, we showed that LCDCAs are extracellular products. Secretion allows LCDCA production to exceed the lipid accumulation capacity of the cell, and promotes their recovery from media after cell harvest at a lower cost. This also can alleviate their inhibitory effect on P450-based hydroxylation reaction due to their competitive binding to P450 [104].

We supplemented the production media with 400 mg/L leucine to take the advantage of its possible stimulatory effect on the metabolite production, under nitrogen-limiting conditions. Supplementation of media with leucine had significant effect on biomass level in shake flask cutures. In addition to that, our preliminary bioreactor fermentation using YNB-Ura, without additional leucine supplementation, resulted in less biomass and LCDCA production from the M strain (Biomass up to 7 g/L and maximum LCDCA of 3 g/L). Similarly, addition of leucine was reported to increase the growth and lipid accumulation in *S. cerevisiae* devoid of active *SNF2* [105].

Although nitrogen-limiting conditions hinder growth, they induce lipid and citric acid production. The resting cell production of LCDCAs has several advantages, including the dedication of a larger proportion of supplied oxygen to metabolism and the targeted P450 reaction [106], the repeated utilization of LCDCA-exporting cells, the use of high cell concentration, particularly for the multi-copy transformants [107], as well as more strain robustness against fermentation media. In fact, resting cells consume nearly half of what they typically utilize when they have access to a nitrogen source [106]. We used this strategy for the *de novo* production of free fatty acids. This strategy also caused our engineered cell factory to produce more than 36 g/L of citric acid as the major bioproduct which was nearly stable during the 4-day bioreactor fermentation. The production of this byproduct, which is extensively used in pharmaceutical industries [108] raises more revenue and offsets the total production cost.

This *Y. lipolytica* platform can be tailored to produce LCDCA of higher specificity with respect to chain length and saturation. This yeast accumulates a higher proportion of unsaturated fatty acids, oleic and linoleic acids, when it is grown on sugar, due to the strong expression of endogenous desaturases [53]. Over expression of these native desaturase genes, including Δ-9 and Δ-12 desaturases, can increase the proportion of desaturated fatty acids in lipid bodies. Together with *FAA1* inactivation, this can result in a larger pool of unsaturated free fatty acids for higher unsaturated LCDCA production. In fact, *Y. lipolytica* predominantly accumulate fatty acids in the form of C16:0 and C18:1. These fatty acids are more susceptible to remobilization [109]. Disruption of their re-activation in the Δ*FAA1* background, enables their availability for the ω-oxidation. Our study confirmed that this oxidation does not change the configuration of double bonds. As a result, LCDCA16 and LCDCA18:1 were the predominant LCDCA products of our platform. Moreover, according to the chromatogram of GC-FID, a smaller portion of LCDCA-18 had two double bonds (less than 10%). These bonds are derived from the linolenic acid, FA 18:2, whose concentration was less than oleic acid. Regarding chain length specificity and selectivity, there is room for improvement. For example, manipulation of *YlYAS* genes [110], and selective over expression of elongase, monooxygenase, and thioesterase can be employed to modulate background activity and enhance the length-specificity of the terminal oxidation.

Y. lipolytica has several advantages as a LCDCA-producing platform. This yeast platform has its own native P450 system that catalyzes terminal oxidation of alkanes and fatty acids [36]. Moreover, this platform possesses an endogenous redox partner (cytochrome P450 reductase), has the capability for large free fatty acid accumulation [111], and can efficiently generates NADPH cofactor mainly through the pentose phosphate pathway (PPP) [112,113]. This pathway is not tightly regulated by nitrogen concentration [114]. Together with glycolysis, it can contribute significantly to the energy supply of resting cells [115]. At the stationary phase, a higher expression of cytochrome P450 monoxygenases [104] and lipid remobilization [40] are also expected. These features enable the large-scale application of engineered *Y. lipolytica* cells without the need for cofactor addition. Regeneration of NADPH is crucial in whole-cell P450-based biotransformation [116], particularly in the NADPH- and adenosine triphosphate (ATP)-consuming process of fatty acid biosynthesis and oxidation. Moreover, whole cells also provide a protected environment for the increased stability of the P450 enzymes [51]. A whole cell biocatalyst is a cheaper option than the isolated or immobilized enzyme [106]. The pathogenic yeast, *C. tropicalis*, has been commonly engineered for whole cell biotransformation of aliphatic feedstock to LCDCAs [14,25]. However, *Y. lipolytica* has safe status, and has been categorized as a GRAS microorganism by the FDA (Food and Drug Administration) [56,88]. This status, as well as its probiotic properties [117], can pave the way toward the biosynthesis of LCDCAs that meet the safety requirements of the food and pharmaceutical industries.

Despite some chemical based-production of LCDCA at large scale, the high price of the LCDCAs, and their limited commercial availability are still major constraints in expanding the spectrum of their application [25]. Therefore, researchers should use various renewable resources for the sustainable production of these multi-purpose building blocks. An alternative sustainable solution is to establish a P450-based biological approach. However, commercial application of a P450-based platform has also its own advantages and limitations. For example, the scalability of modified cytochrome P450 for multi-kilogram scale synthesis remains a challenge [118]. The majority of P450-based biocatalysts have not met a minimum space-time yield of 0.1 g/L·h in previous studies. A solution to this limitation would be the production of high-value pharmaceuticals. In this scenario, a minimum process productivity of 0.001 g/L·h can justify large-scale bioprocessing from an economic standpoint [19]. The LCDCAs have relatively high selling price, depending on their purity, and therefore are outstanding targets for white biotechnology [12]. Our study is the first report on engineering *Y. lipolytica* for the efficient *de novo* production of LCDCAs from glycerol. Results from this study showed a minimum productivity of 0.04 g/L·h, which is more than the foregoing required productivity. However, the provision of adequate agitation and oxygen transfer in larger fermenters will be challenging to maintain this minimum productivity. Our biological-based approach has several advantages over chemically-based approaches, including a broad range of renewable feedstock, milder process conditions, a more environmentally friendly process, and higher selectivity, particularly toward long chain unsaturated fatty acids [10,119].

These monomers have superior properties, promising future, and a large global market that is expected to reach USD $300.3 million by 2025 [120]. These highly demanding monomers are used in manufacturing various high-performance polymeric compounds such as nylon and other polyamides. They have a wide scope of application in the industrial manufacture of automobiles, medicines, fragrances, adhesives, and macrolide antibiotics [10]. However, pharmaceutical applications exhibit the quickest growth in terms of revenue [120].

Compared to the shorter diacids, these building blocks and their LCDCA-based polymers have superior bio-based properties such as improved biodegradability, hydrolytic stability, crystallinity, melting temperature, optical transparency, dispersion, miscibility, material toughness, and flexibility; lower moisture uptake, dielectric constant, and surface tension [5]. These long chain monomers are also the building blocks of naturally occurring polyesters [121]. Additionally, the double bonds of unsaturated monomers, which are maintained during terminal oxidation, show promise for additional functional groups and linkages. These features also allow for the production of novel unsaturated LCDCAs [54] to synthesize cross-linked polymer networks with adjustable properties. Synthesizing these unsaturated long chain monomers is difficult via chemical routes due to the reactivity and participation of the double bonds in unwanted reactions [122].

Our findings suggest that the tunable high-level expression of selected P450 genes, by high copy and stable integration or by high copy number expression plasmids, can foster LCDCA overproduction. Application of a lipid responsive strong promoter may provide additional benefit for P450 gene expression. The multiple-copy integration of a selected gene at random loci is influenced by the integration site underlining the need for an intensive screening afterwards [62,123]. We recommend an intensive screening process for choosing the most productive multiple-copy transformants. We selected the *YlALK5* for multiple-copy integration since it exhibited the highest ω-oxidation activity, among the tested *YlALKSs* toward the endogenously synthesized fatty acids of 16 carbons. The *YlALK5* gene product is a good candidate for protein engineering, allowing higher selectivity and turn-over rates for long chain fatty acids. The combination of the *YlALK5* and *YlFAO1* over expression can create a push and pull of fatty acid precursors and oxygenated fatty acids in the ω-oxidation pathway. Further studies are recommended to identify the major native cytochome monoxygenese member involved in the temrinal oxidation of 18 carbon-fatty acids. We also suggest modulating the ratio of P450 monoxygenase/reductase over expression for better performance. Our engineered cell factory can be further optimized to produce a higher proportion of LCDCAs with specific chain lengths or degrees of desaturation. This can be achieved in two ways: by introducing heterologous thioesterase-coding genes that have long chain fatty acid specificity (e.g., *Cocos nucifera* fatty acyl-ACP thioesterase CnFatB2 and *Iris germanica* acyl-ACP thioesterase AAG43857 [124]) or by manipulating the native desaturase and elongase systems. The *LEU2* gene was disrupted for enabling a successful site-specific multiple gene integration and testing various expression systems. However, we suggest targeting other sites for future multiple-gene integrations, due to the positive role of the *LEU2* gene on biomass and metabolite production. We used a standard minimal medium for production. However, the availability of rich nutrients, heme containing compounds, and low cost substrates, including lignocellulosic hydrolysates and crude glycerol, could enhance production level and economy of this bioprocess. The over expression of key genes from PPP can promote NADPH availability [113], and this may serve the NADPH-consuming process of endogenous fatty acid and LCDCA production. We also recommend the application of immobilized cells and a biofilm reactor for integrated production and recovery and for efficient use of multi-copy transformants. The *de novo* production of LCDCA can be further improved by generating a larger pool of free fatty acids [125], which have stronger stimulatory effect on the induction of the background *ω*-oxidation activity. This could be achieved by over expressing the upstream and downstream targets of the fatty acid biosynthesis pathway. Acetyl-CoA carboxylase (*ACC1*) and thioesterase (*TE*) are examples of these targets. Disruption of other acyl-CoA synthetases (e.g., *YlFat1*) [81] and overexpression of intracellular lipases (e.g., *TGL3* and *TGL4*) [126] may also contribute to a larger pool of free fatty acids. Microscopic examination of resting cells may reveal more information about the status of remaining fatty acids. Moreover, time course qualitative analysis of free fatty acids can also provide further insight into their threshold for the induction of background ω-oxidation activity. Despite the good LCDCA transport efficiency of our platform, the overexpression of potential heterologous transporters such as CmCDR1 [127], or homologous transporter like ABC2 [128] could improve LCDCA export to minimize product inhibition. Optimization of culture conditions was beyond the scope of this study. However, production titer and yield can be enhanced by optimizing production media and developing fed-batch or continuous bioprocess. For example, providing a reduced aeration rate [129] or nitrogen as co-feed [130] might lessen citric acid production in favour of more fatty acid accumulation. Summary of elements for further improvement is shown in Figure 9.

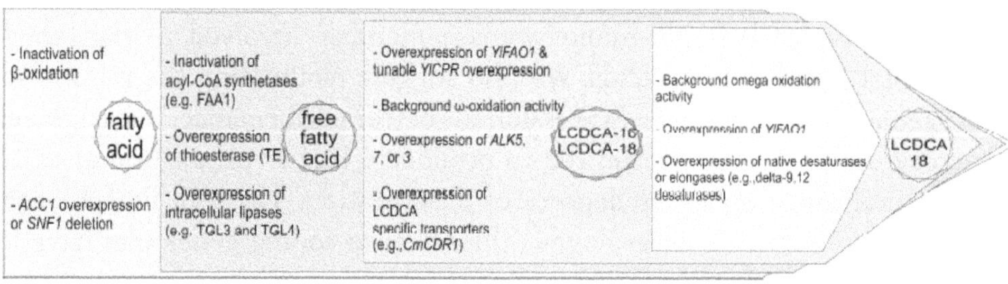

Figure 9. Summary of elements for further improvement.

SUMMARY

Our study presents the construction of *Y. lipolytica* as a cell factory platform for the *de novo* production of LCDCAs from glycerol. We accomplished this by harnessing the native capabilities of this yeast for endogenous fatty acid accumulation and ω-oxidation. Our combinatorial genetic engineering mainly involved inactivating the peroxisomal fatty acid degradation, disrupting cytosolic fatty acid activation, and upregulating fatty acid ω-oxidation. Further enhancement was obtained through fermentation scale-up to the 1-L nitrogen-limited bioreactor culture. Our engineering strategies led *Y. lipolytica* strains to produce about 3.49 g/L LCDCA, with a yield of 0.06 g/g. To the best of our knowledge, this is the highest *de novo* production level ever reported for LCDCA monomers. We envision that this study constitutes a foundational work in developing *Y. lipolytica* as a safe oleaginous yeast platform to upgrade low-value, unrelated feedstock to high-value multi-purpose LCDCAs at a larger scale in the foreseeable future. This work represents a key advance in yeast strain manipulation and fermentation for the biosynthesis of LCDCAs from renewable non-oily feedstock. Last, this research paves the way for an industrial process that can supplement the global supply of LCDCA from a second source, independent of the non-renewable hydrophobic substrates such as crude oil.

REFERENCES

1. Pellis, A.; Herrero Acero, E.; Gardossi, L.; Ferrario, V.; Guebitz, G.M. Renewable building blocks for sustainable polyesters: New biotechnological routes for greener plastics. *Polym. Int.* **2016**, *65*, 861–871. [CrossRef]
2. Liao, J.C.; Mi, L.; Pontrelli, S.; Luo, S. Fuelling the future: Microbial engineering for the production of sustainable biofuels. *Nat. Rev. Microbiol.* **2016**, *14*, 288–304. [CrossRef] [PubMed]
3. Chung, H.; Yang, J.E.; Ha, J.Y.; Chae, T.U.; Shin, J.H.; Gustavsson, M.; Lee, S.Y. Bio-based production of monomers and polymers by metabolically engineered microorganisms. *Curr. Opin. Biotechnol.* **2015**, *36*, 73–84. [CrossRef] [PubMed]
4. Díaz, A.; Katsarava, R.; Puiggalí, J. Synthesis, properties and applications of biodegradable polymers derived from diols and dicarboxylic acids: From polyesters to poly (ester amide) s. *Int. J. Mol. Sci.* **2014**, *15*, 7064–7123. [CrossRef] [PubMed]
5. Beuhler, A. C18 Diacid Market to Grow and Expand into an Array of Novel Products with Superior Properties. 2013. Available online: http://www.elevance.com/images/documents/ Elevance-ODDA-C18-white-paper_20130916_F.pdf (accessed on 22 July 2017).
6. Bowen, C.H.; Bonin, J.; Kogler, A.; Barba-Ostria, C.; Zhang, F. Engineering *Escherichia coli* for conversion of glucose to medium-chain ω-hydroxy fatty acids and α, ω-dicarboxylic acids. *ACS Synth. Biol.* **2015**, 5, 200–206. [CrossRef] [PubMed]
7. Sathesh-Prabu, C.; Lee, S.K. Production of long-chain α, ω-dicarboxylic acids by engineered escherichia coli from renewable fatty acids and plant oils. *J. Agric. Food Chem.* **2015**, *63*, 8199–8208. [CrossRef] [PubMed]
8. Lu, W.; Ness, J.E.; Xie, W.; Zhang, X.; Minshull, J.; Gross, R.A. Biosynthesis of monomers for plastics from renewable oils. *J. Am. Chem. Soc.* **2010**, *132*, 15451–15455. [CrossRef] [PubMed]
9. Ngo, H.L.; Foglia, T.A. Synthesis of long chain unsaturated-α, ω-dicarboxylic acids from renewable materials via olefin metathesis. *J. Am. Oil Chem. Soc.* **2007**, *84*, 777–784. [CrossRef]
10. Huf, S.; Kruegener, S.; Hirth, T.; Rupp, S.; Zibek, S. Biotechnological synthesis of long-chain dicarboxylic acids as building blocks for polymers. *Eur. J. Lipid Sci. Technol.* **2011**, *113*, 548–561. [CrossRef]
11. Cho, C.; Choi, S.Y.; Luo, Z.W.; Lee, S.Y. Recent advances in microbial production

of fuels and chemicals using tools and strategies of systems metabolic engineering. *Biotechnol. Adv.* **2015**, *33*, 1455–1466. [CrossRef] [PubMed]

12. Weiss, A. Selective microbial oxidations in industry: Oxidations of alkanes, fatty acids, heterocyclic compounds, aromatic compounds and glycerol using native or recombinant microorganisms. *Mod. Biooxid. Enzym. React. Appl.* **2007**, 193–209. [CrossRef]
13. Kroha, K. Industrial biotechnology provides opportunities for commercial production of new long-chain dibasic acids. *Inform* **2004**, *15*, 568–571.
14. Picataggio, S.; Rohrer, T.; Deanda, K.; Lanning, D.; Reynolds, R.; Mielenz, J.; Eirich, L.D. Metabolic engineering of candida tropicalis for the production of long-chain dicarboxylic acids. *Nat. Biotechnol.* **1992**, *10*, 894–898. [CrossRef]
15. Maurer, S.C.; Schmid, R.D. Biocatalysts for the epoxidation and hydroxylation of fatty acids and fatty alcohols. *Handb. Ind. Biocatal.* **2005.** [CrossRef]
16. Kogure, T.; Horiuchi, H.; Matsuda, H.; Arie, M.; Takagi, M.; Ohta, A. Enhanced induction of cytochromes p450alk that oxidize methyl-ends of n-alkanes and fatty acids in the long-chain dicarboxylic acid-hyperproducing mutant of candida maltosa. *FEMS Microbiol. Lett.* **2007**, *271*, 106–111. [CrossRef] [PubMed]
17. Kim, D.; Cryle, M.J.; De Voss, J.J.; de Montellano, P.R.O. Functional expression and characterization of cytochrome p450 52a21 from candida albicans. *Arch. Biochem. Biophys.* **2007**, *464*, 213–220. [CrossRef] [PubMed]
18. Scheps, D.; Honda Malca, S.; Richter, S.M.; Marisch, K.; Nestl, B.M.; Hauer, B. Synthesis of ω-hydroxy dodecanoic acid based on an engineered cyp153a fusion construct. *Microb. Biotechnol.* **2013**, *6*, 694–707. [CrossRef] [PubMed]
19. Girhard, M.; Bakkes, P.J.; Mahmoud, O.; Urlacher, V.B. P450 biotechnology. In *Cytochrome p450*; Springer: Berlin, Gemany, 2015; pp. 451–520.
20. Li, H.; Pinot, F.; Sauveplane, V.; Werck-Reichhart, D.; Diehl, P.; Schreiber, L.; Franke, R.; Zhang, P.; Chen, L.; Gao, Y. Cytochrome p450 family member cyp704b2 catalyzes the ω-hydroxylation of fatty acids and is required for anther cutin biosynthesis and pollen exine formation in rice. *Plant Cell* **2010**, *22*, 173–190. [CrossRef] [PubMed]
21. Malca, S.H.; Scheps, D.; Kühnel, L.; Venegas-Venegas, E.; Seifert, A.; Nestl, B.M.; Hauer, B. Bacterial cyp153a monooxygenases for the synthesis of omega-hydroxylated fatty acids. *Chem. Commun.* **2012**, *48*, 5115–5117. [CrossRef] [PubMed]
22. Höfer, R.; Briesen, I.; Beck, M.; Pinot, F.; Schreiber, L.; Franke, R. The arabidopsis cytochrome p450 cyp86a1 encodes a fatty acid ω-hydroxylase involved in suberin monomer biosynthesis. *J. Exp. Bot.* **2008**, *59*, 2347–2360. [CrossRef] [PubMed]

23. Kandel, S.; Sauveplane, V.; Compagnon, V.; Franke, R.; Millet, Y.; Schreiber, L.; Werck-Reichhart, D.; Pinot, F. Characterization of a methyl jasmonate and wounding-responsive cytochrome p450 of arabidopsis thaliana catalyzing dicarboxylic fatty acid formation in vitro. *FEBS J.* **2007**, *274*, 5116–5127. [CrossRef] [PubMed]

24. Benveniste, I.; Tijet, N.; Adas, F.; Philipps, G.; Salaün, J.-P.; Durst, F. Cyp86a1 fromarabidopsis thalianaencodes a cytochrome p450-dependent fatty acid omega-hydroxylase. *Biochem. Biophys. Res. Commun.* **1998**, *243*, 688–693. [CrossRef] [PubMed]

25. Craft, D.L.; Madduri, K.M.; Eshoo, M.; Wilson, C.R. Identification and characterization of the cyp52 family of candida tropicalis atcc 20336, important for the conversion of fatty acids and alkanes to alpha, omega-dicarboxylic acids. *Appl. Environ. Microbiol.* **2003**, *69*, 5983–5991. [CrossRef] [PubMed]

26. Eschenfeldt, W.H.; Zhang, Y.; Samaha, H.; Stols, L.; Eirich, L.D.; Wilson, C.R.; Donnelly, M.I. Transformation of fatty acids catalyzed by cytochrome p450 monooxygenase enzymes of candida tropicalis. *Appl. Environ. Microbiol.* **2003**, *69*, 5992–5999. [CrossRef] [PubMed]

27. Hoffmann, S.M.; Danesh-Azari, H.R.; Spandolf, C.; Weissenborn, M.J.; Grogan, G.; Hauer, B. Structure-guided redesign of CYP153AM. aq for the improved terminal hydroxylation of fatty acids. *ChemCatChem* **2016**, *8*, 3234–3239. [CrossRef]

28. Schreiber, L.; Durst, F.; Pinot, F. Cyp94a5, a new cytochrome p450 from nicotiana tabacum is able to catalyze the oxidation of fatty acids to the v-alcohol and to the corresponding diacid. *FEBS J.* **2001**, *268*, 3083–3090.

29. McLean, K.J.; Luciakova, D.; Belcher, J.; Tee, K.L.; Munro, A.W. Biological diversity of cytochrome p450 redox partner systems. In *Monooxygenase, Peroxidase and Peroxygenase Properties and Mechanisms of Cytochrome p450*; Springer: Berlin, Gemany, 2015; pp. 299–317.

30. Girhard, M.; Tieves, F.; Weber, E.; Smit, M.S.; Urlacher, V.B. Cytochrome p450 reductase from candida apicola: Versatile redox partner for bacterial p450s. *Appl. Microbiol. Biotechnol.* **2013**, *97*, 1625–1635. [CrossRef] [PubMed]

31. Theron, C.W. *Hetrologous Expression of Cytochrome p450 Monoxygenase by the Yeast Yarrowia Lipolytica*; University of the Free State: Bloemfontein, South Africa, 2007.

32. Iida, T.; Sumita, T.; Ohta, A.; Takagi, M. The cytochrome p450alk multigene family of an n-alkane-assimilating yeast, yarrowia lipolytica: Cloning and characterization of genes coding for new CYP52 family members. *Yeast* **2000**, *16*, 1077–1087. [CrossRef]

33. Abghari, A.; Chen, S. *Yarrowia lipolytica* as an oleaginous cell factory platform for the production of fatty acid-based biofuel and bioproducts. *Front. Energy Res.* **2014**, *2*. [CrossRef]

34. Juretzek, T.; Mauersberger, S.; Barth, G. Recombinant Haploid or Diploid Yarrowia Lipolytica Cells for the Functional Heterologous Expression of Cytochrome p450 Systems. Patent WO2000003008 A3, 27 April 2000.

35. Fickers, P.; Benetti, P.H.; Waché, Y.; Marty, A.; Mauersberger, S.; Smit, M.S.; Nicaud, J.M. Hydrophobic substrate utilisation by the yeast yarrowia lipolytica, and its potential applications. *FEMS Yeast Res.* **2005**, *5*, 527–543. [CrossRef] [PubMed]

36. Iwama, R.; Kobayashi, S.; Ishimaru, C.; Ohta, A.; Horiuchi, H.; Fukuda, R. Functional roles and substrate specificities of twelve cytochromes p450 belonging to cyp52 family in n-alkane assimilating yeast yarrowia lipolytica. *Fungal Genet. Biol.* **2016**, *91*, 43–54. [CrossRef] [PubMed]

37. Braun, A.; Geier, M.; Buehler, B.; Schmid, A.; Mauersberger, S.; Glieder, A. Steroid biotransformations in biphasic systems with yarrowia lipolytica expressing human liver cytochrome p450 genes. *Microb. Cell Factories* **2012**, *11*, 106. [CrossRef] [PubMed]

38. Mauersberger, S. Cytochromes p450 of the alkane-utilising yeast yarrowia lipolytica. In *Yarrowia Lipolytica*; Springer: Berlin, Gemany, 2013; pp. 227–262.

39. Takai, H.; Iwama, R.; Kobayashi, S.; Horiuchi, H.; Fukuda, R.; Ohta, A. Construction and characterization of a yarrowia lipolytica mutant lacking genes encoding cytochromes p450 subfamily 52. *Fungal Genet. Biol.* **2012**, *49*, 58–64. [CrossRef] [PubMed]

40. Thevenieau, F.; Beopoulos, A.; Desfougeres, T.; Sabirova, J.; Albertin, K.; Zinjarde, S.S.; Nicaud, J.-M. Uptake and assimilation of hydrophobic substrates by the oleaginous yeast yarrowia lipolytica. In *Handbook of Hydrocarbon and Lipid Microbiology*; Timmis, K.N., Ed.; Springer: Berlin/Heidelberg, Gemany, 2010; pp. 1513–1527.

41. Hanley, K.; Nguyen, L.V.; Khan, F.; Pogue, G.P.; Vojdani, F.; Panda, S.; Pinot, F.; Oriedo, V.B.; Rasochova, L.; Subramanian, M. Development of a plant viral-vector-based gene expression assay for the screening of yeast cytochrome p450 monooxygenases. *Assay Drug Dev. Technol.* **2003**, *1*, 147–160. [CrossRef] [PubMed]

42. Gatter, M.; Förster, A.; Bär, K.; Winter, M.; Otto, C.; Petzsch, P.; Ježková, M.; Bahr, K.; Pfeiffer, M.; Matthäus, F. A newly identified fatty alcohol oxidase gene is mainly responsible for the oxidation of long-chain ω-hydroxy fatty acids in yarrowia lipolytica. *FEMS Yeast Res.* **2014**, *14*, 858–872. [CrossRef] [PubMed]

43. Iwama, R.; Kobayashi, S.; Ohta, A.; Horiuchi, H.; Fukuda, R. Alcohol dehydrogenases and an alcohol oxidase involved in the assimilation of exogenous fatty alcohols in yarrowia lipolytica. *FEMS Yeast Res.* **2015**, *15*, fov014. [CrossRef] [PubMed]
44. Beopoulos, A.; Chardot, T.; Nicaud, J.-M. *Yarrowia lipolytica*: A model and a tool to understand the mechanisms implicated in lipid accumulation. *Biochimie* **2009**, *91*, 692–696. [CrossRef] [PubMed]
45. Smit, M.S.; Mokgoro, M.M.; Setati, E.; Nicaud, J.-M. A, ω-dicarboxylic acid accumulation by acyl-coa oxidase deficient mutants of yarrowia lipolytica. *Biotechnol. Lett.* **2005**, *27*, 859–864. [CrossRef] [PubMed]
46. Wache, Y. *Yarrowia lipolytica* biotechnological applications: Production of dicarboxylic acids and flagrances by yarrowia lipolytica. *Microbiol. Monogr.* **2013**, *25*, 151–170.
47. Nicaud, J.-M.; Thevenieau, F.; Le Dall, M.-T.; Marchal, R. Production of Dicarboxylic Acids by Improved Mutant Strains of Yarrowia Lipolytica. U.S. Patent 20100041115 A1, 18 February 2010.
48. Foo, J.L.; Susanto, A.V.; Keasling, J.D.; Leong, S.S.J.; Chang, M.W. Whole-cell biocatalytic and de novo production of alkanes from free fatty acids in saccharomyces cerevisiae. *Biotechnol. Bioeng.* **2017**, *114*, 232–237. [CrossRef] [PubMed]
49. Ledesma-Amaro, R.; Dulermo, R.; Niehus, X.; Nicaud, J.-M. Combining metabolic engineering and process optimization to improve production and secretion of fatty acids. *Metab. Eng.* **2016**, *38*, 38–46. [CrossRef] [PubMed]
50. Cintolesi, A.; Rodríguez-Moyá, M.; Gonzalez, R. Fatty acid oxidation: Systems analysis and applications.*Wiley Interdiscip. Rev. Syst. Biol. Med.* **2013**, *5*, 575–585. [CrossRef] [PubMed]
51. O'Reilly, E.; Köhler, V.; Flitsch, S.L.; Turner, N.J. Cytochromes p450 as useful biocatalysts: Addressing the limitations. *Chem. Commun.* **2011**, *47*, 2490–2501. [CrossRef] [PubMed]
52. Endoh-Yamagami, S.; Hirakawa, K.; Morioka, D.; Fukuda, R.; Ohta, A. Basic helix-loop-helix transcription factor heterocomplex of yas1p and yas2p regulates cytochrome p450 expression in response to alkanes in the yeast yarrowia lipolytica. *Eukaryot. Cell* **2007**, *6*, 734–743. [CrossRef] [PubMed]
53. Beopoulos, A.; Mrozova, Z.; Thevenieau, F.; Le Dall, M.T.; Hapala, I.; Papanikolaou, S.; Chardot, T.; Nicaud, J.M. Control of lipid accumulation in the yeast yarrowia lipolytica. *Appl. Environ. Microbiol.* **2008**, *74*, 7779–7789. [CrossRef] [PubMed]

54. Goldbach, V.; Roesle, P.; Mecking, S. Catalytic isomerizing ω-functionalization of fatty acids. *ACS Catal.***2015**, *5*, 5951–5972. [CrossRef]
55. Isikgor, F.; Becer, C.R. Lignocellulosic biomass: A sustainable platform for production of bio-based chemicals and polymers. *Polym. Chem.* **2015**, *6*, 4497–4559. [CrossRef]
56. Groenewald, M.; Boekhout, T.; Neuveglise, C.; Gaillardin, C.; Dijck, P.W.; Wyss, M. *Yarrowia lipolytica*: Safety assessment of an oleaginous yeast with a great industrial potential. *Crit. Rev. Microbiol.* **2014**, *40*, 187–206. [CrossRef] [PubMed]
57. Sambrook, J.; Russell, D.W. *Molecular Cloning: A Laboratory Manual*, 3rd ed.; Coldspring-Harbour Laboratory Press: Plymouth, UK, 2001.
58. Tan, M.-J.; Chen, X.; Wang, Y.-K.; Liu, G.-L.; Chi, Z.-M. Enhanced citric acid production by a yeast yarrowia lipolytica over-expressing a pyruvate carboxylase gene. *Bioprocess Biosyst. Eng.* **2016**, *39*, 1–8. [CrossRef] [PubMed]
59. Xie, D.; Jackson, E.N.; Zhu, Q. Sustainable source of omega-3 eicosapentaenoic acid from metabolically engineered yarrowia lipolytica: From fundamental research to commercial production. *Appl. Microbiol. Biotechnol.* **2015**, *99*, 1599–1610. [CrossRef] [PubMed]
60. Chuang, L.-T.; Chen, D.-C.; Nicaud, J.-M.; Madzak, C.; Chen, Y.-H.; Huang, Y.-S. Co-expression of heterologous desaturase genes in yarrowia lipolytica. *New Biotechnol.* **2010**, *27*, 277–282. [CrossRef] [PubMed]
61. Madzak, C.; Nicaud, J.M.; Gaillardin, C. Yarrowia lipolytica. In *Production of Recombinant Proteins: Novel Microbial and Eukaryotic Expression Systems*; Gellissen, G., Ed.; Wiley-Blackwell: Hoboken, NJ, USA, 2005; pp. 163–189.
62. Madzak, C. *Yarrowia lipolytica*: Recent achievements in heterologous protein expression and pathway engineering. *Appl. Microbiol. Biotechnol.* **2015**, *99*, 4559–4577. [CrossRef] [PubMed]
63. Fickers, P.; Le Dall, M.T.; Gaillardin, C.; Thonart, P.; Nicaud, J.M. New disruption cassettes for rapid gene disruption and marker rescue in the yeast yarrowia lipolytica. *J. Microbiol. Methods* **2003**, *55*, 727–737. [CrossRef] [PubMed]
64. Lõoke, M.; Kristjuhan, K.; Kristjuhan, A. Extraction of genomic DNA from yeasts for pcr-based applications.*Biotechniques* **2011**, *50*, 325–328. [CrossRef] [PubMed]
65. O'fallon, J.; Busboom, J.; Nelson, M.; Gaskins, C. A direct method for fatty acid methyl ester synthesis: Application to wet meat tissues, oils, and feedstuffs. *J. Anim. Sci.* **2007**, *85*, 1511–1521. [CrossRef] [PubMed]
66. Abghari, A.; Chen, S. Engineering *yarrowia lipolytica* for enhanced production of lipid and citric acid.*Fermentation* **2017**, *3*, 34. [CrossRef]

67. Tai, M.; Stephanopoulos, G. Engineering the push and pull of lipid biosynthesis in oleaginous yeast yarrowia lipolytica for biofuel production. *Metab. Eng.* **2013**, *15*, 1–9. [CrossRef] [PubMed]

68. Morin, N.; Cescut, J.; Beopoulos, A.; Lelandais, G.; Le Berre, V.; Uribelarrea, J.-L.; Molina-Jouve, C.; Nicaud, J.-M. Transcriptomic analyses during the transition from biomass production to lipid accumulation in the oleaginous yeast yarrowia lipolytica. *PLoS ONE* **2011**, *6*, e27966. [CrossRef] [PubMed]

69. Kirtz, M.; Klebensberger, J.; Otte, K.B.; Richter, S.M.; Hauer, B. Production of ω-hydroxy octanoic acid with escherichia coli. *J. Biotechnol.* **2016**, *230*, 30–33. [CrossRef] [PubMed]

70. Cao, Y.; Cheng, T.; Zhao, G.; Niu, W.; Guo, J.; Xian, M.; Liu, H. Metabolic engineering of *Escherichia coli* for the production of hydroxy fatty acids from glucose. *BMC Biotechnol.* **2016**, *16*, 26. [CrossRef] [PubMed]

71. Clomburg, J.M.; Blankschien, M.D.; Vick, J.E.; Chou, A.; Kim, S.; Gonzalez, R. Integrated engineering of β-oxidation reversal and ω-oxidation pathways for the synthesis of medium chain ω-functionalized carboxylic acids. *Metab. Eng.* **2015**, *28*, 202–212. [CrossRef] [PubMed]

72. Wang, X.; Li, L.; Zheng, Y.; Zou, H.; Cao, Y.; Liu, H.; Liu, W.; Xian, M. Biosynthesis of long chain hydroxyfatty acids from glucose by engineered escherichia coli. *Bioresour. Technol.* **2012**, *114*, 561–566. [CrossRef] [PubMed]

73. Picataggio, S.; Rohrer, T.; Eirich, L.D. Method for Increasing the Omega-Hydroxylase Activity in Candida Tropicalis. U.S. Patent 5620878 A, 15 April 1997.

74. Sumita, T.; Iida, T.; Yamagami, S.; Horiuchi, H.; Takagi, M.; Ohta, A. Ylalk1 encoding the cytochrome p450alk1 in yarrowia lipolytica is transcriptionally induced by n-alkane through two distinct cis-elements on its promoter. *Biochem. Biophys. Res. Commun.* **2002**, *294*, 1071–1078. [CrossRef]

75. Titorenko, V.I.; Rachubinski, R.A. Dynamics of peroxisome assembly and function. *Trends Cell Biol.* **2001**, *11*, 22–29. [CrossRef]

76. Runguphan, W.; Keasling, J.D. Metabolic engineering of saccharomyces cerevisiae for production of fatty acid-derived biofuels and chemicals. *Metab. Eng.* **2014**, *21*, 103–113. [CrossRef] [PubMed]

77. Seip, J.; Jackson, R.; He, H.; Zhu, Q.; Hong, S.-P. Snf1 is a regulator of lipid accumulation in yarrowia lipolytica. *Appl. Environ. Microbiol.* **2013**, *79*, 7360–7370. [CrossRef] [PubMed]

78. Hofbauer, H.F.; Schopf, F.H.; Schleifer, H.; Knittelfelder, O.L.; Pieber, B.; Rechberger, G.N.; Wolinski, H.; Gaspar, M.L.; Kappe, C.O.; Stadlmann, J.

Regulation of gene expression through a transcriptional repressor that senses acyl-chain length in membrane phospholipids. *Dev. Cell* **2014**, *29*, 729–739. [CrossRef] [PubMed]

79. Thevenieau, F.; Le Dall, M.T.; Nthangeni, B.; Mauersberger, S.; Marchal, R.; Nicaud, J.M. Characterization of yarrowia lipolytica mutants affected in hydrophobic substrate utilization. *Fungal Genet. Biol.* **2007**, *44*, 531–542. [CrossRef] [PubMed]
80. Scharnewski, M.; Pongdontri, P.; Mora, G.; Hoppert, M.; Fulda, M. Mutants of saccharomyces cerevisiae deficient in acyl-coa synthetases secrete fatty acids due to interrupted fatty acid recycling. *FEBS J.* **2008**, *275*, 2765–2778. [CrossRef] [PubMed]
81. Dulermo, R.; Gamboa-Meléndez, H.; Ledesma-Amaro, R.; Thévenieau, F.; Nicaud, J.-M. Unraveling fatty acid transport and activation mechanisms in yarrowia lipolytica. *Biochim. Biophys. Acta (BBA)-Mol. Cell Biol. Lipids* **2015**, *1851*, 1202–1217. [CrossRef] [PubMed]
82. Dulermo, R.; Gamboa-Meléndez, H.; Ledesma, R.; Thevenieau, F.; Nicaud, J.-M. *Yarrowia lipolytica* aal genes are involved in peroxisomal fatty acid activation. *Biochim. Biophys. Acta (BBA)-Mol. Cell Biol. Lipids* **2016**, *1861*, 555–565. [CrossRef] [PubMed]
83. Fukuda, R.; Ohta, A. Utilization of hydrophobic substrate by yarrowia lipolytica. In *Yarrowia Lipolytica Genetics, Genomics, and Physiology*; Barth, G., Ed.; Springer: Berlin/Heidelberg, Gemany, 2013; pp. 111–119.
84. Park, J.S.; Iwama, R.; Kobayashi, S.; Ohta, A.; Horiuchi, H.; Fukuda, R. Involvement of acyl-coa synthetase genes in n-alkane assimilation and fatty acid utilization in yeast yarrowia lipolytica. *FEMS Yeast Res.* **2015**, *15*, fov031. [CrossRef]
85. Krivoruchko, A.; Zhang, Y.; Siewers, V.; Chen, Y.; Nielsen, J. Microbial acetyl-coa metabolism and metabolic engineering. *Metab. Eng.* **2015**, *28*, 28–42. [CrossRef] [PubMed]
86. Chen, L.; Zhang, J.; Lee, J.; Chen, W.N. Enhancement of free fatty acid production in saccharomyces cerevisiae by control of fatty acyl-coa metabolism. *Appl. Microbiol. Biotechnol.* **2014**, *98*, 6739–6750. [CrossRef] [PubMed]
87. Petschnigg, J.; Wolinski, H.; Kolb, D.; Zellnig, G.; Kurat, C.F.; Natter, K.; Kohlwein, S.D. Good fat, essential cellular requirements for triacylglycerol synthesis to maintain membrane homeostasis in yeast. *J. Biol. Chem.* **2009**, *284*, 30981–30993. [CrossRef] [PubMed]

88. Stefan, A.; Ugolini, L.; Lazzeri, L.; Conte, E.; Hochkoeppler, A. The expression of the cuphea palustris thioesterase cpfatb2 in yarrowia lipolytica triggers oleic acid accumulation. *Biotechnol. Prog.* **2015**, *32*, 26–35. [CrossRef] [PubMed]

89. Xu, P.; Qiao, K.; Stephanopoulos, G. Engineering oxidative stress defense pathways to build a robust lipid production platform in yarrowia lipolytica. *Biotechnol. Bioeng.* **2017**, *114*, 1521–1530. [CrossRef] [PubMed]

90. Kohlwein, S.D. Triacylglycerol homeostasis: Insights from yeast. *J. Biol. Chem.* **2010**, *285*, 15663–15667. [CrossRef] [PubMed]

91. Hardwick, J.P. Cytochrome p450 omega hydroxylase (cyp4) function in fatty acid metabolism and metabolic diseases. *Biochem. Pharmacol.* **2008**, *75*, 2263–2275. [CrossRef] [PubMed]

92. Björkhem, I. On the mechanism of regulation of omega oxidation of fatty acids. *J. Biol. Chem.* **1976**, *251*, 5259–5266. [PubMed]

93. Wanders, R.J.; Komen, J.; Kemp, S. Fatty acid omega-oxidation as a rescue pathway for fatty acid oxidation disorders in humans. *FEBS J.* **2011**, *278*, 182–194. [CrossRef] [PubMed]

94. Geier, M.; Braun, A.; Emmerstorfer, A.; Pichler, H.; Glieder, A. Production of human cytochrome p450 2d6 drug metabolites with recombinant microbes–a comparative study. *Biotechnol. J.* **2012**, *7*, 1346–1358. [CrossRef] [PubMed]

95. Eirich, L.D.; Craft, D.L.; Steinberg, L.; Asif, A.; Eschenfeldt, W.H.; Stols, L.; Donnelly, M.I.; Wilson, C.R. Cloning and characterization of three fatty alcohol oxidase genes from candida tropicalis strain atcc 20336. *Appl. Environ. Microbiol.* **2004**, *70*, 4872–4879. [CrossRef] [PubMed]

96. Kang, W.-R.; Seo, M.-J.; An, J.-U.; Shin, K.-C.; Oh, D.-K. Production of δ-decalactone from linoleic acid via 13-hydroxy-9 (z)-octadecenoic acid intermediate by one-pot reaction using linoleate 13-hydratase and whole yarrowia lipolytica cells. *Biotechnol. Lett.* **2016**, *38*, 817–823. [CrossRef] [PubMed]

97. Iwama, R.; Kobayashi, S.; Ohta, A.; Horiuchi, H.; Fukuda, R. Fatty aldehyde dehydrogenase multigene family involved in the assimilation of n-alkanes in yarrowia lipolytica. *J. Biol. Chem.* **2014**, *289*, 33275–33286. [CrossRef] [PubMed]

98. Gatter, M.; Matthaus, F.; Barth, G. Yeast Strains and Method for the Production of Omega-Hydroxy Fatty Acids and Dicarboxylic Acids. U.S. Patent 20160304913 A1, 20 October 2016.

99. Vanhanen, S.; West, M.; Kroon, J.T.; Lindner, N.; Casey, J.; Cheng, Q.; Elborough, K.M.; Slabas, A.R. A consensus sequence for long-chain fatty-acid alcohol oxidases from candida identifies a family of genes involved in lipid ω-oxidation in yeast with homologues in plants and bacteria. *J. Biol. Chem.* **2000**, *275*, 4445–4452. [CrossRef] [PubMed]

100. Liu, S.; Li, C.; Fang, X.; Cao, Z.A. Optimal ph control strategy for high-level production of long-chain α, ω-dicarboxylic acid by candida tropicalis. *Enzym. Microb. Technol.* **2004**, *34*, 73–77. [CrossRef]

101. Lin, R.; Cao, Z.; Zhu, T.; Zhang, Z. Secretion in long-chain dicarboxylic acid Fermentation. *Bioprocess Eng.***2000**, *22*, 391–396. [CrossRef]

102. Bankar, A.V.; Kumar, A.R.; Zinjarde, S.S. Environmental and industrial applications of yarrowia lipolytica.*Appl. Microbiol. Biotechnol.* **2009**, *84*, 847–865. [CrossRef] [PubMed]

103. Casal, M.; Paiva, S.; Queirós, O.; Soares-Silva, I. Transport of carboxylic acids in yeasts. *FEMS Microbiol. Rev.***2008**, *32*, 974–994. [CrossRef] [PubMed]

104. Liu, S.; Li, C.; Xie, L.; Cao, Z.A. Intracellular ph and metabolic activity of long-chain dicarDylic acid-producing yeast*Candida tropicalis*. *J. Biosci. Bioeng.* **2003**, *96*, 349–353. [CrossRef]

105. Kamisaka, Y.; Tomita, N.; Kimura, K.; Kainou, K.; Uemura, H. DGA1 (diacylglycerol acyltransferase gene) overexpression and leucine biosynthesis significantly increase lipid accumulation in the Δ*snf* 2 disruptant of saccharomyces cerevisiae. *Biochem. J.* **2007**, *408*, 61–68. [CrossRef] [PubMed]

106. Lundemo, M.T.; Woodley, J.M. Guidelines for development and implementation of biocatalytic p450 processes. *Appl. Microbiol. Biotechnol.* **2015**, *99*, 2465–2483. [CrossRef] [PubMed]

107. Wang, Z.; Zhao, F.; Chen, D.; Li, D. Biotransformation of phytosterol to produce androsta-diene-dione by resting cells of mycobacterium in cloud point system. *Process Biochem.* **2006**, *41*, 557–561. [CrossRef]

108. Vandenberghe, L.P.S.; Soccol, C.R.; Pandey, A.; Lebeault, J.-M. Microbial production of citric acid. *Braz. Arch. Biol. Technol.* **1999**, *42*, 263–276. [CrossRef]

109. Papanikolaou, S.; Aggelis, G. Selective uptake of fatty acids by the yeast yarrowia lipolytica. *Eur. J. Lipid Sci. Technol.* **2003**, *105*, 651–655. [CrossRef]

110. Hirakawa, K.; Kobayashi, S.; Inoue, T.; Endoh-Yamagami, S.; Fukuda, R.; Ohta, A. Yas3p, an opi1 family transcription factor, regulates cytochrome p450 expression in response to n-alkanes in yarrowia lipolytica. *J. Biol. Chem.* **2009**, *284*, 7126–7137. [CrossRef] [PubMed]

111. Beopoulos, A.; Haddouche, R.; Kabran, P.; Dulermo, T.; Chardot, T.; Nicaud, J.-M. Identification and characterization of dga2, an acyltransferase of the dgat1 acyl-coa: Diacylglycerol acyltransferase family in the oleaginous yeast yarrowia lipolytica. New insights into the storage lipid metabolism of oleaginous yeasts. *Appl. Microbiol. Biotechnol.* **2012**, *93*, 1523–1537. [CrossRef] [PubMed]

112. Wasylenko, T.M.; Ahn, W.S.; Stephanopoulos, G. The oxidative pentose phosphate pathway is the primary source of nadph for lipid overproduction from glucose in yarrowia lipolytica. *Metab. Eng.* **2015**, *30*, 27–39. [CrossRef] [PubMed]

113. Silverman, A.M.; Qiao, K.; Xu, P.; Stephanopoulos, G. Functional overexpression and characterization of lipogenesis-related genes in the oleaginous yeast yarrowia lipolytica. *Appl. Microbiol. Biotechnol.* **2016**, *100*, 3781–3798. [CrossRef] [PubMed]

114. Zhang, H.; Wu, C.; Wu, Q.; Dai, J.; Song, Y. Metabolic flux analysis of lipid biosynthesis in the yeast yarrowia lipolytica using 13 c-labled glucose and gas chromatography-mass spectrometry. *PLoS ONE* **2016**, *11*, e0159187. [CrossRef]

115. Fang, F.; Dai, B.; Zhao, G.; Zhao, H.; Sun, C.; Liu, H.; Xian, M. In depth understanding the molecular response to the enhanced secretion of fatty acids in saccharomyces cerevisiae due to one-step gene deletion of acyl-coa synthetases. *Process Biochem.* **2016**, *51*, 1162–1174. [CrossRef]

116. Jin, Z.; Wong, A.; Foo, J.L.; Ng, J.; Cao, Y.X.; Chang, M.W.; Yuan, Y.J. Engineering saccharomyces cerevisiae to produce odd chain-length fatty alcohols. *Biotechnol. Bioeng.* **2016**, *113*, 842–851. [CrossRef] [PubMed]

117. Vohra, A.; Syal, P.; Madan, A. Probiotic yeasts in livestock sector. *Anim. Feed Sci. Technol.* **2016**, *219*, 31–47. [CrossRef]

118. Martinez, C.A.; Rupashinghe, S.G. Cytochrome p450 bioreactors in the pharmaceutical industry: Challenges and opportunities. *Curr. Top. Med. Chem.* **2013**, *13*, 1470–1490. [CrossRef] [PubMed]

119. Holtmann, D.; Hollmann, F. The oxygen dilemma: A severe challenge for the application of monooxygenases? *ChemBioChem* **2016**, *17*, 1391–1398. [CrossRef] [PubMed]

120. Grand View Research, Inc. *Long Chain Dicarboxylic Acid Market Analysis*; Grand View Research: Maharashtra, India, 2016.

121. Stempfle, F.; Ortmann, P.; Mecking, S. Long-chain aliphatic polymers to bridge the gap between semicrystalline polyolefins and traditional polycondensates. *Chem. Rev.* **2016**, *116*, 4597–4641. [CrossRef] [PubMed]

122. De Montellano, P.R.O. *Cytochrome p450: Structure, Mechanism, and Biochemistry*; Springer Science & Business Media: Berlin, Gemany, 2005.

123. Cambon, E.; Piamtongkam, R.; Bordes, F.; Duquesne, S.; Laguerre, S.; Nicaud, J.-M.; Marty, A. A new yarrowia lipolytica expression system: An efficient tool for rapid and reliable kinetic analysis of improved enzymes. *Enzym. Microb. Technol.* **2010**, *47*, 91–96. [CrossRef]

124. Xu, P.; Gu, Q.; Wang, W.; Wong, L.; Bower, A.G.W.; Collins, C.H.; Koffas, M.A.G. Modular optimization of multi-gene pathways for fatty acids production in *E. coli*. *Nat. Commun.* **2013**, *4*, 1409. [CrossRef] [PubMed]

125. Teixeira, P.G.; Ferreira, R.; Zhou, Y.J.; Siewers, V.; Nielsen, J. Dynamic regulation of fatty acid pools for improved production of fatty alcohols in saccharomyces cerevisiae. *Microb. Cell Factories* **2017**, *16*, 45. [CrossRef] [PubMed]

126. Dulermo, T.; Tréton, B.; Beopoulos, A.; Gnankon, A.P.K.; Haddouche, R.; Nicaud, J.-M. Characterization of the two intracellular lipases of y. Lipolytica encoded by tgl3 and tgl4 genes: New insights into the role of intracellular lipases and lipid body organisation. *Biochim. Biophys. Acta (BBA)-Mol. Cell Biol. Lipids* **2013**, *1831*, 1486–1495. [CrossRef] [PubMed]

127. Sagehashi, Y.; Horiuchi, H.; Fukuda, R.; Ohta, A. Identification and characterization of a gene encoding an abc transporter expressed in the dicarboxylic acid-producing yeast candida maltosa. *Biosci. Biotechnol. Biochem.* **2013**, *77*, 2502–2504. [CrossRef] [PubMed]

128. Chen, B.; Ling, H.; Chang, M.W. Transporter engineering for improved tolerance against alkane biofuels in saccharomyces cerevisiae. *Biotechnol. Biofuels* **2013**, *6*, 21. [CrossRef] [PubMed]

129. Kavšcˇek, M.; Bhutada, G.; Madl, T.; Natter, K. Optimization of lipid production with a genome-scale model of yarrowia lipolytica. *BMC Syst. Biol.* **2015**, *9*, 72. [CrossRef] [PubMed]

130. Cescut, J. *Accumulation D'acylglycérols par des Espèces Levuriennes à Usage Carburant Aéronautique: Physiologie et Performances de Procédés*; INSA Toulouse: Toulouse, France, 2009. (In French) Available online: http://www.theses.fr/2009ISAT0022 (accessed on 22 July 2017).

CHAPTER-4

INVESTIGATING THE PRODUCTION OF CITRIC ACID BY YARROWIA LIPOLYTICA YEAST USING VARIOUS RENEWABLE RAW MATERIALS

INTRODUCTION: AN OVERVIEW

Embarking on the frontier of sustainable bioproduction, this study delves into the remarkable versatility of Yarrowia lipolytica yeast in the synthesis of citric acid using diverse renewable raw materials. As the global demand for eco-friendly alternatives intensifies, exploring Yarrowia lipolytica's proficiency in citric acid production from various renewable sources represents a crucial stride toward sustainable industrial practices. In this exploration, we delve into the adaptability of Yarrowia lipolytica, unraveling its potential as a potent catalyst for citric acid synthesis across a spectrum of renewable raw materials.

The world market of citric acid (CA) is one of the largest and fastest growing markets in the biotechnological industry. Microbiological processes for CA production have usually used the mycelial fungi *Aspergillus niger* as a producer and molasses as a carbon source. In this paper, we propose methods for CA production from renewable carbon substrates (rapeseed oil, glucose, glycerol, ethanol, glycerol-containing waste of biodiesel industry and glucose-containing aspen waste) by the mutant strain *Yarrowia lipolytica* NG40/UV5. It was revealed that *Y. lipolytica* grew and synthesized CA using all tested raw materials. The obtained results are sufficient for industrial use of most of the raw materials studied for CA production. Using rapeseed oil, ethanol and raw glycerol (which is an important feedstock of biodiesel production), a high CA production (100–140 g L^{-1}) was achieved.

Citric acid (CA) and its salts are widely used as an acidulate, favouring agent and antioxidant in the production of beverages and confectionery, in infant formula, as well as in the chemical, pharmaceutical, electronic, defense, and other industries. The volume of citric acid globally exceeds two million tons per year and its production is annually increased by 5% [1].

Modern technologies of citric acid (CA) production are based on using various mutant strains of the mycelial fungi *Aspergillus niger* as a producer and molasses as a raw material. CA production by fungi is a complicated and environmentally unsafe process; as a result of its implementation, a large number of both liquid effluents containing mineral acids, ballast organic substances, cyanides, and solid wastes, primarily gypsum, are accumulated. Moreover, *A. niger* is an opportunistic pathogenic fungi and can cause allergic diseases and aspergillosis [2,3].

Over the past 40 years, the interest of researchers has focused on yeast as a producer of CA; the yeast *Yarrowia lipolytica* has been the most used CA producer [1–3]. Initially, this kind of yeast attracted the attention of researchers due to its ability to grow and synthesize CA in media with n alkanes—an available and cheap substrate [2]. However, due to changes in the world oil market, the use of this substrate has become economically un-profitable. In this regard, it is of interest to use other types of raw materials.

The choice of raw materials for developing CA biotechnology is determined by factors such as renewability, ability of the producer to assimilate the substrate with a high conversion rate, consumption value and cost price of the target product. To carry out fermentation processes, in addition to very expensive food raw materials, such as glucose [4–7] and plant oils [8–10], much cheaper substrates which are waste products of various industries, such as glycerol-containing waste of the biodiesel industry [9–17], glucose-containing wood hydrolysates [18,19], olive mill wastewater [20],and inulin [21] are used. Ethanol, a water-soluble individual compound which ensures the formation of a pure product and facilitates the isolation process, is also of great importance [22,23]. In the works of the above-mentioned researchers it was shown that all these substrates are promising for CA production and the use of glycerol-containing waste of biodiesel industry and glucose-containing wood hydrolyzates can increase the profitability of CA production process.

The aim of this work was a comparative study of CA production by the yeast *Y. lipolytica* on different types of renewable raw materials.

MATERIALS AND METHODS

Microorganisms and Chemicals

The mutant strain *Y. lipolytica* NG40/UV5 was obtained as described previously [9,24].Chemicals, the manufacturer of the carbon sources, and their characteristics were presented in our published articles [7,9,15,17].

Media and Cultivation Conditions

All experiments were done using the same equipment, nutrient medium and cultivation conditions. Strain *Y. lipolytica* NG40/UV5 was cultivated in a 10-L ANKUM-2M fermenter (Pushchino, Russia) with an initial volume of 5 L. The medium contained (g L^{-1}): carbon source: $(NH_4)_2SO_4$, 6; $MgSO_4 \cdot 7H_2O$, 1.4; NaCl, 0.5; $Ca(NO_3)_2$, 0.8; KH_2PO_4, 2.0; K_2HPO_4, 0.2; Difco yeast extract (BD Diagnostic Systems, Sparks, MD, USA), 1.0; trace elements (mg L^{-1}): $FeSO_4 \times 7H_2O$, 14.9; $MnSO_4 \times 4H_2O$, 0.2; $ZnSO_4 \times 7H_2O$, 8.1; $CuSO_4 \times 5H_2O$, 3.9. The fermentation conditions were maintained automatically at a constant level: Temperature 28-0.5 C; pH 4.5.0-0.1; pO_2 50% (of air saturation); agitation rate of 800 rpm. Pulsed addition of carbon source (by 2–20 g L^{-1}) depending on the carbon source used was performed as the pO_2 value changed by 10%. Cultivation was continued for 6 days.

Assays

Biomass, concentration of CA, isocitric acid (ICA) and other organic acids were determined as described previously [7].

Calculations

Earlier, it was found that the mass yield of CA production (Y_{CA}), expressed in g of CA per g of carbon source, and the fermented productivity, expressed in g $(L \cdot h)^{-1}$ were influenced by the medium dilution due to the addition of NaOH solution for maintaining a constant pH value [9,17]. In this regard, the total amount of CA in the culture broth at the end of the fermentation was used to calculate Y_{CA} and fermented productivity. Formulas for calculation of Y_{CA} value and fermented productivity were described earlier [9,17].

All the data presented are the mean values of three experiments and two measurements for each experiment; standard deviations were calculated (S.D. < 10%).

RESULTS AND DISCUSSION

The dynamics of nitrogen consumption and the accumulation of biomass and CA by *Y. lipolytica* NG40/UV5 grown on rapeseed oil are shown in Figure 1a, while the logarithmic growth curve (μ) and the specific rate of biosynthesis of CA (q_p) are shown in Figure 1b. As it can be seen in the latter figure, the growth curve had an exponential phase (phase I) lasting for 12 h, growth retardation phase (phase II) lasting from 12 to 36 h of cultivation, and stationary phase (phase III) lasting from 36 h to the end of cultivation. The retardation of growth coincided with the exhaustion of nitrogen from the medium. The specific growth rate attained a maximum (μ_{max} = 0.360 h^{-1}) in the exponential growth phase (12 h of cultivation). This value of μ_{max} was more than two times higher than that of the other CA-producing strain

Y. lipolytica (0.17–0.22 h⁻1) [4,8]. After 12 h of cultivation, μ gradually decreased to zero after 48 h of cultivation. The excretion of CA did not occur in the exponential growth phase but became active in the growth retardation and stationary phases. Within this cultivation period, the specific rate of CA production (qp) was between 0.065 and 0.104 g CA/g·h. At the end of cultivation (144 h), the strain produced 140 g L⁻1 CA and 5.3 g L⁻1 ICA (data not shown) with CA:ICA ratio of 26.4:1. The CA production yield Y_{CA} was 1.5 g g⁻1; the fermented productivity was calculated to be 1.46 g (L·h)⁻1 with account for the dilution factor.

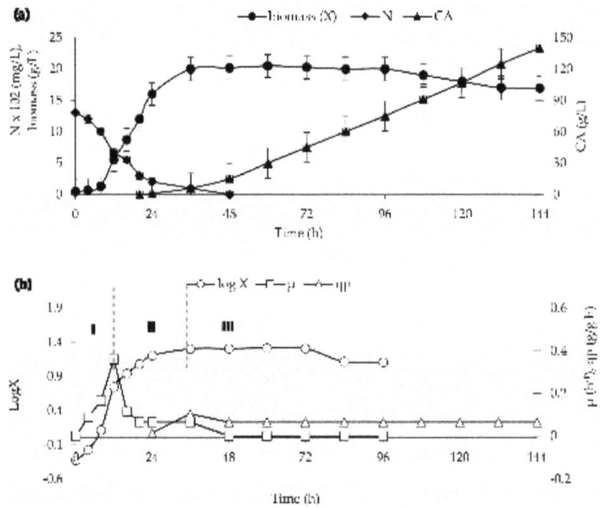

Figure 1. Time courses of nitrogen consumption, biomass accumulation, and citric acid production in *Y. lipolytica* grown on rapeseed oil (a) and calculated parameters of the process (b): I—the exponential cell growth; II—the cell growth retardation; III—the stationary phase.

The data of the accumulation of biomass and CA by *Y. lipolytica* NG40/UV5 grown on other substrates compared to rapeseed oil are shown in Table 1.

Table 1. Citric acid production by Y. lipolytica on various carbon sources.

Substrates	Biomass (g L-1)	CA (g L-1)	ICA (g L-1)	CA:ICA	Productivity (g (L-h)-1)	YCA (g g-1)
Rapeseed oil	17.0 ± 1.1	140.0 ± 5.0	5.3 ± 0.8	26.2:1	1.46	1.50
Glucose	18.7 ± 1.3	100.8 ± 9.2	4.9 ± 0.9	20.6:1	1.05	0.80
Glucose-containing aspen waste	5.6 ± 0.8	31.2 ± 2.1	7.84 ± 0.9	4:1	0.325	0.50
Glycerol	16.8 ± 1.1	87 ± 6.4	13 ± 1.1	6.7:1	0.906	0.64
Glycerol waste of biodiesel industry	20.0 ± 1.8	100 ± 3.4	15 ± 1.2	7.7:1	1.04	0.90
Ethanol	15.3 ± 1.4	106.7 ± 2.7	15 ± 1.4	7.1:1	1.32	0.87

As seen in Table 1, at the end of cultivation (144 h), *Y. lipolytica* NG40/UV5 produced 100.8 g L^{-1} CA and 4.9 g L^{-1} ICA with CA:ICA ratio of 20.6:1 in the medium containing glucose. The CA production yield Y_{CA} was 0.80 g g^{-1}; the fermented productivity was calculated to be 1.05 g $(L \cdot h)^{-1}$.

As seen in Table 1, *Y. lipolytica* NG40/UV5 only produced 31.2 g L^{-1} CA and 7.84 g L^{-1} ICA with CA:ICA ratio of 4:1 in the medium containing glucose-containing aspen waste. The CA production yield Y_{CA} was 0.50 g g^{-1}; the fermented productivity was calculated to be 0.325 g $(L \cdot h)^{-1}$.

As it can be seen from the data in Table 1, the mutant grows perfectly and synthesizes CA both in a medium with pure glycerol and in a medium with biodiesel-derived glycerol. *Y. lipolytica* NG40/UV5 produced 87 g L^{-1} CA with a ratio of CA to ICA of 6.7:1. The application of waste glycerol for *Y. lipolytica* NG40/UV5 cultivation increased CA production by 15% (up to 100 g L^{-1}) compared to that obtained from pure glycerol; the CA to ICA ratio was 7.7:1. The fermented productivity was high and reached 0.906 and 1.04 (g $(L \cdot h)^{-1}$) in the media with pure- and biodiesel-derived glycerol, respectively. The mass yield (Y_{CA}) reached 0.64 and 0.9 g/g in the media with pure- and biodiesel-derived glycerol, respectively.

As seen in Table 1, *Y. lipolytica* NG40/UV5 produced 106.7 g L^{-1} CA and 15 g L^{-1} ICA with CA:ICA ratio of 7.1:1 in the medium containing ethanol. The CA production yield Y_{CA} was 0.87 g g^{-1}; the fermented productivity was calculated to be 1.32 g/L·h.

Comparative data on the most efficient processes of CA production by yeasts *Y. lipolytica* from various carbohydrate-containing substrates are given in Table 2. As seen in this table, in the experiments with wild and mutant strains, *Y. lipolytica* produced CA in industrially sufficient amounts. For instance, in the mutant strain *Saccharomycopsis lipolytica* NTG9 grown on rapeseed oil, the CA concentration reached 152.3 g L^{-1} with the yield (Y_{CA}) of 1.5 g g^{-1} [25]. Aurich et al. [26] obtained a CA concentration of 198 g L^{-1} with the yield (Y_{CA}) of 1.16 g g^{-1}, which was achieved after a 300 h fed-batch cultivation of the wild strain *Y. lipolytica* H181. The wild strains *Y. lipolytica* H222 and *Y. lipolytica* W29, grown on glucose, produced 41 g L^{-1} [4] and 49 g L^{-1} of CA, respectively [27]. Recently, we found that the wild strain *Y. lipolytica* VKM Y 2373, cultivated in a medium with glucose under cell growth limitation using nitrogen, phosphorus and sulfur, produced CA at a level of 80–85 g L^{-1} with a yield of 0.70–0.75 g g^{-1} [7]. The over expression of gene *PYR* encoding pyruvate carboxylase in *Y. lipolytica*, resulted in the production of CA at a level of 95–111.1 g L^{-1} with the yield (Y_{CA}) of 0.75–0.93 g g^{-1} [5,6]. Strain *Y. lipolytica* ACA-DS 50109 cultivated on glucose and olive mill wastewaters produced CA (28.9 g L^{-1}) with the product yield (Y_{CA}) of 0.53 g g^{-1} [28]. Later, the authors of the last article improved the process of CA production up to 52.0 g L^{-1} with the product yield (Y_{CA}) of 0.64 g g^{-1} using strain *Y. lipolytica* ACA-YC 5033, which was also able to remove harmful phenolic compounds from olive mill wastewaters [29]. The glycerol-grown yeast *Y. lipolytica* NRRL YB-423 produced21.6 g L^{-1} of CA with mass yield of 0.54 g g^{-1} [12]; strain *Y. lipolytica* ACA-DC 50109 synthesized62.5 g L^{-1} of CA with mass yield of 0.56 g g^{-1} from raw glycerol [20], while the recombinant strain *lipolytica* NCYC3825 was able to produce 58 g L^{-1} of CA [30]. Earlier, we indicated that the other mutant *Y. lipolytica* NG40/UV7 synthesized, CA (122.2 g L^{-1}) with the yield of 0.95 g g^{-1}. The high CA production (up to 140 g L^{-1}) has also been reported for acetate-negative mutant *Y. lipolytica* Wratislavia AWG7, grown on crude glycerol [13] and recombinant strain *Y. lipolytica* H222-S4 (p67ICL1), harboring the invertase encoding *ScSUC2* gene of *Saccharomyces cerevisiae* under inducible XPR2 promoter control and multiple ICL1 copies, cultivated on sucrose [31].

It should be noted that plant raw materials, such as wood, straw, and agricultural products processing waste, are inexpensive, accessible, renewable, and environmentally friendly substrates for microbiological synthesis of practically valuable compounds. However, the effective conversion of these substrates into easily assimilable carbohydrates (glucose, xylose, and higher glucose-containing polymers) is a difficult task. The traditional technologies of hydrolysis of plant raw materials with the use of strong acids and alkalis are associated with the formation of by-products that inhibit the growth of microorganisms and biosynthesis of target substances. Therefore, recombinant producers, effectively assimilating plant raw materials, were developed. For instance, Ledesmo-Amaro et al. (2016) engineered *Y. lipolytica* able to metabolize xylose to produce CA and lipids. The over expression of xylose reductase, xylitol dehydrogenase and xylulokinase resulted in a production of 80 g L^{-1} of CA from xylose by mutant strain *Y. lipolytica* [19].

Table 2. Comparative data of the processes of CA production from various substrates using Y. lipolytica strains.

Strain	Substrate	Characteristics of Strain CA YCA (g L^{-1}) (g g^{-1})		References
S. lipolytica NTG9	canola oil	mutant/ nitrosoguanidine 152	.3 1.50	[25]
Y. lipolytica H181	sunflower oil	wild type 198	.0 1.16	[26]
Y. lipolytica H222		wild type 41.	0 0.55	[4]
Y. lipolytica W29l		wild type 49.	0 0.85	[27]
Y. lipolytica VKM Y 2373	glucose	wild type 80–85 0.70–0.75		[7]
Y. lipolytica PG86		*PYC* gene expression 95.	0 0.75	[5]

Strain	Substrate	Characteristics of Strain CA YCA (g L⁻1) (g g⁻1)		References
Y. lipolytica PR32		*PYC* gene expression 111	.1 0.93	[6]
Y. lipolytica ACA-DS 50109	glucose +	wild type 28.	9 0.53	[28]
Y. lipolytica ACA-YC 5033	olive mill wastewaters	wild type 52.	0 0.64	[29]
Y. lipolytica NRRL YB-423	glycerol	wild type 21.	6 0.54	[12]
Y. lipolytica ACA-DC 50109	raw glycerol	wild type 62.	5 0.56	[20]
Y. lipolytica NCYC 3825	raw glycerol	multigene expression 58.	8 0.17	[30]
Y. lipolytica NG40/UV7	raw glycerol	mutant/ nitrosoguanidine/UV 122	.2 0.95	[17]
Y. lipolytica Wratislavia AWG7	raw glycerol	mutant/acetate⁻ 139	.0 0.70	[13]
Y. lipolytica H222-S4 (p67ICL1)	sucrose	ScSUC2/ICL1 127–140 0.75–0.82		[31]
Y. lipolytica XYL+	xylose	*XYL* gene expression 80.	0 0.53	[19]
Y. lipolytica Wratislavia K1	inulin	*INU1* gene expression 105	.2 0.53	[21]
Y. lipolytica NG40/UV5	rapeseed oil	mutant/ nitrosoguanidine/UV 140	.0 1.5	Present study

Rakicka et al. (2016) reported the production from inulin by engineered strain *Y. lipolytica* Wratislavia K1. The overexpression of the *INU1* gene from *Kluyveromyces marxianus* coding inulinase resulted in the effective hydrolysis of inulin by mutant and the production of a high amount of CA (105.2 g L⁻1 from 200 g L⁻1 inulin) [21].

SUMMARY

The results of the experiments indicated that the mutant *Y. lipolytica* NG40/UV5 was able to grow and synthesize CA on media containing all types of the investigated renewable raw materials. However, it should be noted that the accumulation of the by-product of fermentation—ICA—was even very high in four of the six substrates investigated (glucose-containing aspen waste, glycerol, glycerol waste of biodiesel industry and ethanol). The best results were obtained using rapeseed oil (140 g L⁻1 CA; CA:ICA ratio of 26.2:1; mass yield of CA production (Y_{CA}) of 1.5 g g⁻1, and fermented productivity of 1.46 g (L·h)⁻1). However, economic considerations may lead to the fact that using a cheaper and less pure substrate (glycerol waste of biodiesel industry; 100 g L⁻1 CA; CA:ICA ratio of 7.7:1; mass yield of CA production

(YCA) of 0.9 g $g^{-}1$, and fermented productivity of 1.04 g $(L \cdot h)^{-}1$) would be preferable on a production scale.

REFERENCES

1. Cavallo, E.; Charreau, H.; Cerrutti, P.; Foresti, M.L. *Yarrowia lipolytica*: A model yeast for citric acid production.*FEMS Yeast Res.* **2017**, *17*. [CrossRef] [PubMed]
2. Anastassiadis, S.; Morgunov, I.G.; Kamzolova, S.V.; Finogenova, T.V. Citric acid production patent review.*Recent Pat. Biotechnol.* **2008**, *2*, 107–123. [CrossRef] [PubMed]
3. Finogenova, T.V.; Morgunov, I.G.; Kamzolova, S.V.; Chernyavskaya, O.G. Organic acid production by the yeast *Yarrowia lipolytica*: A review of prospects. *Appl. Biochem. Microbiol.* **2005**, *41*, 418–425. [CrossRef]
4. Moeller, L.; Strehlitz, B.; Aurich, A.; Zehnsdorf, A.; Bley, T. Optimization of citric acid production from glucose by *Yarrowia lipolytica*. *Eng. Life Sci.* **2007**, *7*, 504–511. [CrossRef]
5. Tan, M.J.; Chen, X.; Wang, Y.K.; Liu, G.L.; Chi, Z.M. Enhanced citric acid production by a yeast *Yarrowia lipolytica*over-expressing a pyruvate carboxylase gene. *Bioprocess Biosyst. Eng.* **2016**, *39*, 1289–1296. [CrossRef] [PubMed]
6. Fu, G.Y.; Lu, Y.; Chi, Z.; Liu, G.L.; Zhao, S.F.; Jiang, H.; Chi, Z.M. Cloning and characterization of a pyruvate carboxylase from *Penicillium rubens* and overexpression of the gene in the yeast *Yarrowia lipolytica* for enhanced citric acid production. *Mar. Biotechnol.* **2016**, *18*, 1–14. [CrossRef] [PubMed]
7. Kamzolova, S.V.; Morgunov, I.G. Metabolic peculiarities of the citric acid overproduction from glucose in yeasts *Yarrowia lipolytica*. *Bioresour. Technol.* **2017**, *243*, 433–440. [CrossRef] [PubMed]
8. Kamzolova, S.V.; Morgunov, I.G.; Aurich, A.; Perevoznikova, O.A.; Shishkanova, N.V.; Stottmeister, U.; Finogenova, T.V. Lipase secretion and citric acid production in *Yarrowia lipolytica* yeast grown on animal and vegetable fat. *Food Technol. Biotechnol.* **2005**, *43*, 113–122.
9. Kamzolova, S.V.; Lunina, J.N.; Morgunov, I.G. Biochemistry of citric acid production from rapeseed oil by*Yarrowia lipolytica* yeast. *J. Am. Oil Chem. Soc.* **2011**, *88*, 1965–1976. [CrossRef]
10. Aurich, A.; Specht, R.; Müller, R.A.; Stottmeister, U.; Yovkova, V.; Otto, C.; Holz, M.; Barth, G.; Heretsch, P.; Thomas, F.A.; et al. Microbiologically Produced Carboxylic Acids Used as Building Blocks in Organic Synthesis. In *Reprogramming Microbial Metabolic Pathways*; Wang, X., Chen, J., Quinn, P., Eds.; Springer: Dordrecht, The Netherlands, 2012; pp. 391–424. Available online: https://link.springer.com/chapter/10. 1007%2F978-94-007-5055-5_19 (accessed on 28 September 2012).

11. Rymowicz, W.; Rywin´ska, A.; Z˙arowska, B.; Juszczyk, P. Citric acid production from raw glycerol by acetatemutants of *Yarrowia lipolytica*. *Chem. Pap.* **2006**, *60*, 391–395. [CrossRef]

12. Levinson, W.E.; Kurtzman, C.P.; Kuo, T.M. Characterization of *Yarrowia lipolytica* and related species for citric acid production from glycerol. *Enzym. Microb. Technol.* **2007**, *41*, 292–295. [CrossRef]

13. Rywin´ska, A.; Rymowicz, W.; Z´arowska, B.; Wojtatowicz, M. Biosynthesis of citric acid from glycerol byacetate mutants of *Yarrowia lipolytica* in fed-batch fermentation. *Food Technol. Biotechnol.* **2009**, *47*, 1–6.

14. Makri, A.; Fakas, S.; Aggelis, G. Metabolic activities of biotechnological interest in *Yarrowia lipolytica* grown on glycerol in repeated batch cultures. *Bioresour. Technol.* **2010**, *101*, 2351–2358. [CrossRef] [PubMed]

15. Morgunov, I.G.; Kamzolova, S.V.; Lunina, J.N. The citric acid production from raw glycerol by *Yarrowia lipolyticayeast* and its regulation. *Appl. Microbiol. Biotechnol.* **2013**, *97*, 7387–7397. [CrossRef] [PubMed]

16. Rywinska, A.; Juszczyk, P.; Wojtatowicz, M.; Robak, M.; Lazar, Z.; Tomaszewska, L.; Rymowicz, W. Glycerol as a promising substrate for *Yarrowia lipolytica* biotechnological applications. *Biomass Bioenergy* **2013**, *48*, 148–166. [CrossRef]

17. Morgunov, I.G.; Kamzolova, S.V. Physiologo-biochemical characteristics of citrate-producing yeast *Yarrowia lipolytica* grown on glycerol-containing waste of biodiesel industry. *Appl. Microbiol. Biotechnol.* **2015**, *99*, 6443–6450. [CrossRef] [PubMed]

18. Wojtatowicz, M.; Rymowicz, W.; Kautola, H. Comparison of different strains of the yeast *Yarrowia lipolytica* for citric acid production from glucose hydrol. *Appl. Biochem. Biotechnol.* **1991**, *31*, 165–174. [CrossRef] [PubMed]

19. Ledesma-Amaro, R.; Lazar, Z.; Rakicka, M.; Guo, Z.; Fouchard, F.; Coq, A.C.; Nicaud, J.M. Metabolic engineering of *Yarrowia lipolytica* to produce chemicals and fuels from xylose. *Metab. Eng.* **2016**, *38*, 115–124. [CrossRef] [PubMed]

20. Papanikolaou, S.; Fakas, S.; Fick, M.; Chevalot, I.; Galiotou-Panayotou, M.; Komaitis, M.; Marc, I.; Aggelis, G. Biotechnological valorisation of raw glycerol discharged after biodiesel (fatty acid methyl-esters) manufacturing process: Production of 1,3-propanediol, citric acid and single oil. *Biomass Bioengergy* **2008**, *32*, 60–71. [CrossRef]

21. Rakicka, M.; Lazar, Z.; Rywinska, A.; Rymowicz, W. Efficient utilization of inulin and glycerol as fermentation substrates in erythritol and citric acid production using expressing inulinase. *Chem. Pap.* **2016**, *70*, 1452–1459. [CrossRef]

22. Stephanopoulos, G. Challenges in engineering microbes for biofuels production. *Science* **2007**, *315*, 801–804. [CrossRef] [PubMed]

23. Weusthuis, R.A.; Aarts, J.M.M.J.G.; Sanders, J.P.M. From biofuel to bioproduct: Is bioethanol a suitable fermentation feedstock for synthesis of bulk chemicals? *Biofuels Bioprod. Biorefin.* **2011**, *5*, 486–494. [CrossRef]
24. Finogenova, T.V.; Puntus, I.F.; Kamzolova, S.V.; Lunina, I.N.; Monastyrskaia, S.E.; Morgunov, I.G.; Boronin, A.M. Obtaining of the mutant *Yarrowia lipolytica* strains producing citric acid from glucose. *Prikl. Biokhimiia Mikrobiol.* **2008**, *44*, 219–224.
25. Good, D.W.; Droniuk, R.; Lawford, R.G.; Fein, J.E. Isolation and characterization of a *Saccharomycopsis lipolytica* mutant showing increased production of citric acid from canola oil. *Can. J. Microbiol.* **1985**, *31*, 436–440. [CrossRef]
26. Aurich, A.; Förster, A.; Mauesberger, S.; Barth, G.; Stottmeister, U. Citric acid production from renewable resources by *Yarrowia lipolytica. Biotechnol. Adv.* **2003**, *21*, 454–455.
27. Papanikolaou, S.; Chatzifragkou, A.; Fakas, S.; Galiotou-Panayotou, M.; Komaitis, M.; Nicaud, J.M.; Aggelis, G. Biosynthesis of lipids and organic acids by *Yarrowia lipolytica* strains cultivated on glucose. *Eur. J. Lipid Sci. Technol.* **2009**, *111*, 1221–1232. [CrossRef]
28. Papanikolaou, S.; Galiotou-Panayotou, M.; Fakas, S.; Komaitis, M.; Aggelis, G. Citric acid production by *Yarrowia lipolytica* cultivated on olive-mill wastewater-based media. *Bioresour. Technol.* **2008**, *99*, 2419–2428. [CrossRef] [PubMed]
29. Sarris, D.; Stoforos, N.G.; Mallouchos, A.; Kookos, I.K.; Koutinas, A.A.; Aggelis, G.; Papanikolaou, S. Production of added-value metabolites by *Yarrowia lipolytica* growing in olive mill wastewater-based media under aseptic and non-aseptic conditions. *Eng. Life Sci.* **2017**, *17*, 695–709. [CrossRef]
30. Celin′ ska, E.; Grajek, W. A novel multigene expression construct for modification of glycerol metabolism in *Yarrowia lipolytica. Microb. Cell Fact.* **2013**, *12*, 102. [CrossRef] [PubMed]
31. Förster, A.; Aurich, A.; Mauersberger, S.; Barth, G. Citric acid production from sucrose using a recombinant strain of the yeast *Yarrowia lipolytica. Appl. Microbiol. Biotechnol.* **2007**, *75*, 1409–1417. [CrossRef] [PubMed]

CHAPTER-5

DEVELOPING A RECOMBINANT DIPLOID STRAIN OF SACCHAROMYCES CEREVISIAE FOR EFFICIENT CO-FERMENTATION OF GLUCOSE AND XYLOSE

INTRODUCTION: AN OVERVIEW

In the pursuit of advancing biofuel and biorefinery technologies, significant efforts are dedicated to the development of innovative yeast strains capable of efficient co-fermentation of glucose and xylose. In this context, a key stride is the creation of a recombinant diploid strain of Saccharomyces cerevisiae, tailored to address the challenges associated with the simultaneous utilization of these two essential sugars. This engineered strain is designed with precision to enhance its ability to rapidly convert both glucose and xylose into valuable biofuel precursors. By leveraging the genetic potential of Saccharomyces cerevisiae, this research aims to contribute to the optimization of bioprocessing strategies, unlocking new possibilities for sustainable and economically viable bioenergy production.

Cost-effective production of cellulosic ethanol requires robust microorganisms for rapid co-fermentation of glucose and xylose. This study aims to develop a recombinant diploid xylose-fermenting *Saccharomyces cerevisiae* strain for efficient conversion of lignocellulosic biomass sugars to ethanol. Episomal plasmids harboring codon-optimized *Piromyces* sp. E2 xylose isomerase (*PirXylA*) and *Orpinomyces* sp. ukk1 xylose (*OrpXylA*) genes were constructed and transformed into

S. cerevisiae. The strain harboring plasmids with tandem *PirXylA* was favourable for xylose utilization when xylose was used as the sole carbon source, while the strain harboring plasmids with tandem *OrpXylA* was beneficial for glucose and xylose cofermentation. *PirXylA* and *OrpXylA* genes were also individually integrated into the genome of yeast strains in multiple copies. Such integration was beneficial for xylose alcoholic fermentation. The respiration-deficient strain carrying episomal or integrated *OrpXylA* genes exhibited the best performance for glucose and xylose co-fermentation. This was partly attributed to the high expression levels and activities of xylose isomerase. Mating a respiration-efficient strain carrying the integrated *PirXylA* gene with a respiration-deficient strain harboring integrated *OrpXylA* generated a diploid recombinant xylose-fermenting yeast strain STXQ with enhanced cell growth and xylose fermentation. Co-fermentation of 162 g L^{-1} glucose and 95 g L^{-1} xylose generated 120.6 g L^{-1} ethanol in 23 h, with sugar conversion higher than 99%, ethanol yield of 0.47 g g^{-1}, and ethanol productivity of 5.26 g $L^{-1} \cdot h^{-1}$.

Ethanol can be produced from renewable resources such as crops or agricultural waste. It is therefore a sustainable and clean fuel. Further growth in bioethanol production largely depends on the effective conversion of lignocellulosic feedstock such as agricultural and forestry wastes to bioethanol because they are the most abundant polymers of fermentable sugars [1–3].

Saccharomyces cerevisiae is the most effective microorganism for fermenting sugars to ethanol due to its rapid sugar consumption rate, high sugar and ethanol tolerance, and resistance to biomass-derived inhibitors [4,5]. Much research has been done to genetically engineer *S. cerevisiae* strains for xylose fermentation [6–10].

Two xylose-assimilating pathways were heterologously engineered in *S. cerevisiae* for xylose-fermenting yeast construction. One focused on the D-xylose isomerase (XI) pathway [11,12], the other focused on the D-xylose reductase (XR) and xylitol dehydrogenase (XDH) pathway [10,13–15]. In the XI pathway, xylose is first isomerized into xylulose by XI and xylulose was then phosphorylated into xylulose 5-phosphate by xylulokinase. Xylulose is subsequently metabolized to ethanol through glycolysis in the pentose phosphate pathway [11]. As no xylitol is produced in this pathway, much research focused on the XI pathway construction in xylose-fermenting yeast [12].

However, for all the engineered strains developed, rapid glucose and xylose co-fermentation is still challenging, in particular when sugar concentration is high [4,8–10]. Because fermentation time on mixed-substrate hydrolysates is still not cost-effective, strategies in evolutionary engineering were used to improve fermentation kinetics [16–18] and much research focused on the search for new or engineered sugar transporters [19–21]. More recently, robust diploid *S. cerevisiae* strains were developed for rapid xylose-fermentation [22–25].

This study aims to develop a recombinant *S. cerevisiae* strain for rapid glucose and xylose co-fermentation through metabolic engineering, evolutionary engineering and strain mating. Firstly, four episomal plasmids containing the two-copy codon-optimized *Piromyces* sp. E2 XI gene (*PirXylA*, GenBank accession number AJ249909.1), one-copy codon-optimized *Orpinomyces* sp. ukk1 XI gene (*OrpXylA*, GenBank accession number EU411046), one-copy *PirXylA* and *OrpXylA* in tandem, and two-copy *OrpXylA*, were individually constructed. They were subsequently transformed to an evolved respiration-deficient yeast strain. Four engineered strains with episomal XI genes were generated and optimal XI functional expression was identified. Afterwards, *PirXylA* and *OrpXylA* were separately integrated into the genome of two evolved *S. cerevisiae* strains (one respiration-efficient and the other respiration-deficient) in multiple copies according the methods described previously [26,27]. Four engineered yeasts with integrated XI genes were generated and screened for xylose fermentation or glucose/xylose co-fermentation. In the end, a diploid recombinant xylose-fermenting *S. cerevisiae* was constructed by mating a respiration-efficient haploid strain with a respiration-deficient haploid strain. To the best of our knowledge, this is the first report on diploid xylose-fermenting yeast strain construction by such strain mating. The resulted diploid yeast strain displayed superior glucose and xylose co-fermentation performance, which far exceeded that by engineered *S. cerevisiae* reported to-date [28,29].

MATERIALS AND METHODS

Plasmid Construction

All plasmids used in this work are listed in Table 1. All primers used in this study are listed in Table S1. *E. coli* strain DH5α (Life Technologies, Rockville, MD, USA) were used as the transformation host for plasmid construction. *E. coli* were grown in LB medium containing 100 μg/mL ampicillin at 30°C or 37°C for plasmid maintenance [25]. The *PGK1* promoter was amplified from genomic DNA of *S. cerevisiae* strain ATCC 24860 and was ligated to pYES2 (Thermo-Fischer Scientific, Singapore) to replace the *GAL1* promoter, resulting in pPY1. *PirXylA* and *OrpXylA* were synthesized by Integrated DNA Technologies Singapore). Cassettes *PGK1*p-*OrpXylA*-*CYC1*t, *PGK1*p-*PirXylA*-*CYC1*t-*PGK1*p-*OrpXylA*-*CYC1*t, *PGK1*p-*OrpXylA*-*CYC1*t-*PGK1*p-*OrpXylA*-*CYC1*t and *PGK1*p-*PirXylA*-*CYC1*t-*PGK1*p-*PirXylA*-*CYC1*t were cloned into pPY1 individually, resulting in plasmids pPYXo, pPYXpXo, pPYXoXo and pPYXpXp (Figure 1A–D).

NTS2-2 partial fragment (pNTS) was obtained by overlap extension polymerase chain reaction (OE-PCR) of *S. cerevisiae* ATCC 24860 genomic DNA over 274 bp upstream and 245 bp downstream homologous regions of NTS2-2. The XXUN plasmid (Figure 1E) is an integrating yeast plasmid based on pPYXpXp whereby the 2 μ origin was replaced with pNTS. The *loxP-KanMX4-loxP* cassette was amplified from the plasmid pUG6 [30]. Cassettes *PGK1*p-*OrpXylA*-*CYC1*t and *loxP-KanMX4-loxP*-

pNTS were obtained by OE-PCR and then subsequently constructed into plasmid pUC19 [12], resulting in plasmid XoNK (Figure 1F). Both plasmids were digested with *Swa*I for XI genome integration using NTS2-2 homologous recombinant arms.

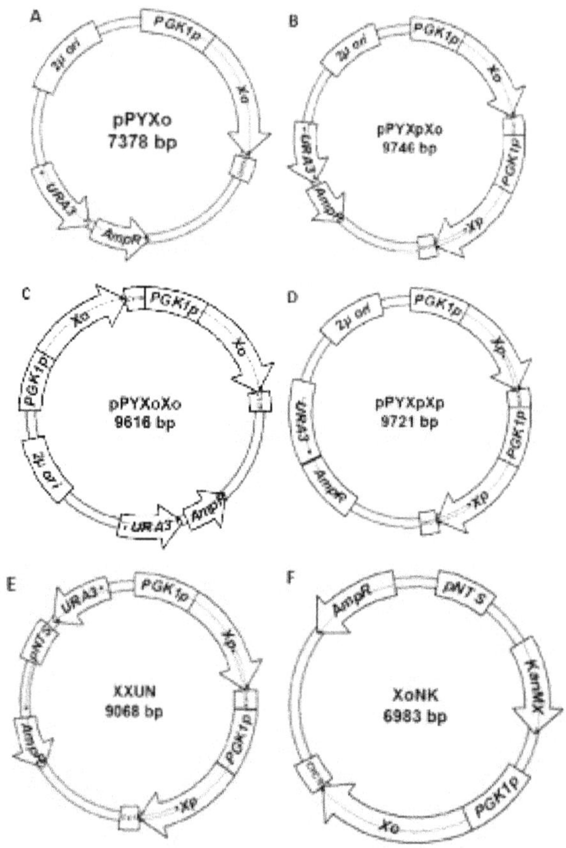

Figure 1. Map of plasmids. (A) pPYXo; (B) pPYXpXo; (C) pPYXoXo; (D) pPYXpXp; (E) XXUN; and (F) XoNK.

Strain Construction and Adaptive Evolution

All strains used in this work are listed in Table 1. *S. cerevisiae* haploid strains JUK36□ and JUK39a were isolated from the diploid strain *S. cerevisiae* ATCC 24860. They were both overexpressed with the non-oxidative pentose phosphate pathway (PPP) genes and xylulokinase gene, *XKS1*. *URA3* and *GRE3* genes were disrupted in both haploid strains. In addition, the *CYC3* gene, encoding cytochrome c heme lyase, was knocked out in strain JUK39a to eliminate respiration [12].

Plasmid pPYXpXp was transformed into strain JUK39a using the LiAc/SS carrier DNA/PEG method [31], resulting in recombinant *S. cerevisiae* 39aXpXp. The respiration-deficient strain 39aXpXp was evolved by continuous transfer and cultivation in a synthetic medium (SM) containing 6.7 g L^{-1} yeast nitrogen base without amino acids (YNB) (Difco Laboratories Inc., Detroit, MI, USA) and 20 g L^{-1} xylose (SMX) under oxygen-limited conditions according to the protocols described in our earlier report [12]. After 75-day continuous transfer, cell doubling time was reduced from 15.9 h to 6.4 h. Samples were taken on day 75 and streaked on SMX plates containing 20 g L^{-1} xylose and 20 g L^{-1} agar. Fifteen large single colonies were selected. They were then incubated in 50 mL SMX medium in 100 mL Erlenmeyer flasks capped with rubber stoppers, shaken at 200 rpm and 30°C. Weight loss of the cultures from the 15 colonies was individually measured on Day 4. The best ethanol-producing strain was indicated by the highest weight loss [18]. This strain was denoted 39aXpXp2415, and plasmid pPYXpXp in this strain was removed by streaking the culture on 5-FOA plates [12]. This generated the background strain 39a2 (Table 1). On the other hand, strain JUK51a_2 (Table 1) was evolved an aerobically in a chemostat with an increase in the dilution rate from 0.01 to 0.14 h^{-1} on xylose using the method described in our earlier report [12]. The fastest growing strain was selected, and the plasmid pJFX11 was removed according to the above-described method on 5-FOA plates. The background strain 36α2 was later obtained (Table 1).

Table 1. Strains and plasmids used in this study.

Strains/Plasmid	Genotype/Phenotype	References
Saccharomyces cerevisiae strains		
Saccharomyces cerevisiae ATCC 24860	Obtained from American Type Culture Collection (ATCC)	
JUK36α	*S. cerevisiae* ATCC 24860 segregant; *MAT*; *ura3::loxP*; *TKL1::RKIt-RKI1-ADH1p-RPE1t-RPE1-TPI1p-loxP-XKS1t-XKS1-PGK1p-* *PDC1p-TAL1-TAL1t-FBA1p*; *gre3::loxP*	[12]
JUK39a	*S. cerevisiae* ATCC 24860 segregant; *MATa*; *ura3::loxP*; *TKL1::RKIt-RKI1-ADH1p-RPE1t-RPE1-TPI1p-loxP-XKS1t-XKS1-PGK1p-* *PDC1p-TAL1-TAL1t-FBA1p*; *gre3::loxP*; *cyc3::loxP*	[12]
JUK51a_2	JUK36α derivative; {pJFX11}/(*BvuXylA*, *XK*, *PPP*, *gre3Δ*)	[12]
39aXpXp	JUK39a derivative; {pPYXpXp}/(two-copy *PirXylA*, *XK*, *PPP*, *gre3Δ*, *cyc3Δ*)	This work
39aXpXp2415	39aXpXp derivative; {pPYXpXp}/(two-copy *PirXylA*, *XK*, *PPP*, *gre3Δ*, *cyc3Δ*, AE)	This work
36α2	Isolate from chemostat anaerobic and adaptive evolution at a dilution rate of 0.15 h-1 on xylose of JUK51a_2 and loss of plasmid pJFX11	This work

Strains/Plasmid	Genotype/Phenotype	References
39a2	Isolate from 39aXpXp2415 and loss of plasmid pPYXpXp	This work
39a2XpXp	39a2 derivative; {pPYXpXp}/(two-copy *PirXylA*, *XK*, *PPP*, *gre3*Δ, *cyc3*Δ)	This work
39a2XpXo	39a2 derivative; {pPYXpXo}/(*OrpXylA*, *PirXylA*, *XK*, *PPP*, *gre3*Δ, *cyc3*Δ)	This work
39a2Xo	39a2 derivative; {pPYXo}/(*OrpXylA*, *XK*, *PPP*, *gre3*Δ, *cyc3*Δ)	This work
39a2XoXo	39a2 derivative; {pPYXoXo}/}/(two-copy *OrpXylA*, *XK*, *PPP*, *gre3*Δ, *cyc3*Δ)	This work
36α2XpXpUN	36α2 derivative;*NTS2-2*::two-copy *PirXylA*, *ura3*, *XK*, *PPP*, *gre3*Δ	This work
36α2XoNK	36α2 derivative;*NTS2-2*::*OrpXylA-KanMX4*, *XK*, *PPP*, *gre3*Δ, *ura3*Δ	This work
39a2XpXpUN	39a2 derivative; *NTS2-2*::two-copy *PirXylA*, *ura3*, *XK*, *PPP*, *gre3*Δ, *cyc3*Δ	This work
39a2XoNK	39a2 derivative; *NTS2-2*::*OrpXylA-KanMX4*, *XK*, *PPP*, *gre3*Δ, *cyc3*Δ, *ura3*Δ	This work
STXQ	Isolate from mating of 36α2XpXpUN with 39a2XoNK	This work
Plasmids		
pUG6	*E. coli* plasmid with segment *loxP–KanMX4–loxP*	[30]
pJFX11	YEp, *TEF1*p-*BvuXylA*-*CYC1t*	[12]
pPY1	pPYES2; *GAL1*p replaced by *PGK1*p	This work
pPYXo	pPY1; *PGK1*p-*OrpXylA*-*CYC1*t	This work
pPYXpXp	pPY1; 2 copies of *PGK1*p-*PirXylA*-*CYC1*t in tandem	This work
pPYXpXo	pPY1; *PGK1*p-*PirXylA*-*CYC1*t-*PGK1*p-*OrpXylA*-*CYC1*t	This work
pPYXoXo	pPY1; 2 copies of *PGK1p-OrpXylA-CYC1t* in tandem	This work
XXUN	pPYXpXp-based yeast integration plasmid; 2 μ and *ura3* were replaced with *ura3* and *NTS2-2* partial fragment	This work
XoNK	pUC19-based yeast integration plasmid; *loxP-KanMX4-loxP-pNTS-PGK1*p-*OrpXylA-CYC1*t	This work

Plasmids pPYXo, pPYXpXo, pPYXoXo and pPYXpXp were individually transformed into 39a2, resulting in strains 39a2Xo, 39a2XpXo, 39a2XoXo and 39a2XpXp, respectively. Plasmids XoNK and XXUN were digested with *Swa*I and linearized. The linearized fragments were integrated into the genome of 39a2 and 36α2 at the NTS2-2 site, individually, resulting in recombinant strains 39a2XoNK, 39a2XpXpUN, 36α2XoNK and 36α2XpXpUN. For each plasmid transformation, a pool of transformants were generated. The best transformant was isolated based on its cell growth and ethanol production in xylose medium as described in the isolation of 39aXpXp2415. Recombinant strains, 36α2XpXpUN and 39a2XoNK, were later mated to obtain strain STXQ through screening on SMX agar plates containing 20 g L^{-1} xylose at pH 6 followed by fermentation in SMX medium containing up to 250 g L^{-1} xylose.

Enzyme Activity Assay

Cells were grown to the exponential phase in SMX medium containing 20 g L^{-1} xylose. After centrifugation, cells were washed twice with chilled distilled water, and then lysed in chilled extraction buffer (100 mM Tris-HCl, 10 mM $MgSO_4$, pH 7.5) by vortex mixing using a Vortex Mixer (Mixer UZUSIO, Tokyo, Japan) with 0.5 mg of 0.5 mm glass beads (Sigma-Aldirich, Singapore). Protease inhibitor cocktail set V (Merck Millipore, Singapore) was added to inhibit serine and cysteine generated in the lysis process. Cell debris was removed by centrifugation (Micro-centrifuge D3024, DR. LAB Technology Hong Kong, Hong Kong, China) at 4000× *g* for 10 min at 4 °C, and the crude extract was stored for enzyme activity assay. The protein concentration of the cell extract was determined by the Bradford Assay using a Coomassie Protein Assay Kit (Thermo Scientific, Singapore), and bovine serum albumin (BSA) was used as the standard. Extraction of raw proteins from the yeast strains was performed in duplicate.

The in-vitro XI activity was determined by measuring NADH absorbance using sorbitol dehydrogenase (SDH) (Sigma Aldrich, Singapore). The assay mixture (1 mL) contained extraction buffer, 0.15 mM NADH, 1 U SDH, and 50 μL crude extract. It was equilibrated at 30°C for 2 min. The reaction was started by the addition of D-xylose to a final concentration of 500 mM. The change of NADH concentration within 3 min was detected using a UV-visible spectrophotometer (Shimadzu, Tokyo, Japan) at wavelength 340 nm, and the specific activity of XI in the recombinant strains was determined [32]. A molar extinction coefficient of 6.25 $(mM\ cm)^{-1}$ at 340 nm for NADH was used to calculate specific activity. Specific activity was expressed as units per mg protein. One unit of enzyme activity was defined as the amount of enzyme required to oxidize 1 μmol of coenzyme/min, under the specified reaction conditions [12].

Glucose and Xylose Fermentation by the Recombinant Strains

The preculture of the evolved engineered strains was prepared by growing them in 40 mL SMX medium containing 20 g L^{-1} xylose in 100 mL Erlenmeyer flaks at 200 rpm and 30°C for 24 h. The oxygen-limited conditions in the flasks was maintained by capping the flasks with rubber stoppers pierced with a needle to allow the release of CO_2. Cells in the exponential phase were harvested by centrifugation

(Microcentrifuge D3024, DR. LAB Technology Hong Kong, Hong Kong, China) at 14,000× *g* for 1 min. Cell pellets were washed twice and were then inoculated into SM medium supplemented with 20 g L^{-1} xylose with (SMGX) or without 20 g L^{-1} glucose (SMX). The initial optical cell density at 600 nm (OD600) of the culture was about 2 unless otherwise stated. Fermentation was conducted in 100 mL Erlenmeyer shaking flasks under oxygen-limited conditions with a working volume of 40 mL at 200 rpm and 30°C. The pH value was adjusted at 5.0–6.0 using 3 M NaOH during fermentation. All fermentation experiments were performed in duplicate. Samples were taken periodically to measure OD600, sugar and metabolite concentration.

Analytical Methods

Cell densities (OD600) were determined using a UV-visible spectrophotometer (Shimadzu, Tokyo, Japan). Fifty-mL cell cultures with varying OD600 (1–5) were filtered with 0.22 μm glass fiber filter membrane (Merck Millipore, Singapore) using Aspirator A-3S (Fisher Scientific, Tokyo, Japan). Cells were washed twice with distilled water, dried at 105°C in an oven for 24 h, and then weighed. One OD600 unit corresponded to 0.241 g L^{-1} dry cell weight (DCW). Concentrations of glucose, xylose, xylitol, acetate, glycerol and ethanol produced in fermentation were determined by Agilent 1200 series HPLC system (Agilent Technologies, Santa Clara, CA, USA) equipped with a refractive index detector RID-10A using an Aminex HPX-87H ion exchange column (Bio-Rad Laboratories, Woodinville, WA, USA). The column was eluted at 60 °C with 5 mM of sulfuric acid as the mobile phase at a flow rate of 0.6 mL min^{-1}.

Quantitative Reserve Transcription Polymerase Chain Reaction (RT-PCR)

Recombinant yeast strains 39a2Xo, 39a2XpXo, 39a2XoXo and 39a2XpXp were individually cultivated in 40 mL SMX medium containing 20 g L^{-1} xylose under oxygen-limited conditions at 200 rpm and 30°C. The expression of XI gene transcripts was determined by quantitative reverse transcription polymerase chain reaction (qRT-PCR). Primers for RT-PCR are listed in Table S1. Samples were taken at 24 h, and cells were harvested by centrifuging 2-mL culture at 14,000× *g* and 4°C for 1 min. Cell pellets were washed twice using double distilled water and total RNA was isolated by using the E.Z.N.A™ Yeast RNA Kit (Omega Bio-tek, Norcross, GA, USA). First-strand cDNA was obtained by using the SuperScript® First-Strand Synthesis System for RT-PCR (Invitrogen, Carlsbad, CA, USA). Such cDNA was then used as the template for qRT-PCR using iCycler iQ™ Real-time PCR Detection System (Bio-Rad Laboratories, Woodinville, WA, USA) and FastStart Universal SYBR Green Master (Roche Applied Science, Penzberg, Germany). The cycle threshold values were calculated with the Optical System Software Version 3.1 (Bio-Rad Laboratories,

Woodinville, WA, USA), and the detection threshold over the cycle range was set at 2 to 10. Each PCR was carried out in duplicate. All kits were used under conditions recommended by the manufacturers. The $2^{-\Delta\Delta Ct}$ method [33] was used to analyze the relative changes in gene expression using the housekeeping *ACT1* gene as the reference.

Biomass Hydrolysate Fermentation Using Strain S. cerevisiae STXQ

The diploid recombinant *S. cerevisiae* strain STXQ (Table 1) was inoculated into YP medium (10 g L^{-1} yeast extract, 20 g L^{-1} peptone, pH 5.0) containing 132 g L^{-1} xylose or mixture of 162 g L^{-1} glucose and 95 g L^{-1} xylose at an initial OD600 of about 13. Fermentation was conducted in 40 mL YP medium in 100 mL shaking flasks under oxygen limited conditions at 200 rpm and 30°C.

Oil palm empty fruit bunch (OPEFB) Hydrolysate was obtained using crude cellulase from *Trichoderma reesei* Rut-C30 according to the protocols described in our earlier report [34]. OPEFB hydrolysate was sterilized using 0. 22 μm filter membrane (Merck Millipore, Singapore) and was supplemented with 7 g L^{-1} yeast extract 2 g L^{-1} peptone, 2 g L^{-1} $(NH_4)_2SO_4$, 2.05 g L^{-1} KH_2PO_4, and 0.25 g L^{-1} Na_2HPO_4. The diploid recombinant *S. cerevisiae* strain STXQ was inoculated into the above OPEFB hydrolysate medium with an initial OD600 about 5. Fermentation was carried out in 40 mL fermentation medium in 100 mL shaking flasks under oxygen-limited conditions at 200 rpm and 30 °C with an initial pH of 4.48. Samples were taken periodically for OD600, sugar and metabolite analysis. Experiments were conducted in duplicate.

RESULTS

Expression of XIs with Various Combinations

XI activities were assayed for 39a2 recombinant strains with episomal XI genes (Table 2). XI activity of 39a2XoXo was three times higher than that of 39a2XpXp and 2.5 times higher than that of 39a2Xo. Strain 39a2XoNK presented the highest XI activity among all the yeast strains with integrated XI genes, followed by 39a2XpXpUN. Quantitative RT-PCR showed that the transcription level of XI gene in the engineered strains significantly increased compared to that in the parent strain 39a2 (Table 3). In 36α2XpXpUN and 39a2XoNK, respective 1.04-fold and 1.41-fold increases in XI gene transcription levels were observed compared to those in 39a2XpXp and 39a2XoXo. Inconsistent XI activity and transcription level were observed.

Table 2. Specific activity of xylose isomerase in the recombinant strains.

Strains	Specific Activity (U mg-1 Protein)
39a2XpXp	0.10 ± 0.003
39a2XpXo	0.11 ± 0.019
39a2Xo	0.12 ± 0.041
39a2XoXo	0.30 ± 0.079
36α2XpXpUN	0.11 ± 0.007
36α2 XoNK	0.04 ± 0.005
39a2XpXpUN	0.26 ± 0.004
39a2XoNK	0.72 ± 0.006

The results represent the mean ± standard deviation of duplicate independent experiments.

Table 3. Fold-change in xylose isomerase (XI) mRNA expression.

Strain	Fold-Change a	
39a2XpXp	50.21 (47.81–52.74)	nil
39a2XpXo	29.86 (28.43–31.36)	11.71 (10.62–12.92)
39a2Xo	nil	59.71 (51.98–68.59)
39a2XoXo	nil	59.71 (55.72–64)
36α2XpXpUN	51.98 (49.50–54.60)	nil
39a2XoNK	nil	84.45 (59.71–119.43)

a Fold-change of XI mRNA level in the recombinant yeast strains compared to the parent strain 39a2 . Results were based on duplicate independent real-time RT-PCR reactions.

Fermentation of Glucose and Xylose by Engineered 39a2 Strains with Episomal XI Genes.

Fermentation performance of 39a2 recombinant strains with episomal XI genes was tested in SMX medium containing 20 g L^{-1} xylose under oxygen-limited conditions. Apparently, strain 39a2XpXp displayed the fastest xylose utilization rate, followed by 39a2XoXo (Figure 2). Strain 39a2XpXo utilized xylose more slowly than 39a2XpXp, though it demonstrated almost the same XI activity (Table 2). On the other hand, strain 39a2XoXo showed a faster xylose utilization rate than strain 39a2Xo; however, the former displayed much higher activity than the latter. On the contrary, xylose utilization results accorded quite well with the results of qRT-PCR analysis showing that strain 39a2XpXp had higher XI gene transcription levels than strain 39a2XpXo (Table 3) and a faster xylose utilization rate (Figure 2). However, strains 39a2XoXo and 39a2Xo displayed identical XI gene transcription levels (Table 3), almost the same xylose consumption rate (Figure 2) and xylose conversion (Table 4). Interestingly, strain 39a2XpXp also exhibited the maximal ethanol yield of 0.472 g g^{-1} (Table 4). Such results suggest that expression of two-copy *PirXylA* is favourable for xylose alcoholic fermentation when xylose is used as the sole carbon source.

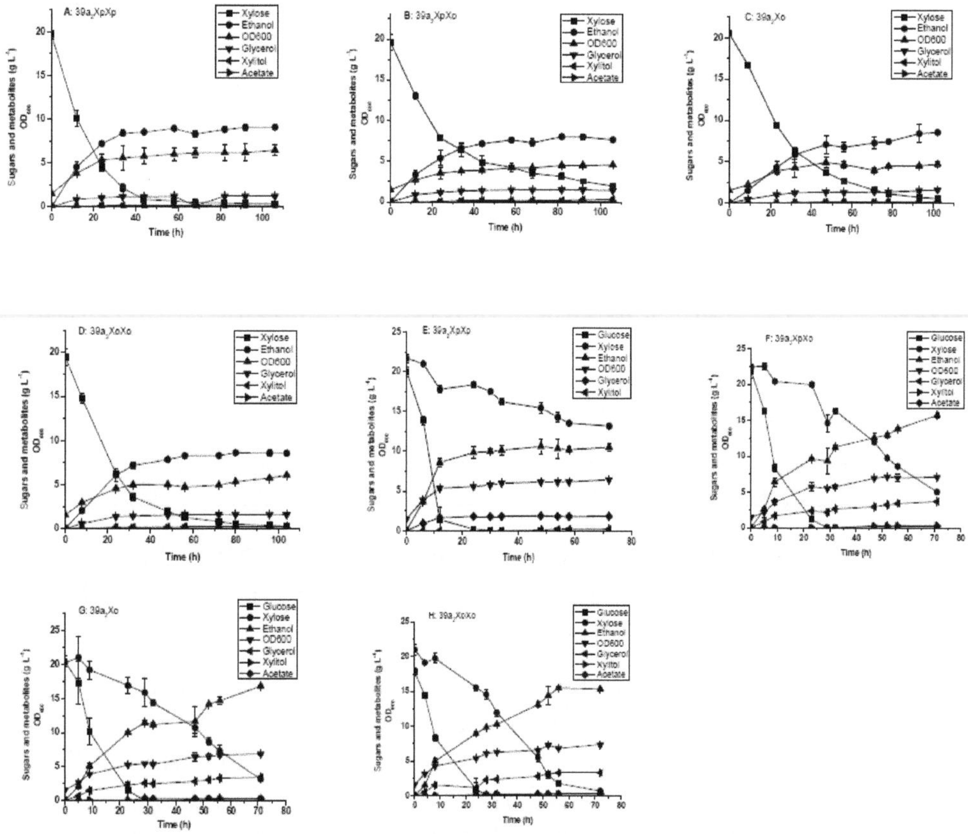

Figure 2. Sugar fermentation under oxygen-limited conditions by the engineered *S. cerevisiae* strains in SM medium containing 20 g L^{-1} xylose (SMX) or 20 g L^{-1} glucose and 20 g L^{-1} xylose (SMGX). (A) 39a2XpXp SMX; (B) 39a2XpXo SMX; (C) 39a2Xo SMX; (D) 39a2XoXo SMX; (E) 39a2XpXpSMGX; (F) 39a2XpXo SMGX; (G) 39a2Xo SMGX; (H) 39a2XoXo SMGX. The results represent the mean ± standard deviation of duplicate independent experiments ($p < 0.01$).

Table 4. Glucose and xylose fermentation by recombinant xylose-fermenting yeast strains.

39a2xpXp [a]	-	20	98% ± 0.004	0.472 ± 0.013	0.098 ± 0.004	0.014 ± 0.001
39a2XpXo [a]	-	20	89.8% ± 0.011	0.434 ± 0.017	0.098 ± 0.004	0.011 ± 5.89 × 10^{-6}
39a2Xo [a]	-	20	97.3% ± 0.0138	0.428 ± 0.026	0.084 ± 0.001	0.011 ± 0.0003
39a2XoXo [a]	-	20	98.5% ± 0.0057	0.449 ± 0.038	0.108 ± 0.001	0.013 ± 2.29 × 10^{-5}

36α2XpXpUN [a]	-	20	86.4% ± 0.052	0.318 ± 0.01	0.069 ± 0.007	0.014 ± 0.0003
36α2XoNK [a]	-	20	6.42% ± 0.057	ND	ND	0.003 ± 0.0006
39a2XpXpUN[a]	-	20	67.0% ± 0.089	0.398 ± 0.048	0.067 ± 0.001	0.013 ± 0.0002
39a2XoNK [a]	-	20	99.0% ± 0.002	0.368 ± 0.029	0.107 ± 0.007	0.0098 ± 0.0003
39a2XpXp [a]	20	20	22.0% ± 0.003	0.444 ± 0.029	0.221 ± 0.020	0.020 ± 0.0007
39a2XpXo [a]	20	20	77.5% ± 0.017	0.392 ± 0.001	0.081 ± 0.006	0.022 ± 0.0001
39a2Xo [a]	20	20	84.3% ± 0.0007	0.428 ± 0.01	0.238 ± 0.003	0.021 ± 0.0004
39a2XoXo [a]	20	20	96.7% ± 0.011	0.449 ± 0.022	0.213 ± 0.004	0.021 ± 0.001
36α2XpXpUN [a]	20	20	73.8% ± 0.011	0.348 ± 0.049	0.225 ± 0.007	0.029 ± 0.0006
36α2XoNK [a]	20	20	2.62% ± 0.000	0.390 ± 0.043	0.222 ± 0.003	0.016 ± 0.0013
39a2XpXpUN [a]	20	20	25.6% ± 0.097	0.405 ± 0.049	0.213 ± 0.002	0.013 ± 0.0002
39a2XoNK [a]	20	20	97.3% ± 0.006	0.387 ± 0.002	0.243 ± 0.003	0.020 ± 0.0005
36α2XpXpUN [b]	-	40	75.9% ± 0.041	0.384 ± 0.033	0.108 ± 0.011	0.013 ± 0.0004
39a2XoNK [b]	-	40	81.6% ± 0.004	0.421 ± 0.004	0.131 ± 0.000	0.010 ± 0.0003
STXQ [b]	-	40	93.3% ± 0.019	0.393 ± 0.024	0.146 ± 0.009	0.017 ± 0.0002
STXQ [b]	-	132	100%	0.498 ± 0.006	1.13 ± 0.01	0.014 ± 0.0004
STXQ [c]	162	95	99.27% ± 0.002	0.475 ± 0.01	5.24 ± 0.02	0.024 ± 0.0001

[a] Fermentation in SM medium in 72 h; [b] Fermentation in YP medium in 102 h; [c] Fermentation in YP medium. The results represent the mean ± standard deviation of duplicate independent experiments.

Co-fermentation of 20 $g\ L^{-1}$ glucose and 20 $g\ L^{-1}$ xylose was carried out in SM medium under oxygen-limited conditions. It can be seen that for all strains, glucose was almost completely consumed at 24 h (Figure 2). However, xylose utilization varied greatly for the four engineered 39a2 strains. Noticeably, in the presence of 20 $g\ L^{-1}$ glucose, strain 39a2XpXp utilized xylose quite slowly and about 22% xylose was consumed at 72 h (Table 4). On the other hand, xylose utilization was improved to 77.5% by strain 39a2XpXo. Moreover, xylose consumption was improved to 84.3% by strain 39a2Xo and it was further improved to 97% by 39a2XoXo (Table 4). It is worth noting that strain 39a2XoXo presented the highest ethanol yield (0.449 $g\ g^{-1}$). The above results suggest that the expression of two-copy of *OrpXylA* is beneficial for glucose and xylose co-fermentation.

Fermentation of Glucose and Xylose by Yeast Strains Engineered with Integrated XI Genes

Fermentation performance of 39a2 recombinant strains with integrated XI genes was tested in SM medium containing 20 $g\ L^{-1}$ xylose (SMX) or 20 $g\ L^{-1}$ glucose and 20 $g\ L^{-1}$ xylose (SMGX) under oxygen-limited conditions (Figure 3). In SMX fermentation, the specific growth rate of 36α2XpXpUN on xylose was 0.014 h⁻1, much higher than that of 36α2XoNK (0.003 h⁻1), whereas xylose conversion by strain 39a2XoNK was 99.0%, much higher than that by 39a2XpXpUN (67.0%) (Table 4). In SMGX fermentation, the specific growth rate of 36α2XpXpUN on xylose was 0.029 h⁻1, much higher than that of 36α2XoNK (0.016 h⁻1), whereas xylose conversion by strain 39a2XoNK was 97.30%, much higher than that by 39a2XpXpUN (73.8%) (Table 4). In both SMX and SMGX fermentation, strain 36α2XpXpUN presented the best cell growth, whereas strain 39a2XoNK displayed the best xylose conversion. Furthermore, strain 39a2XoNK exhibited the highest ethanol productivity in both SMX and SMGX fermentation. Despite the possible XI gene copy number variation in these strains, the above results demonstrated that the elimination of respiration was favourable for xylose fermentation, which corroborated quite well with previous reports [12,14]. Strains 36α2XpXpUN and 39a2XoNK were therefore selected for mating to generate the diploid recombinant strain STXQ to attain good cell growth and xylose fermentation.

Fermentation of Glucose and Xylose by the Recombinant Diploid Strain STXQ

Fermentation performance of STXQ was tested in YP medium containing 40 g L⁻1 xylose under oxygen-limited conditions. As expected, strain STXQ inherited the advantages of both parent strains, 36α2XpXpUN and 39a2XoNK. It presented 93.3% xylose conversion and a specific growth rate of0.017 h⁻1; both were higher than those of its parent strains (Table 4). In addition, ethanol productivitywas greatly improved (Figure 3, Table 4). A very minimum amount of glycerol was produced by STXQ, although glycerol production was notable for the respiration-deficient parent strain 39a2XoNK (Figure 3). Moreover, production of xylitol and acetate was almost undetectable.

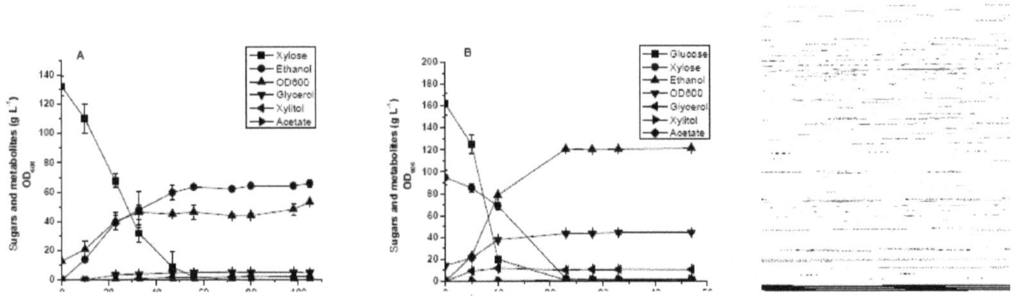

Figure 3. Xylose fermentation in YPX medium containing 40 g L⁻1 xylose under oxygen-limited conditions. (A) 36α2XpXpUN; (B) 39a2XoNK; (C) STXQ. The results represent the mean ± standard deviation of duplicate independent experiments ($p < 0.01$).

Fermentation performance of strain STXQ was further invested in YP medium containing high-concentration sugar. With an initial OD600 of about 13, strain STXQ consumed 100% xylose in fermenting 132 g L^{-1} xylose and produced 65.8 g L^{-1} ethanol at 46 h. The corresponding ethanol yield was 0.498 g g^{-1} (Figure 4A, Table 4). In fermentation of 162 g L^{-1} glucose and 95 g L^{-1} xylose with about the same initial OD600, more than 99% glucose and xylose were co-utilized within 23 h (Figure 4B, Table 4). Ethanol titer reached 120.6 g L^{-1} corresponding to an ethanol volumetric productivity of 5.26 g L^{-1} h^{-1}. These are so far the highest values compared to those reported in the literature. The above results indicate that the strain development strategy elaborated in this study is efficient in generating a robust *S. cerevisiae* strain with improved xylose fermentation capabilities.

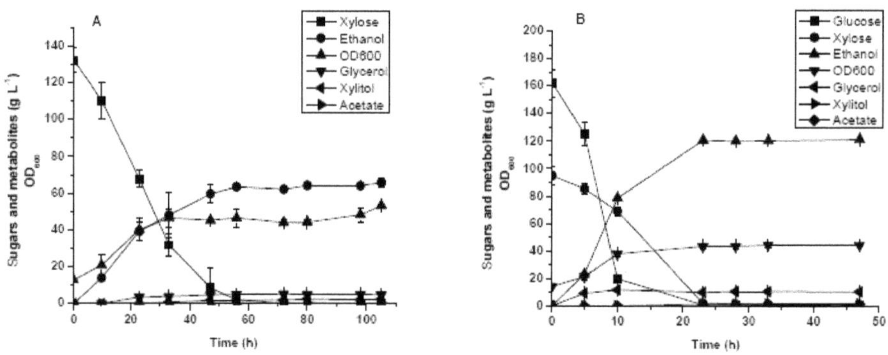

Figure 4. High-titer sugar fermentation under oxygen-limited conditions by strain STXQ in YP media. (A) Xylose fermentation in YPX medium containing 132 g L-1 xylose; (B) Glucose and xylose co-fermentation in YPGX medium containing 162 g L-1 glucose and 95 g L-1 xylose. The results represent the mean ± standard deviation of duplicate independent experiments ($p < 0.01$).

Oil Palm Empty Fruit Bunch Hydrolysate Fermentation by Diploid Recombinant Strain STXQ

Fermentation performance of strain STXQ was further tested in OPEFB hydrolysate containing41.81 g L^{-1} glucose, 30.00 g L^{-1} xylose, 7 g L^{-1} yeast extract, 2 g L^{-1} peptone, 2 g L^{-1} $(NH_4)_2SO_4$,2.05 g L^{-1} KH_2PO_4, and 0.25 g L^{-1} Na_2HPO_4 under oxygen-limited conditions with an initial OD600of about 10. Strain STXQ consumed 95.3% glucose and 88.9% xylose (Figure 5). The pH value decreased from 4.48 to 4.00 within 72 h. Strain STXQ consumed 94.0% total sugar from the OPEFB hydrolysate without detoxification. The specific cell growth rate (μmax) reached 0.013 h^{-1} and the ethanol yield was0.420 g g^{-1}. Glucose was quickly consumed, and glucose-xylose co-fermentation was clearly observed within 24 h. Subsequently, ethanol concentration kept increasing with almost the sole consumption of xylose until it reached the final titer of 28.4 g L¯1 at 72 h.

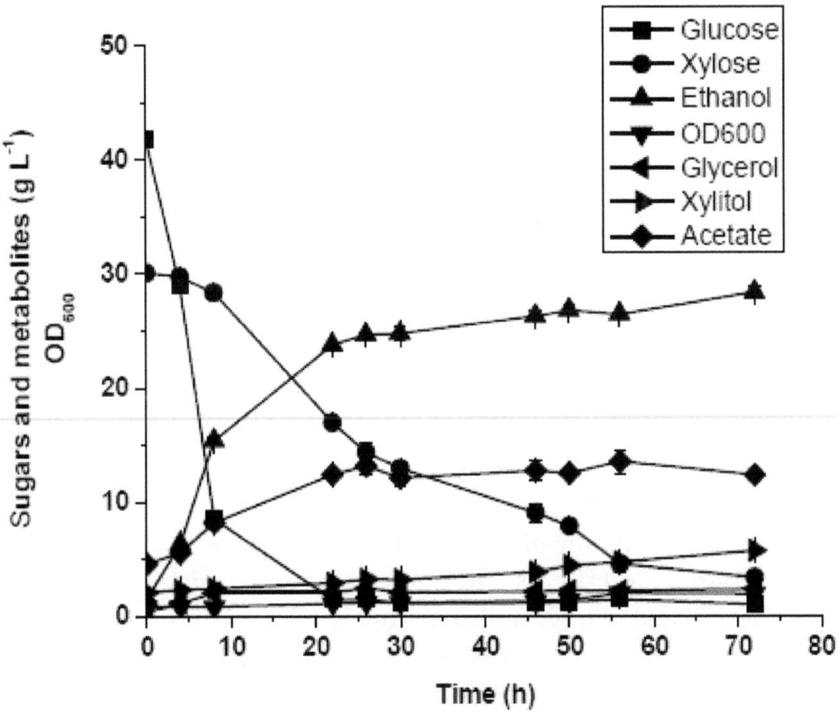

Figure 5. Sugar fermentation under oxygen-limited conditions by strain STXQ in oil palm empty fruit bunch hydrolysate supplemented with 7 g L-1 yeast extract, 2 g L-1 peptone, 2 g -1 (NH4)2SO4, 2.05 g L-1 KH2PO4, and 0.25 g L-1 Na2HPO4. The results represent the mean ± standard deviation of duplicate independent experiments (p < 0.01).

DISCUSSION

Episomal plasmids containing *OrpXylA*, *PirXylA* and *OrpXylA* in tandem, two-copy *PirXylA* and two-copy *OrpXylA* expression cassettes (Figure 1) were transformed into the background strain 39a2 individually. Among all the 39a2 strains harboring episomal XI genes, strain 39a2XpXp presented the fastest rate of xylose utilization when xylose was used as the sole carbon source. However, it did not exhibit the highest XI activity and transcription level (Tables 2 and 3). Inconsistent XI activity and transcription level in recombinant xylose-fermenting *S. cerevisiae* strain was reported and it could be due to rearrangement of pentose phosphate pathway (PPP) genes, decreased glycolysis activity, repressed respiration activity, and enhanced gluconeogenesis [35]. Besides XI activity, the enhanced xylose utilization and fermentation could be associated with the elevated expression of sugar transporter genes, non-oxidative pentose phosphate pathway (PPP) genes such as *TAL1*, *TKL1*, *RKI1*, and *RPE1* and xylulokinase gene, *XKS1* [19,20,25,36].

For glucose and xylose co-fermentation, strain 39a2XoXo presented the best glucose and xylose co-utilization (Figure 2) and exhibited the highest xylose conversion (96.7%) (Table 4). *S. Cerevisiae* does not contain specific xylose transporters. Xylose was therefore transported by glucose transporters. Xylose affinity for the glucose transporters was very low at a high glucose concentration; however, it could increase to a similar level of glucose affinity at low glucose concentration [19,20]. As a result, diauxic lag is still a practical problem associated with mixed sugar utilization by xylose-fermenting yeasts. However, the diauxic growth was not significant for strains expressing *OrpXylA*, 39a2Xo and 39a2XoXo (Figure 2). The above results and analysis suggest that expression of *OrpXylA* is beneficial for glucose and xylose co-fermentation. This is consistent with a previous report [37].

Chromosomal integration of the XI gene into the *S. cerevisiae* genome has received a significant amount of attention in recent years [25,26]. It allows the recombinant strain to retain its physiological characteristics in non-selective medium. In the present study, multiple copies of XI genes were integrated at the 18S rDNA sites based on random homologous recombination. Such genome integration led to stable recombinant yeast strains (Figure S1) and boosted XI activity to 0.72 U mg^{-1} protein in 39a2XoNK (Table 2). Such improvement might be associated with the multiple copies of XI integrated in the yeast genome [25]. This further led to high XI transcription levels (Table 3), high xylose conversion and ethanol production (Table 4). Such results accorded quite well with earlier reports [25,26,36]. Notably, among these strains, 39a2 strains with integrated XI genes presented higher XI activity and transcription level than 36α2 strains (Tables 2 and 3). This further confirmed that the elimination of respiration was favourable for xylose fermentation [12,14].

S. cerevisiae strains are regarded as industrial working horses for ethanol production owing to their high ethanol titer and sugar tolerance. Mating is one of the traditional yeast breeding methods to develop improved *S. cerevisiae* strains without genetic modifications [38]. Through strain mating, robust diploid *S. cerevisiae* strains were developed for enhanced xylose-fermentation and inhibitor resistance [23,24]. In the present work, a recombinant diploid *S. cerevisiae* strain STXQ was obtained by mating respiration-dependent strain 36α2XpXpUN with the respiration-deficient strain 39a2XoNK. Such strain mating enabled strain STXQ to present improved cell growth, xylose utilization and ethanol production (Figure 4, Table 4).

For glucose-xylose co-fermentation by engineered laboratory *S. cerevisiae* strains, higher ethanol concentration of ~60 g L¯1 was reported by Ho et al. [39] (Table 5). An ethanol titer of 47.5 g L¯1 was obtained by recombinant *S. cerevisiae* RWB218 expressing *PirXylA* [40]. About 53 g L¯1 ethanol titer was reported by Diao and his colleagues using the diploid recombinant *S. cerevisiae* strain CIBTS0735 [22]. Demeke et al. obtained an inhibitor-resistant recombinant *S. cerevisiae* through metabolic engineering and adaptive evolution [18]. It utilized glucose and xylose rapidly with ethanol titer up to 46 g L¯1 and ethanol productivity of 2.58 g L¯1·h¯1. This is the highest volumetric ethanol productivity reported to-date in the literature. More recently, about 58 g L¯1 ethanol titer was obtained by a diploid recombinant yeast strain LF1 developed from a wild-type *S. cerevisiae* strain in YP medium [25]. Strain STXQ yielded 65.8 g L¯1 ethanol with an ethanol yield of 0.50 g g¯1 at 56 h in YPX medium containing 132 g L¯1 xylose (Figure 4A, Table 4). Further glucose-xylose co-fermentation produced 120.6 g L¯1 ethanol with an ethanol yield of 0.48 g g¯1 at 23 h in YP medium containing 162 g L¯1 glucose and 95 g L¯1 xylose. The corresponding volumetric ethanol productivity reached 5.24 g L¯1·h¯1. (Figure 4B, Table 4). Both ethanol titer and volumetric productivity far exceeded the results in previous reports.

Table 5. Comparison of fermentation performance of engineered xylose-fermenting *S. cerevisiae* Strain

Initial Glucose (g L¯1)	Initial Xylose (g L¯1)	Final Ethanol (g L¯1)	Ethanol Volumetric Yield (g Ethanol Reference g-1) Productivity(g L-1 ·h¯1)		
	40	17.47	0.44	1.09	[18]
Rich medium, initial inoculum size at 0.63 g DCW L¯1 80	40	53	0.45	2.22	
RWB218 *PirXylA*; *XKS1*; PPP; *gre3Δ*; adaptive defined synthetic medium, initialevolution) inoculum size at 1.1 g DCW L¯1 100	25	47.5	0.38	1.98	[40]
1400(pLNH32) *XR*; *XDH*; *XK*; adaptive evolution Rich medium; OD 40–45 50		24	0.48	0.52	[39]
Rich medium; OD 40–45 80	40	60	0.45	1.3	
H31-A3-ALCS*PirXylA*;*XKS1*;PPP;*gre3Δ*; adaptive Defined medium;initial inoculum sizeevolution at 0.05 g DCW L¯1	40	16.4	0.41	0.55	[28]

Initial Glucose (g L^{-1})	Initial Xylose (g L^{-1})	Final Ethanol (g L^{-1})	Ethanol Volumetric Yield (g Ethanol Reference g-1) Productivity(g L-1 ·h^{-1})		
cpXylA; *XKS1*; PPP; *HXT7*; *AraT*; *AraA*; Rich medium; initial inoculum size at 1.3 gGS1.11- *AraB*; *AraD*; *TAL2*; *TKL2* mutagenesis; DCW L^{-1} 36genome shuffling; adaptive evolution	37	33.6	0.46	2.58	[18]
N360F LF1 *Ru-XylA*, *XKS1*, PPP, *gre3::MGT05196* , Rich medium, initial inoculum size at 1.00 adaptive evolution g DCW L^{-1} 80	42	58.0	0.47	3.60	[25]
STXQ *OrpXylA*, *PirXylA*, *XKS1*, PPP, *gre3Δ*; Rich medium, initial inoculum size at 3.13 adaptive evolution g DCW L^{-1}	132	65.8	0.50	1.13	This study
Rich medium, initial inoculum size at 3.43 g DCW L^{-1} 162	95	120.6	0.48	5.24	This study

The OPEFB hydrolysate fermentation result demonstrated that strain STXQ could co-ferment glucose and xylose without detoxification, with 95.3% glucose conversion and 88.9% xylose conversion at 72 h (Figure 5). Ethanol titer reached 28.4 g L^{-1} at 72 h with an ethanol yield of 0.420 g g^{-1}. Strain STXQ has potential in the conversion of lignocellulosic biomass hydrolysate to ethanol. These results suggest that the strain development protocols outlined in this study are effective in obtaining robust xylose-fermenting yeast strains for industrial applications.

SUMMARY

A recombinant haploid strain containing plasmids harboring two-copy *PirXylA* presented the best xylose utilization among the engineered yeast strains harboring episomal XI genes when xylose was used as the sole carbon source. On the other hand, the strain containing plasmids with two-copy *OrpXylA* exhibited the best glucose and xylose co-fermentation. Respiration-deficient 39a2 strains harboring the *OrpXylA* gene were favourable for xylose fermentation and glucose-xylose co-fermentation in the engineered yeast strains. Chromosome integration of XI genes in *S. cerevisiae* resulted in high XI activity, high XI transcription levels, and improved xylose fermentation. Mating the respiration-efficient strain 36α2XpXpUN with the respiration-deficient strain 39a2XoNK resulted in a diploid recombinant *S. cerevisiae* strain STXQ with enhanced cell growth and xylose fermentation. Strain STXQ demonstrated superior glucose and xylose co-fermentation performance.

It produced 120.6 g L^{-1} ethanol with a volumetric productivity of 5.24 g L^{-1} h^{-1}, the highest among those reported to-date. Such superior performance by strain STXQ is largely associated with its development process, in particular with strain adaptive evolution, XI gene chromosome integration and strain mating.

REFERENCES

1. Somerville, C.; Youngs, H.; Taylor, C.; Davis, S.C.; Long, S.P. Feedstocks for lignocellulosic biofuels. *Science***2010**, *329*, 790–792. [CrossRef] [PubMed]
2. Miao, Z.; Grift, T.E.; Hansen, A.C.; Ting, K.C. An overview of lignocellulosic biomass feedstock harvest, processing and supply for biofuel production. *Biofuels* **2013**, *4*, 5–8. [CrossRef]
3. Fatma, S.; Hameed, A.; Noman, M.; Ahmed, T.; Shahid, M.; Tariq, M.; Sohail, I.; Tabassum, R. Lignocellulosic biomass: A sustainable bioenergy source for the future. *Protein Pept. Lett.* **2018**, *25*, 148–163. [CrossRef] [PubMed]
4. Moyses, D.N.; Reis, V.C.; de Almeida, J.R.; de Moraes, L.M.; Torres, F.A. Xylose fermentation by *Saccharomyces cerevisiae*: Challenges and prospects. *Int. J. Mol. Sci.* **2016**, *17*, 207. [CrossRef] [PubMed]
5. Li, Y.C.; Gou, Z.X.; Zhang, Y.; Xia, Z.Y.; Tang, Y.Q.; Kida, K. Inhibitor tolerance of a recombinant flocculating industrial *Saccharomyces cerevisiae* strain during glucose and xylose co-fermentation. *Braz. J. Microbiol.* **2017**, *48*, 798–800. [CrossRef] [PubMed]
6. Hahn-Hägerdal, B.; Karhumaa, K.; Jeppsson, M.; Gorwa-Grauslund, M.F. Metabolic engineering for pentose utilization in *Saccharomyces cerevisiae. Adv. Biochem. Eng. Biotechnol.* **2007**, *108*, 147–177. [CrossRef] [PubMed]
7. Madhavan, A.; Srivastava, A.; Kondo, A.; Bisaria, V.S. Bioconversion of lignocellulose-derived sugars to ethanol by engineered *Saccharomyces cerevisiae. Crit. Rev. Biotechnol.* **2012**, *32*, 22–48. [CrossRef] [PubMed]
8. Zhang, G.C.; Liu, J.J.; Kong, I.I.; Kwak, S.; Jin, Y.S. Combining C6 and C5 sugar metabolism for enhancing microbial bioconversion. *Curr. Opin. Chem. Biol.* **2015**, *29*, 49–57. [CrossRef] [PubMed]
9. Kwak, S.; Jin, Y.S. Production of fuels and chemicals from xylose by engineered *Saccharomyces cerevisiae*: A review and perspective. *Microb. Cell Factories* **2017**, *16*, 82. [CrossRef] [PubMed]
10. Hou, J.; Qiu, C.; Shen, Y.; Li, H.; Bao, X. Engineering of *Saccharomyces cerevisiae* for the efficient co-utilization of glucose and xylose. *FEMS Yeast Res.* **2017**, *17*. [CrossRef] [PubMed]
11. Kuyper, M.; Hartog, M.M.; Toirkens, M.J.; Almering, M.J.; Winkler, A.A.; van Dijken, J.P.; Pronk, J.T. Metabolic engineering of a xylose-isomerase-expressing

Saccharomyces cerevisiae strain for rapid anaerobic xylose fermentation. *FEMS Yeast Res.* **2005**, *5*, 399–409. [CrossRef] [PubMed]

12. Peng, B.; Huang, S.; Liu, T.; Geng, A. Bacterial xylose isomerases from the mammal gut Bacteroidetes cluster function in *Saccharomyces cerevisiae* for effective xylose fermentation. *Microb. Cell Factories* **2015**, *14*, 70. [CrossRef] [PubMed]
13. Johansson, B.; Hahn-Hägerdal, B. The non-oxidative pentose phosphate pathway controls the fermentation rate of xylulose but not of xylose in *Saccharomyces cerevisiae* TMB3001. *FEMS Yeast Res.* **2002**, *2*, 277–282. [PubMed]
14. Peng, B.; Shen, Y.; Li, X.; Chen, X.; Hou, J.; Bao, X. Improvement of xylose fermentation in respiratory-deficient xylose-fermenting *Saccharomyces cerevisiae. Metab. Eng.* **2012**, *14*, 9–18. [CrossRef] [PubMed]
15. Lee, S.H.; Kodaki, T.; Park, Y.C.; Seo, J.H. Effects of NADH-preferring xylose reductase expression on ethanol production from xylose in xylose-metabolizing recombinant *Saccharomyces cerevisiae. J. Biotechnol.* **2012**, *158*, 184–191. [CrossRef] [PubMed]
16. Sonderegger, M.; Sauer, U. Evolutionary engineering of Saccharomyces cerevisiae for anaerobic growth on xylose. *Appl. Environ. Microbiol.* **2003**, *69*, 1990–1998. [CrossRef] [PubMed]
17. Scalcinati, G.; Otero, J.M.; Vleet, J.R.; Jeffries, T.W.; Olsson, L.; Nielsen, J. Evolutionary engineering of *Saccharomyces cerevisiae* for efficient aerobic xylose consumption. *FEMS Yeast Res.* **2012**, *12*, 582–597. [CrossRef] [PubMed]
18. Demeke, M.M.; Dietz, H.; Li, Y.; Foulquie-Moreno, M.R.; Mutturi, S.; Deprez, S.; Den Abt, T.; Bonini, B.M.; Liden, G.; Dumortier, F.; et al. Development of a D-xylose fermenting and inhibitor tolerant industrial *Saccharomyces cerevisiae* strain with high performance in lignocellulose hydrolysates using metabolic and evolutionary engineering. *Biotechnol. Biofuels* **2013**, *6*, 89. [CrossRef] [PubMed]
19. Sedlak, M.; Ho, N.W. Characterization of the effectiveness of hexose transporters for transporting xylose during glucose and xylose co-fermentation by a recombinant *Saccharomyces* yeast. *Yeast* **2004**, *21*, 671–684. [CrossRef] [PubMed]
20. Runquist, D.; Hahn-Hagerdal, B.; Radstrom, P. Comparison of heterologous xylose transporters in recombinant *Saccharomyces cerevisiae. Biotechnol. Biofuels* **2010**, *17*, 3–5. [CrossRef] [PubMed]
21. Goncalves, D.L.; Matsushika, A.; de Sales, B.B.; Goshima, T.; Bon, E.P.; Stambuk, B.U. Xylose and xylose/glucose co-fermentation by recombinant *Saccharomyces*

cerevisiae strains expressing individual hexose transporters. *Enzyme Microb. Technol.* **2014**, *63*, 13–20. [CrossRef] [PubMed]

22. Diao, L.; Liu, Y.; Qian, F.; Yang, J.; Jiang, Y.; Yang, S. Construction of fast xylose-fermenting yeast based on industrial ethanol-producing diploid *Saccharomyces cerevisiae* by rational design and adaptive evolution. *BMC Biotechnol.* **2013**, *13*, 110. [CrossRef] [PubMed]

23. Demeke, M.M.; Dumortier, F.; Li, Y.; Broeckx, T.; Foulquié-Moreno, M.R.; Thevelein, J.M. Combining inhibitor tolerance and D-xylose fermentation in industrial *Saccharomyces cerevisiae* for efficient lignocellulose-based bioethanol production. *Biotechnol. Biofuels* **2013**, *6*, 120. [CrossRef] [PubMed]

24. Kim, S.R.; Lee, K.S.; Kong, I.I.; Lesmana, A.; Lee, W.H.; Seo, J.H.; Kweon, D.H.; Jin, Y.S. Construction of an efficient xylose-fermenting diploid *Saccharomyces cerevisiae* strain through mating of two engineered haploid strains capable of xylose assimilation. *J. Biotechnol.* **2013**, *164*, 105–111. [CrossRef] [PubMed]

25. Li, H.; Shen, Y.; Wu, M.; Hou, J.; Jiao, C.; Li, Z.; Liu, X.; Bao, X. Engineering a wild-type diploid *Saccharomyces cerevisiae* strain for second-generation bioethanol production. *Bioresour. Bioprocess.* **2016**, 3, 51. [CrossRef] [PubMed]

26. Tanino, T.; Hotta, A.; Ito, T.; Ishii, J.; Yamada, R.; Hasunuma, T.; Ogino, C.; Ohmura, N.; Ohshima, T.; Kondo, A. Construction of a xylose-metabolizing yeast by genome integration of xylose isomerase gene and investigation of the effect of xylitol on fermentation. *Appl. Microbiol. Biotechnol.* **2010**, *88*, 1215–1221. [CrossRef] [PubMed]

27. Bamba, T.; Hasunuma, T.; Kondo, A. Disruption of PHO13 improves ethanol production via the xylose isomerase pathway. *AMB Express* **2016**, *6*, 4. [CrossRef] [PubMed]

28. Zhou, H.; Cheng, J.S.; Wang, B.L.; Fink, G.R.; Stephanopoulos, G. Xylose isomerase overexpression along with engineering of the pentose phosphate pathway and evolutionary engineering enable rapid xylose utilization and ethanol production by *Saccharomyces cerevisiae*. *Metab. Eng.* **2012**, *14*, 611–622. [CrossRef] [PubMed]

29. Ko, J.K.; Um, Y.; Woo, H.M.; Kim, K.H.; Lee, S.M. Ethanol production from lignocellulosic hydrolysates using engineered *Saccharomyces cerevisiae* harboring xylose isomerase-based pathway. *Bioresour. Technol.* **2016**, *209*, 290–296. [CrossRef] [PubMed]

30. Güldener, U.; Heck, S.; Fiedler, T.; Beinhauer, J.; Hegemann, J.H. A new efficient gene disruption cassette for repeated use in budding yeast. *Nucleic Acids Res.* **1996**, *24*, 2519–2524. [CrossRef] [PubMed]

31. Gietz, R.D.; Schiestl, R.H. Large-scale high-efficiency yeast transformation using the LiAc/SS carrier DNA/PEG method. *Nat. Protoc.* **2007**, *2*, 38–41. [CrossRef] [PubMed]

32. Kersters-Hilderson, H.; Callens, M.; Opstal, O.V.; Vangrysperre, W.; Bruyne, C.K.D. Kinetic characterization of d-xylose isomerases by enzymatic assays using d-sorbitol dehydrogenase. *Enzyme Microb. Technol.* **1987**, *9*, 145–148. [CrossRef]

33. Livak, K.J.; Schmittgen, T.D. Analysis of relative gene expression data using real-time quantitative PCR and the 2–ΔΔCT method. *Methods* **2001**, *25*, 402–408. [CrossRef] [PubMed]

34. Wang, Z.; Ong, H.X.; Geng, A. Cellulase production and oil palm empty fruit bunch saccharification by a new isolate of *Trichoderma koningii* D-64. *Proc. Biochem.* **2012**, *47*, 1564–1571. [CrossRef]

35. Qi, X.; Zha, J.; Liu, G.G.; Zhang, W.; Li, B.Z.; Yuan, Y.J. Heterologous xylose isomerase pathway and evolutionary engineering improve xylose utilization in *Saccharomyces cerevisiae*. *Front. Microbiol.* **2015**, *6*, 1165. [CrossRef] [PubMed]

36. Vilela Lde, F.; de Araujo, V.P.; Paredes Rde, S.; Bon, E.P.; Torres, F.A.; Neves, B.C.; Eleutherio, E.C. Enhanced xylose fermentation and ethanol production by engineered *Saccharomyces cerevisiae* strain. *AMB Express* **2015**, *5*, 16. [CrossRef] [PubMed]

37. Madhavan, A.; Tamalampudi, S.; Srivastava, A.; Fukuda, H.; Bisaria, V.S.; Kondo, A. Alcoholic fermentation of xylose and mixed sugars using recombinant *Saccharomyces cerevisiae* engineered for xylose utilization. *Appl. Microbiol. Biotechnol.* **2009**, *82*, 1037–1047. [CrossRef] [PubMed]

38. Perez-Traves, L.; Lopes, C.A.; Barrio, E.; Querol, A. Evaluation of different genetic procedures for the generation of artificial hybrids in *Saccharomyces* genus for winemaking. *Int. J. Food Microbiol.* **2012**, *156*, 102–111. [CrossRef] [PubMed]

39. Ho, N.W.; Chen, Z.; Brainard, A.P. Genetically engineered *Saccharomyces* yeast capable of effective cofermentation of glucose and xylose. *Appl. Environ. Microbiol.* **1998**, *64*, 1852–1859. [PubMed]

40. Kuyper, M.; Harhangi, H.R.; Stave, A.K.; Winkler, A.A.; Jetten, M.S.; de Laat, W.T.; den Ridder, J.J.; Op den Camp, H.J.; van Dijken, J.P.; et al. High-level functional expression of a fungal xylose isomerase: The key to efficient ethanolic fermentation of xylose by *Saccharomyces cerevisiae*? *FEMS Yeast Res.* **2003**, *4*, 69–78. [CrossRef]

CHAPTER-6

RESISTANCE TO FLUORINATED PHENYLALANINE PRECURSORS IN YEAST

INTRODUCTION: AN OVERVIEW

In the intricate landscape of yeast biology, the phenomenon of resistance to fluorinated phenylalanine precursors emerges as a focal point of exploration. This study delves into the intriguing dynamics of yeast resistance mechanisms against fluorinated phenylalanine precursors, shedding light on the molecular intricacies that govern this intriguing phenomenon. As we embark on this scientific journey, the aim is to unravel the genetic and biochemical underpinnings that confer resilience to these fluorinated compounds within the yeast cellular framework.

Development of a counter-selection method for phenylalanine auxotrophy could be a useful tool in the repertoire of yeast genetics. Fluorinated and sulfurated precursors of phenylalanine were tested for toxicity in *Saccharomyces cerevisiae*. One such precursor, 4-fluorophenylpyruvate (FPP), was found to be toxic to several strains from the *Saccharomyces* and *Candida* genera. Toxicity was partially dependent on *ARO8* and *ARO9*, and correlated with a strain's ability to convert FPP into 4-fluorophenylalanine (FPA). Thus, strains with deletions in *ARO8* and *ARO9*, having a mild phenylalanine auxotrophy, could be separated from a culture of wild-type strains using FPP. Tetrad analysis suggests FPP resistance in one strain is due to two genes. Strains resistant to FPA have previously been shown to exhibit increased phenylethanol production. However, FPP resistant isolates did not follow this trend. These results suggest that FPP could effectively be used for counter-selection but not for enhanced phenylethanol production.

The phenylalanine biosynthetic pathway is a route to production of the high-value chemical 2-phenylethanol that is used for its rosy scent in cosmetics, foods, and cleaning supplies [1]. Phenylethanol is produced mainly by synthetic chemical processes but also naturally by yeast, as a degradation product of phenylalanine, moving through the intermediate metabolites of phenylpyruvate then phenylacetaldehyde (Figure 1). In *Saccharomyces cerevisiae,* one method of increasing the production of these chemicals is to select for resistance to the fluorinated phenylalanine analog 4-fluorophenylalanine (FPA) [2]. One facet of the mechanism of resistance is enhanced endogenous production of phenylalanine, thus requiring less dependence on exogenously provided phenylalanine [3]. Hence, FPA can be used to increase production of phenylethanol.

There are numerous counter-selection methods for a variety of genes. One method, the use of toxic metabolites to counter-select for auxotrophies such as uracil or tryptophan, has been a boon in yeast genetics, enabling numerous techniques such as the plasmid shuffle and Synthetic Genetic Array analysis [4–6]. These techniques rely on adding a fluorine to a precursor of the final biosynthetic pathway product, such that use of the fluorinated precursor results in a fluorinated end-product, which is toxic to the cell. There are many potential mechanisms of resistance to toxic precursors. One such mechanism is the prohibition of toxic metabolite import, as in the case of canavanine resistance via abrogation of the arginine transporter [4]. Another is to decrease endogenous production of the precursor while relying on exogenous end-product for growth, as in the case of deletion of *URA3* and reliance upon external uracil for 5-fluoroorotic acid resistance. In the case of resistance to a toxic end-product, one mechanism of resistance is via increasing endogenous production of the end-product so as to decrease exogenous nutrient dependence, such as in the case of increased phenylalanine production in response to fluorophenylalanine [3].

We sought to identify toxic precursors to the phenylalanine biosynthetic pathway that could be used for auxotrophic counter-selection, and that may lead to increased production of phenylethanol.

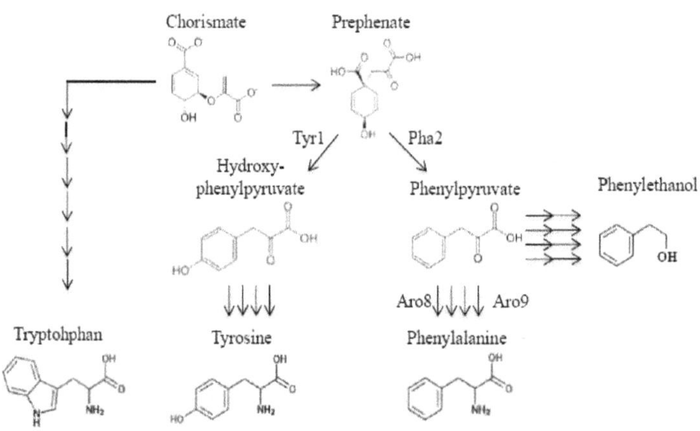

Figure 1. Diagram of the yeast pathway for biosynthesis of tryptophan, tyrosine, and phenylalanine. Number of enzymes and steps are indicated by number of arrows.

MATERIALS AND METHODS

Media and Growth Conditions

Rich medium, or YPD, was made with BactoTM Yeast Extract at 10 g L^{-1} (Becton Dickinson, Franklin Lakes, NJ, USA), BactoTM Peptone at 20 g L^{-1} (Becton Dickinson), and dextrose at 20 g L^{-1} (Fisher Scientific, Waltham, MA, USA). Minimal medium contains 1.8 g L^{-1} yeast nitrogen base without amino acid or ammonium sulfate and 6.2 g L^{-1} ammonium sulphate and dextrose at20 g L^{-1}. Synthetic complete medium contains minimal medium with 20 mg L^{-1} of all the following compounds: uracil, lysine, tyrosine, tryptophan, phenylalanine, adenine, leucine, histidine, and methionine. For clarification, the phenylalanine-related chemicals used in this paper are described in Table 1, their molecular structures are shown in Figure 2.

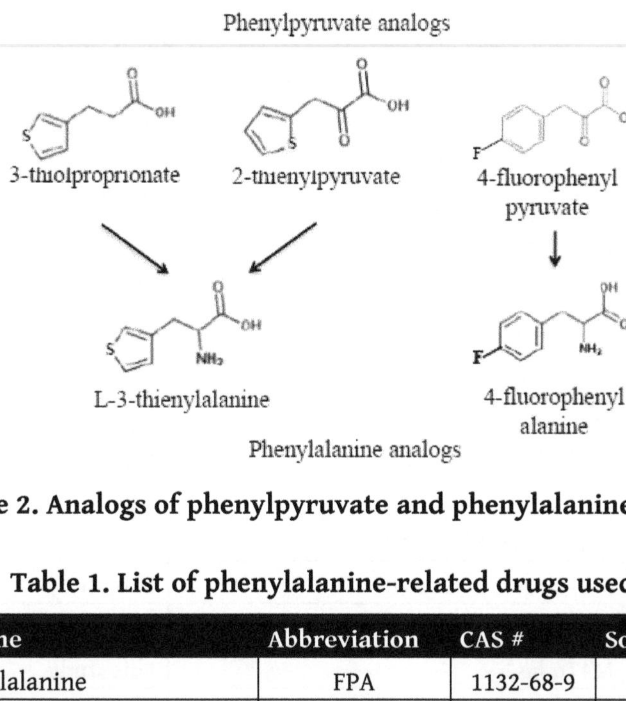

Figure 2. Analogs of phenylpyruvate and phenylalanine tested.

Table 1. List of phenylalanine-related drugs used.

Chemical Name	Abbreviation	CAS #	Source
4-fluorophenylalanine	FPA	1132-68-9	Chem-Impex
4-fluorophenylpyruvate	FFP	7761-30-0	Enamine
2-thienylpyruvate	2TP	15504-41-3	Enamine
3-thiolproprionate	3TP	16378-06-6	Enamine
L-3-thienylalnine	L3T	3685-51-6	Sigma

Chem-Impex (Wood Dale, IL, USA);Enamine (Monmouth Jct., NJ, USA); Sigma (St. Louis, MO, USA). CAS# refers to Chemical Abstracts Service registry number for indicated compound.

Experiments using FPA, FPP, 2TP, 3TP, or L3T were performed in MM2 medium, which consists of minimal medium with 2 mg L^{-1} of all the following compounds: uracil, tyrosine, phenylalanine, leucine, histidine, and methionine. Chemicals FPA, 2TP, 3TP, or L3T were suspended in water, while FPP and FPA were suspended in dimethyl sulfoxide (DMSO).

Culture density was determined by a spectrophotometer reading of the optical density at 600 nm, which was correlated to cell density via hemocytometer counts. Strains used are listed in Table 2. Strains mentioned with a dash (e.g., S288c-3) are derivatives of the parental strain named preceding the dash. Mutation cultures were obtained by exposure to ultraviolet light at 230 nm for 10 s that resulted in viability between 30–70% as measured by methylene blue.

Table 2. List of yeast strains used.

Strain	Genotype	Source
26704c	*MATα ura3 aro8-2 aro9-1*	[7]
BY4700	*MATα S288c ura3Δ0*	ATCC
BY4742	*MATα S288c his3Δ1 leu2Δ0 lys2Δ0 ura3Δ0*	ATCC
A364A	*MATa ade1 ade2 ura1 his7 lys2 tyr1 gal1*	ATCC
AW051	*MATa ura3 aro8-2 aro9-1*	This study
AW052	*MATa ura3 aro8-2 aro9-1*	This study
AW063	*MATα ade2 pha2 arg8 lys2*	This study
AW077	*MATα ade2 ura1*	This study
AW081	MATa ura1 his7	This study
AW083	*MATα lys2 ade1 ade2*	This study
AW102	*MATa/MATα wild-type*	This Study
AW108	*MATα S288c pha2::G418*	This study
AW109	*MATα S288c pha2::G418*	This study
AW258	*MATα ura1*	This study
AW259	*MATα his7*	This study
AW261	*MATα met15*	This study
IM006	*MATα lys2*	This study
Calb	*Candida albicans ATCC 10231*	Gift from Darlene Miller
Ctrop	*Candida tropicalis*	Gift from Darlene Miller
S288c	*MATα SUC2 mal mel CUP1 flo1 flo8-1 hap1*	ATCC
SPara	*lys2 Saccharomyces paradoxus*	Gift from Doug Koshland
SBaya	*ura3 Saccharomyces bayanus*	Gift from Doug Koshland

American Type Culture Collection (ATCC)

Drug resistance was measured by calculating the relative viability of cells, which is the number of viable cells resulting from a drug treated culture divided by the number of viable cells resulting from a carrier treated (water or DMSO) culture. Viability was measured as colony forming units on a petri dish by directly plating cells to a plate or by first serially diluting in water, then pinning to drug and carrier plates (frogging method) or by using the Tadpoling method, discussed below. For time course experiments, relative viability was calculated by using the number of viable cells at time point zero as the denominator in all subsequent time points.

Briefly, the Tadpoling method consists of inoculating 20 µL of yeast containing culture into 180 µL YPD in a well of a 96-well plate and serially diluting 20 µL from this well to subsequent wells [8]. This plate is incubated until individual colonies can be counted in the most dilute well, yielding a quantifiable number used to calculate cell viability in the original well by multiplying by the dilution factor.

$$hours \times log2\left(\frac{ODin}{OD\square n}\right)$$

Growth rate was calculated using the following expression: where ODin is the initial A600, ODfin is the final A600, hours is the number of hours in between A600 readings.

Determination of Phenylethanol and Phenylalanine

After growth in 2 mg L^{-1} uracil, tyrosine, phenylalanine, leucine, histidine, and methionine with 1.8 g L^{-1} yeast nitrogen base without amino acid or ammonium sulfate with 20 g L^{-1} dextrose and 6 g L^{-1} phenylalanine for 6–8 days at 30°C, culture supernatant was removed and selected ion monitoring (SIM) was performed to detect phenylethanol and phenylalanine using liquid chromatography-tandem mass spectrometry (LC-MS/MS). A gradient was run from 100% 50 mM ammonium acetate to 100% methanol with 0.1% formic acid over the course of 20 min. Phenylethanol retention time was 12.5 min while that of phenylalanine was 4.5.

Determination of Fluorophenylalanine

Yeast samples were grown in medium containing FPP at a concentration of 200 mg L^{-1} for 16 h then yeast samples were resuspended in 70% ethanol, boiled for 10 min, then the supernatant was transferred to a new tube and after evaporated in a spinning-vacuum centrifuge. Samples were fully dissolved in 1 mL of 3:1:1 water: acetronitrile: isopropanol, with 2 min of vortexing and 20 min sonication. The resulting solutions were 5× diluted (200/1000) in 10 mM ammonium formate in water (mobile phase A) and was injected in the LC-MS/MS system, an AB Sciex QTRAP 5500 Triple-Quadrupole mass spectrometer, equipped with a Turbospray ESI source. A binary gradient separation program was employed using a reverse phase HPLC column (Dionex Acclaim 120 C18 Column, 250 × 2.1 mm, 5 μm). Sample injection (20 μL) and LC separation was performed by a Shimadzu Prominence LC-20AD Ultra-Fast Liquid Chromatography.

RESULTS

First, we sought to identify a toxic phenylalanine analog and work backward to design toxic phenylalanine precursors. Two promising toxic phenylalanine analogs were identified from the literature: L-3-thienylalanine (L3T), consisting of phenylalanine with the benzene ring substituted for thiophene, and fluorophenylalanine (FPA), consisting of phenylalanine with a fluorine attached to the benzene ring (Figure 2, Table 1). A 2.6 mg L^{-1} concentration of L-3-thienylalanine has been shown to inhibit growth of *S. cerevisiae* by 50% [9]. First, the toxicity of L3-thienylalanine was tested by plating wild-type strains S288c and AW077 to medium with L3-thienylalanine at 0 or 300 mg L^{-1} and determining the relative viability. L3T killed all AW077 cells and nearly all of S288c cells (Figure 3a). Next the toxicity of FPA was tested by plating strains S288c and AW077 onto medium containing FPA at 0 or 300 mg L^{-1} and determining the relative viability (Figure 3b). FPA at both of these concentrations killed almost all cells of S288c and AW077. Thus, we reconfirmed that L3T and FPA are toxic to *S. cerevisiae.*

Next, we sought to design a molecule that may be enzymatically converted to a toxic phenylalanine analog. By identifying a toxic phenylalanine precursor, we could use this compound to select for strains that have mutations in the phenylalanine biosynthetic pathway. We considered several precursors of phenylalanine; chorismate (this precursor is used to synthesize not only phenylalanine, but also tryptophan, tyrosine, folate, and ubiquinone). Thus, a strain with a chorismate deletion would have numerous other auxotrophies besides phenylalanine. Next, prephenate—however, this is not a good candidate for this particular intervention as adding fluorine or sulfur on the cyclohexadiene ring could interfere with transformation to a benzene ring. Additionally, there seems to be at least two mechanisms to convert prephenate to phenylalanine, as we found that deletion of *PHA2* does not confer complete phenylalanine requirement. To test this, strains lacking *PHA2* were inoculated alongside a strain with functioning *PHA2* in minimal medium with or without phenylalanine added (Figure 4, Table 3). The A600 was measured over four days, and maximum growth rate (lowest doubling time) was calculated. Strains AW108 and AW109 grew much better with phenylalanine added (Table 3), showing the dependency on phenylalanine; however, they still grew over time. This is in contrast to previously described observations of strains lacking *PHA2*. However, it is likely that the extremely slow growth rate explains this conclusion (Figure 4) [10]. The next candidate is phenylpyruvate, which does not require modification of its benzene ring prior to becoming phenylalanine, and it is the substrate for only two essentials molecules: phenylalanine and tyrosine. Use of phenylpyruvate has one drawback: there are two primary enzymes that can complete the transamination reaction converting phenylpyruvate to phenylalanine (*ARO8* and *ARO9*) and several others that are predicted to be capable of performing this reaction (Bat1, Bat2, His5) [11].

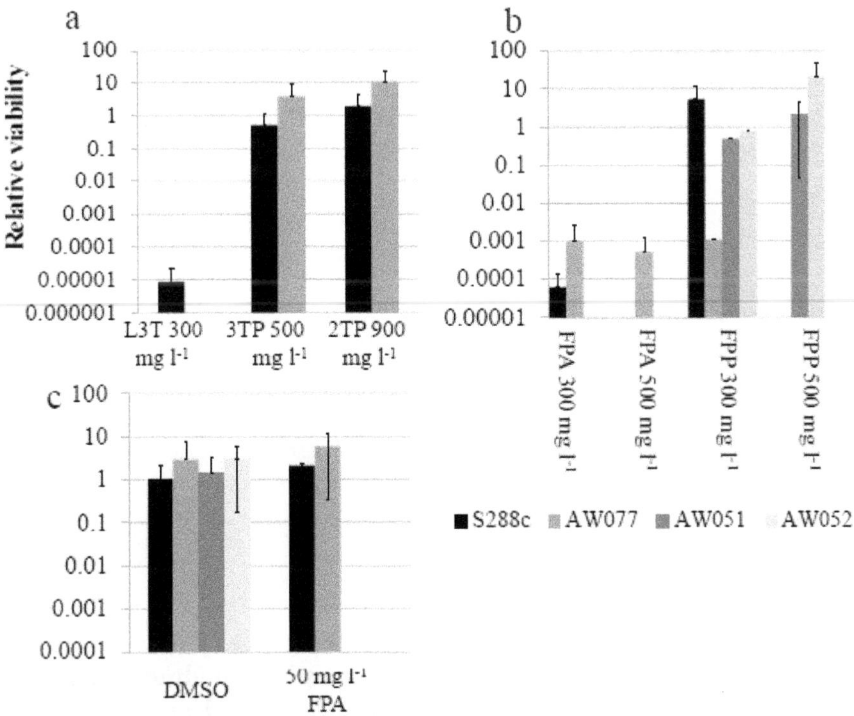

Figure 3. Viability of yeast in response to toxic phenylalanine or phenylpyruvate analogs. Cultures of yeast strains were serially diluted in water then plated onto medium containing either (a) L3T, 3TP, 2TP or (b) FPA, FPP, or (c) dimethyl sulfoxide, FPA. Relative viability was obtained by dividing the number of colony forming units (CFUs) from drug plates by CFUs from YPD plates. Plates were incubated for 2–3 days and CFUs counted. The average and standard deviation of three independent trials is shown.

Table 3. Density of two-day-old cultures.

Strain	Genotype	MM + U a	MM + UP b
AW108	pha2	0.06	3.8
AW109	pha2	0.08	3.6
AW051	aro8 aro9 ura3	0.37	1.1
AW052	aro8 aro9 ura3	1.2	2.2
BY4700	ura3	3.6	3.7

MM: Minimal Medium [a] Minimal medium with 20 mg L^{-1} uracil, [b] Minimal medium with 20 mg L^{-1} of uracil and phenylalanine.

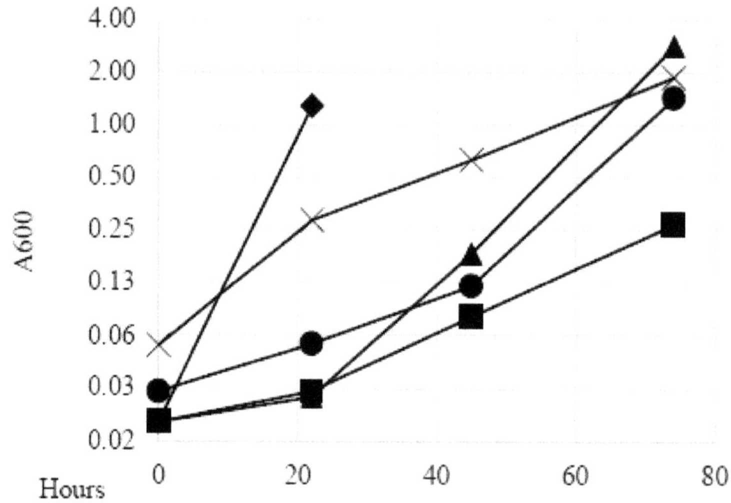

Strain	Genotype	Max Growth (hrs/doubling)
AW108	MATα S288c pha2::G418	16.1
AW109	MATα S288c pha2::G418	8.1
AW051	MATa ura3 aro8-2 aro9-1	6.8
AW052	MATa ura3 aro8-2 aro9-1	5.3
BY4700	MATα S288c ura3Δ0	3.4

Figure 4. Growth rate of strains in medium lacking phenylalanine. The indicated strains were grown overnight in synthetic complete medium then washed and inoculated to minimal medium with uracil and rotated at 23°C. The A600 was measured over subsequent hours. Max growth was the lowest growth rate calculated for all points tested.

This presents a problem for selection, as strains would need to have deletions in two genes, rather than just one, in the case of uracil (URA3), lysine (LYS2), or arginine (CAN1) [6,12,13]. To test if *ARO8* and *ARO9* were required for phenylalanine production, strains lacking *ARO8* and *ARO9* (AW051, AW052) were constructed and inoculated alongside a wild-type strain in minimal medium with or without phenylalanine added (Figure 4, Table 3).

The A600 was measured over four days, and maximum growth rate (lowest doubling time) was calculated. Strains AW051 and AW052 grew much better with phenylalanine added (Table 3), showing their dependency on phenylalanine. However, they still grew over time. These strains do grow faster than the *PHA2* mutants but are not completely rescued solely by phenylalanine, as they also require tyrosine, hence the lower A600 on the second day of growth with phenylalanine added.

Furthermore, to verify there were no residual nutrients in the minimal medium, strains with various individual auxotrophies were inoculated to minimal medium and were found to not grow (Table 4).

Table 4. Density of three-day-old cultures.

Strain	Genotype	MM	MM + Auxa
AW108	pha2	3.0	5.8
AW109	pha2	2.8	5.6
AW258	ura1	0.1	3.6
AW259	his7	0.04	0.3
AW261	met15	0.08	5.2
IM006	lys2	0.07	1.4

a Minimal medium with 20 mg L⁻1 of auxotrophy required for strain.

Nonetheless, multiple enzymes producing phenylalanine could be viewed as a boon, as cells are less likely to develop spontaneous resistance to a toxic phenylpyruvate analog, as they would have to delete multiple processing enzymes. Thus we set out to design phenylpyruvate-like derivatives that may be catalytically converted into toxic phenylalanine analogs so that these can be used for counter-selection of phenylalanine auxotrophies.

The first two phenylpyruvate-like derivatives of L-3-thienylalanine tested were 3-thiolproprionate (3TP) and 2-thienylpyruvate (2TP). Wild-type yeast strains S288c and AW077 were pinned to plates containing these analogs at 500 mg L^{-1} and 900 mg L^{-1}, respectively, and relative viability was calculated (Figure 3a). Despite higher concentrations of the phenylpyruvate derivatives compared to L3T, S288c and AW077 strains exhibited no decrease in viability on 3TP and 2TP containing medium. Thus, while the phenylalanine analog was toxic to these strains, the phenylpyruvate analogs were not. Differences in toxicity could be due to adsorption to the cell wall, decreased uptake, or incompatibility with enzymes.

The toxicity of the FPA derivative fluorophenylpyruvate (FPP) was next examined. Strains AW077 and S288c were serially diluted in water, then pinned to plates containing 0, 300 mg L^{-1}, or 500 mg L^{-1} FPP and relative viability was calculated (Figure 3b). At 300 mg L^{-1} FPP strain S288c showed mild growth inhibition but no lethality, while strain AW077 was more growth-inhibited. At 500 mg L^{-1}, no growth was detected from S288c and AW077 strains (Figure 3b). These results indicate that the phenylpyruvate analog FPP is toxic to S288c and AW077 strains.

Phenylpyruvate is converted to phenylalanine primarily by the proteins Aro8p and Aro9p. However, there are other proteins that can catalyze this reaction. To test if the mechanism of FPP toxicity was due to its conversion to FPA, the toxicity of FPP was tested on strains lacking genes *ARO8* and *ARO9* (Figure 3b). Strains lacking *ARO8* and *ARO9* (AW051, AW052) are able to grow on this medium, showing approximately equal growth on YPD. However, these strains are hypersensitive to FPA, which may reflect the cells increased import of phenylalanine, due to a phenylalanine auxotrophy (Figure 3c). This result suggests that *ARO8* and *ARO9* are primarily responsible for affecting the toxicity of FPP.

For FPP to be of use as an auxotrophic marker, the natural resistance for it should be low. For example, the frequency of spontaneous resistance to 5-fluoroanthranilic acid or 5-fluoroorotic acid, for selection of tryptophan or uracil auxotrophs, respectively, was 6×10^{-7} [5]. To test the natural resistance frequency of S288c and AW077 strains to FPP, we plated 5×10^6 cells on plates containing 500 mg L^{-1} FPP. We found no resistant cells grew on this medium, indicating natural resistance of wild-type strains to FPP is below a frequency of 2×10^{-7}. As a comparison to this, we also plated these strains to medium containing 500 mg L^{-1} FPA. Strains S288c and AW077 formed 240 and 86 colonies giving a frequency of resistance of 3×10^{-5} and 5×10^{-5}, respectively. This indicates that the natural frequency of resistance to FPP is quite low, while that of FPA is higher. This indicates that resistance to FPP toxicity requires either more mutations or more specific mutations, as compared to FPA, and that perhaps these molecules do not induce toxicity via the same mechanism.

To further investigate the difference between FPP and FPA resistance, strains that were FPA resistant were tested for FPP resistance. Isolates that were resistant to FPA were plated onto medium with or without FPP and relative viability was calculated (Figure 5). Three isolates derived from strain AW083 and one from AW077 showed resistance to FPA, but no resistance to FPP. This indicates that FPA resistance is not sufficient for FPP resistance. However, an isolate from a wild-type yeast wine strain (AW102) and an S288c mutant isolate (S288c-3) were recovered that showed resistance to FPP and FPA. When the strain S288c-3 was mated to a non-resistant strain (AW081) and sporulated, the resulting haploid segregants generally showed a pattern of 1:3 resistance to FPP, while resistance to FPA showed a pattern of 2:2 (Table 5).

Table 5. Tetrad analysis of FPP and FPA resistance.

Segregation	FPA Resistance	FPP Resistance
2:2	20	5
1:3	6	21

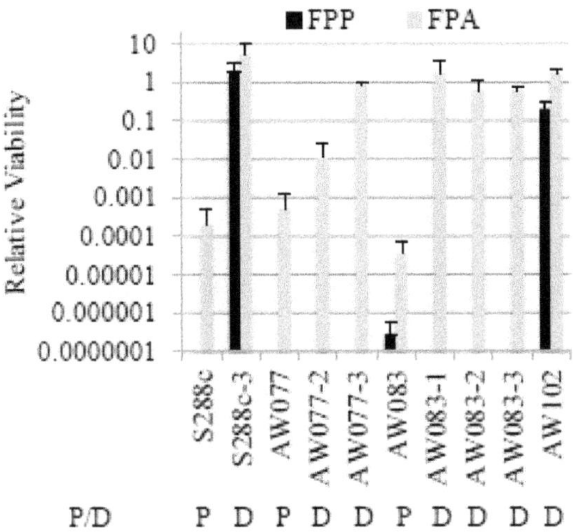

Figure 5. Viability of selected strains in response to FPP or FPA. Cultures of parental and derivative yeast strains were serially diluted in water then plated onto medium containing either 500 mg L⁻1 of FPA, FPP. Plates were incubated for 2–3 days and colony forming units (CFUs) counted. Relative viability was obtained by dividing the number of CFUs from drug plates by CFUs from YPD plates. P indicates the parental strain while D indicates a derivative of that strain. The average and standard deviation of three independent trials is shown.

A predominance of 1:3 segregation (tetratype) indicates that two genes are segregating independently, following the pattern of 1:1:4 PD:NPD:TT, while a 2:2 pattern indicates a single gene. All haploid segregants that were resistant to FPP were also resistant to FPA. The haploid segregants that were only resistant to FPA generally showed slow growth on FPA. One interpretation of these results is that detoxification of FPP requires the activity of two genes, either of which will suffice to detoxify FPA.

To determine if cells converted FPP into FPA or some other compound, we evaluated the presence of FPA in cells treated with FPP. Strains were inoculated to medium containing FPP and allowed to incubate for > 12 h, enzymatically digested to recover individual amino acids, and then tested for the presence of FPA using LC-MS/MS (Figure 6). Strains sensitive to FPP, AW077, and S288c were observed to accumulate the highest amounts of FPA, while the less sensitive strain AW051 accumulated less. The modest decrease in FPP conversion of AW051 is not too surprising, as there are other enzymes capable of converting phenylpyruvate to phenylalanine (Bat1, Bat2, His5), and this strain is capable of growth without phenylalanine supplementation. Strains S288c-3 and AW102, both of which are resistant to FPP, had the lowest levels of FPA. This suggests that their mechanism of resistance involves decreased conversion of FPP to FPA. Thus cells can convert FPP into FPA, and this may be the cause of some of the toxicity of FPP.

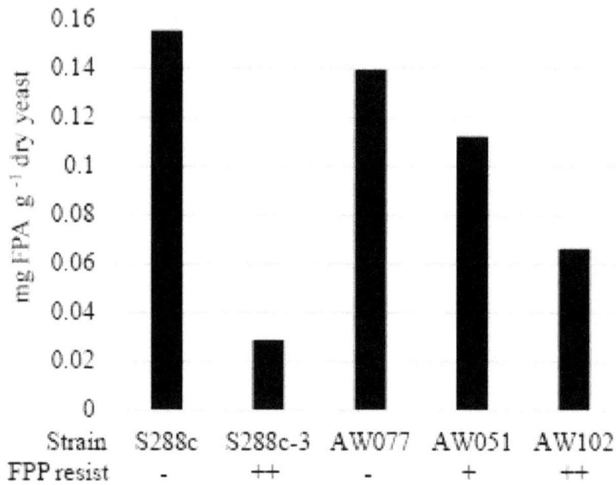

s

Figure 6. FPP is converted by cells to FPA. Yeast cells were inoculated to Synthetic Complete medium overnight at 30 °C, then inoculated to MM2 medium with 200 mg L⁻1 FPP and incubated at 30 °C for 16 h. Cultures were isolated, dried, weighed, and analyzed for the quantity of FPA. FPP resistance is indicated below each strain corresponding to a qualitative measure of their resistance to FPP with - signifying little or no resistance, + indicating moderate resistance, and ++ indicating strong resistance.

For FPP to be an effective selective agent, it should exert its toxic effect amongst diverse genre of yeast. To test the evolutionary conservation of FPP toxicity, strains from diverse species such as *Candida tropicalis, Candida albicans, Saccharomyces bayanus*, and *Saccharomyces paradoxus* were exposed to 0 or 500 mg L⁻1 FPP or FPA and their viability was measured over several days, and relative viabilities were calculated (Figure 7). To test these strains, the FPP and FPA sensitivity test was performed by inoculating cells into media with said compounds and evaluating culture viability using the Tadpoling method [8]. All the strains tested retained viability or slightly increased in cell number when inoculated into media without FPP or FPA, except for AW083. When inoculated to media containing FPA, *C. tropicalis* exhibited high resistance while most of the other strains lost viability over the course of four days. Both *C. albicans* and *S. bayanus* exhibited enhanced sensitivity to FPA, as these lost almost all culture viability within one day. Surprisingly, in media containing FPP, all strains showed greater sensitivity than in FPA and lost at least four orders of magnitude culture viability within two days. This result would suggest that FPP may be transported more effectively than FPA, or that FPP can exert its toxicity through a mechanism distinct from that of FPA. *S. paradoxus* and

AW051 (*aro8-2 aro9-1*) showed the least sensitivity to FPP of all the strains tested, but even these cultures succumbed after four days. This data shows that FPP can act through diverse yeast genre to exert its toxic effect and may be a more effective fungicide than FPA.

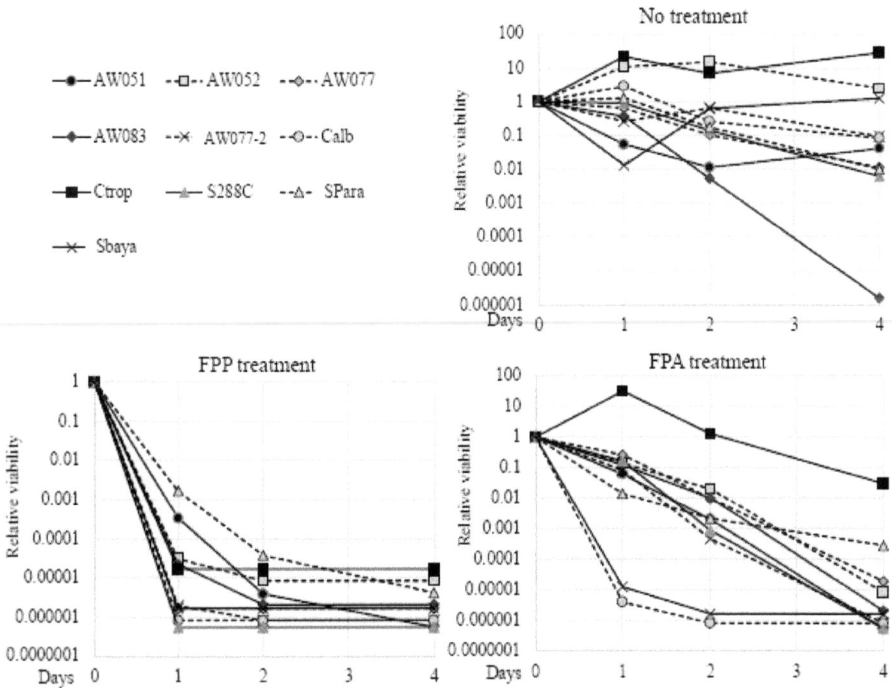

Figure 7. Toxicity of FPP and FPA are well conserved. Yeast cells were inoculated to Synthetic Complete medium overnight, then inoculated to MM2 medium with 500 mg L^{-1} FPP, FPA, or no drug, and incubated at 37°C in an immobile 96-well plate. Culture viability was assessed over time using the Tadpoling method. Relative viability was assessed by dividing the number of CFUs on day 0, 1, 2, or 5 by the number of CFUs on day 0. The average of three independent trials is shown.

One potential mechanism of FPP toxicity is to affect the mitochondria. Phenylpyruvate acts as an inhibitor of pyruvate transport into the mitochondria where pyruvate would normally be decarboxylated to acetate [14]. Indeed, accumulation of phenylpyruvate in phenylketonuria patients becomes toxic as it may disrupt oxidative metabolism by preventing normal pyruvate movement into the mitochondria. To test if phenylpyruvate alone may exert a toxic effect on yeast, exponential phase cells were inoculated to 500 mg L^{-1} phenylpyruvate and culture viability was measured over several days using the Tadpoling method. The ratio of CFUs of cells exposed to phenylpyruvate with control cells was calculated over the course of several days (Figure 8). There were no significant ($p < 0.05$) deviations from the ideal 1 ratio for all strains tested, indicating that phenylpyruvate does not significantly affect culture viability. Nonetheless, there were notable deviations in decreased viability of S288c cells exposed to phenylpyruvate and an increase in viability of AW052 cells exposed to phenylpyruvate. Thus, it is likely that fluorine is required for fluorophenylalanine toxicity.

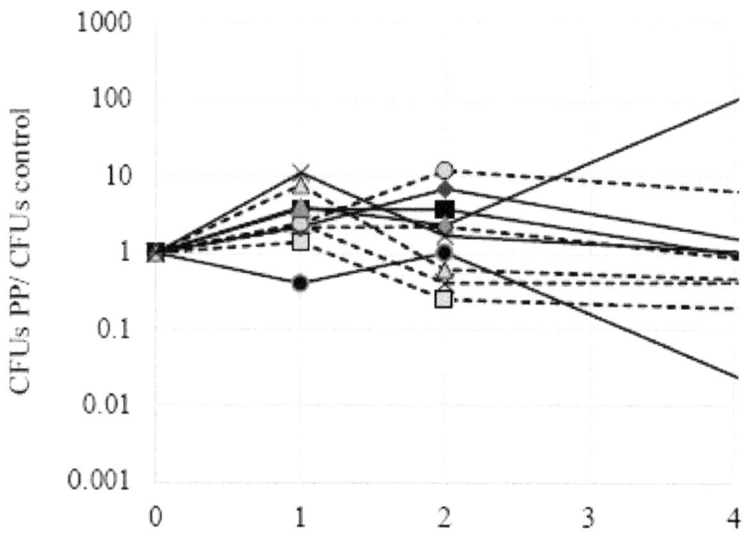

Figure 8. Phenylpyruvate does not affect culture viability. Yeast cells were inoculated to Synthetic Complete medium overnight, then inoculated to MM2 medium with 500 mg L^{-1} phenylpyruvate, or no drug, and incubated at 37 °C in an immobile 96-well plate. Culture viability was assessed over time using the Tadpoling method. To calculate the CFUs PP/CFUs control, the number of CFUs in phenylpyruvate treated cultures was divided by the number of CFUs in the untreated culture for each day. The average of three independent trials is shown. Markers are the same as used in Figure 7.

In *S. cerevisiae*, one method of increasing the production of the valuable chemical phenylethanol is to select for resistance to FPA [2]. We wanted to test whether FPA resistant strains obtained herein would produce high amounts of phenylethanol. To do this, we inoculated strains into medium containing phenylalanine as the source of nitrogen and measured the phenylethanol and phenylalanine concentration in the medium after three days of growth using GC-MS (Figure 9A).

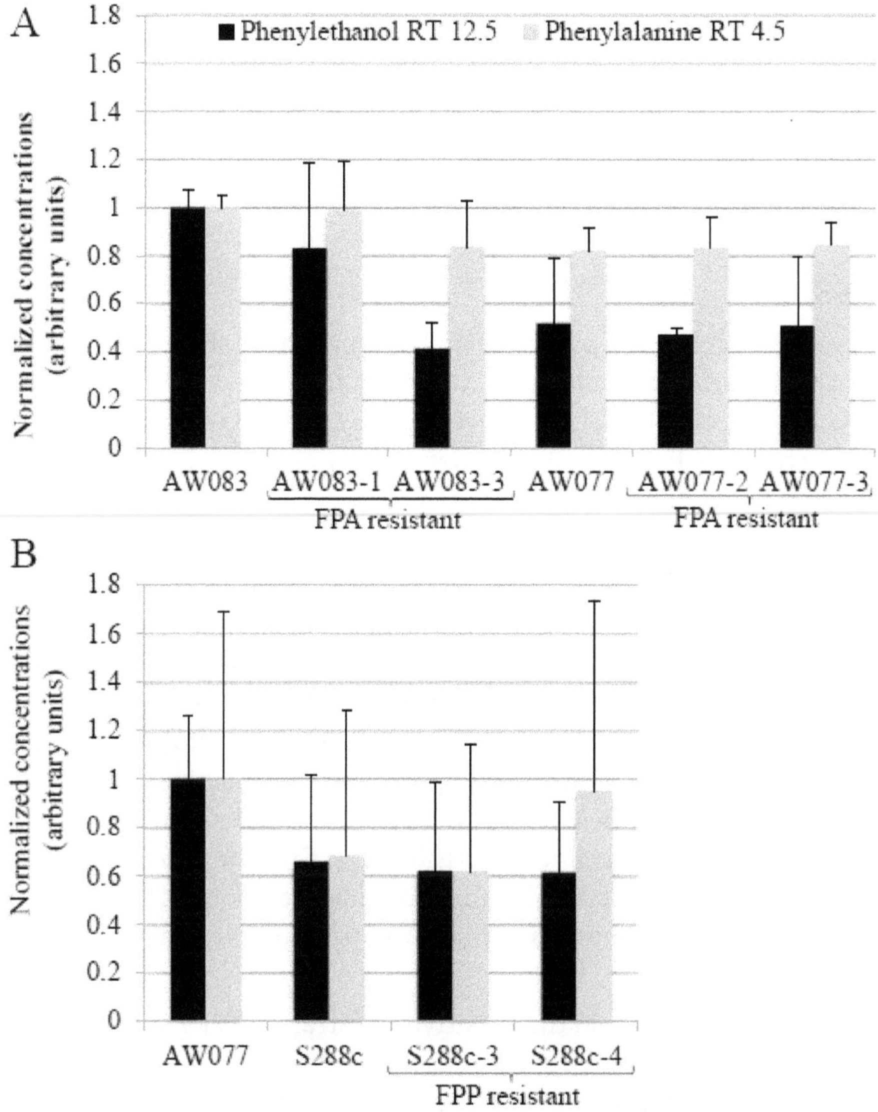

Figure 9. FPP or FPA resistance is not sufficient for increased phenylethanol production. Strains were incubated in medium containing phenylalanine as the source of nitrogen for three days, and then phenylethanol and phenylalanine concentration in the medium were measured using GC-MS. Each strain's production of phenylethanol or phenylalanine was normalized to 1 using (A) AW083 strain or (B) AW077 strain. The average and standard deviation of three independent trials is shown.

Resistance to FPA did not induce higher phenylethanol production in the four strains that were tested. This is not surprising, as previously, the frequency of FPA-resistant mutants with high levels of phenylethanol production was approximately 20% [15]. Thus, if we had tested five or more isolates we may have expected to obtain at least one. However, because of the low frequency of obtaining FPA-resistant strains we did not pursue this further. Because we had seen that FPP resistance was more difficult to attain than FPA, we predicted that these mutants may produce phenylalanine at a greater rate than FPA-resistant mutants and thus may have higher phenylethanol production. However, upon testing FPP-resistant strains, no increase in phenylethanol production was noted (Figure 9B). Thus, FPA and FPP resistance is not sufficient to induce high levels of phenylethanol production.

DISCUSSION

Here, we report that FPP can act as a toxic agent in diverse strains of yeast, and this effect is at least partially mediated by *ARO8* and *ARO9* conversion to FPA. Whether FPP exerts its toxic effect solely by conversion to FPA or via another route requires further investigation. FPP exhibited higher toxicity than FPA, suggesting a separate mechanism of toxicity, however this could be due to enhanced adsorption to cell wall, decreased cell uptake, or other enzymatic activities that decrease the effective concentration of FPA.

The mechanism of FPP toxicity still remains somewhat of a mystery. It is possible that there is preferred transport of phenylpyruvate over phenylalanine although evidence for this is absent. Another hypothesis is that FPP is converted into some other compound that FPA is not, such as fluoro-phenylethanol. Phenylalanine degradation follows this pathway: phenylalanine > phenylpyruvate > phenylacetaldehyde > phenylethanol. Phenylpyruvate is one enzymatic step closer to phenylethanol, and it is possible the fluorine moiety decreases its ability to convert to fluoro-phenylalanine, but instead allows it to become fluoro-phenylethanol. Phenylethanol has been shown to be toxic to *C. albicans* at 20 mg L^{-1} [16].

We observed differences in the sensitivity of strains to FPP and FPA depending on the method of delivery—solid or liquid medium. Phenylpyruvate shows increased solubility relative to phenylalanine (112 g $L^{-}1$ versus 27 g $L^{-}1$) that may explain its increased toxicity in liquid medium [17]. This observation may be explained by the increased exposure of cells to all toxic molecules in a liquid, as opposed to a solid medium in which cells are only exposed to toxins within close physical proximity. Previous groups have reported similar enhanced drug potency in liquid media [18].

The effect of fluorinated phenylalanine metabolites has been observed previously in other fungal species. Fluorophenylalanine at concentrations as low as 2.5 mM can inhibit growth of fungal species *Cladosporium cucumerinum* and *Colletotrichum lagenarium* [19]. We found that FPP was toxic to a variety of yeast species at 2.74 mM indicating that it is a more potent fungal inhibitor than FPA.

Interestingly, we also found that *PHA2* was not required for slow growth. This is in contrast to several other works that indicate that *PHA2* causes an absolute requirement for phenylalanine in the medium [10,20]. However, strain growth is typically assayed over a period of up to three days on solid medium, while in our studies, liquid medium was used which allows for faster growth. *PHA2* converts prephenate to phenylpyruvate, but an alternative method to converting prephenate to phenylalanine is via the arogenate dehydratase enzyme. This enzyme has not been reported in *S. cerevisiae*. The closest species having a putative arogenate dehydratase is *Neurospora crassa*, which shares the same phylum with *S. cerevisiae*. Further investigation is needed to determine the mechanism of slow phenylalanine generation in strains lacking *PHA2*.

These observations should be helpful in better understanding phenylalanine metabolism and may be of interest in identifying potential secondary targets of FPP.

REFERENCES

1. Etschmann, M.; Bluemke, W.; Sell, D.; Schrader, J. Biotechnological production of 2-phenylethanol.*Appl. Microbiol. Biotechnol.* **2002**, *59*, 1–8. [PubMed]
2. Akita, O.; Ida, T.; Obata, T.; Hara, S. Mutants of Saccharomyces cerevisiae producing a large quantity of β-phenethyl alcohol and β-phenethyl acetate. *J. Ferment. Bioeng.* **1990**, *69*, 125–128. [CrossRef]
3. Fukuda, K.; Watanabe, M.; Asano, K. Altered regulation of aromatic amino acid biosynthesis in β-phenylethyl-alcohol-overproducing mutants of sake yeast Saccharomyces cerevisiae. *Agric. Biol. Chem.* **1990**, *54*, 3151–3156. [CrossRef]
4. Forsburg, S.L. The art and design of genetic screens: Yeast. *Nat. Rev. Genet.* **2001**, *2*, 659–668. [CrossRef] [PubMed]
5. Toyn, J.H.; Gunyuzlu, P.L.; Hunter White, W.; Thompson, L.A.; Hollis, G.F. A counter selection for the tryptophan pathway in yeast: 5-fluoroanthranilic acid resistance. *Yeast* **2000**, *16*, 553–560. [CrossRef]
6. Boeke, J.D.; La Croute, F.; Fink, G.R. A positive selection for mutants lacking orotidine-5t-phosphatedecarboxylase activity in yeast: 5-fluoro-orotic acid resistance. *Mol. Gen. Genet. MGG* **1984**, *197*, 345–346. [CrossRef] [PubMed]
7. Iraqui, I.; Vissers, S.; Cartiaux, M.; Urrestarazu, A. Characterisation of Saccharomyces cerevisiae ARO8 and ARO9 genes encoding aromatic aminotransferases I and II reveals a new aminotransferase subfamily. *Mol. Gen. Genet. MGG* **1998**, *257*, 238–248. [CrossRef] [PubMed]
8. Welch, A.Z.; Koshland, D.E. A simple colony-formation assay in liquid medium, termed tadpoling, provides a sensitive measure of Saccharomyces cerevisiae culture viability. *Yeast* **2013**, *30*, 501–509. [CrossRef] [PubMed]

9. Dittmer, K. The Synthesis and Microbiological Properties of β-3-Thienylalanine, a New anti-PhenylalanineJ. *Am. Chem. Soc.* **1949**, *71*, 1205–1207. [CrossRef] [PubMed]

10. Bross, C.D.; Corea, O.R.A.; Kaldis, A.; Menassa, R.; Bernards, M.A.; Kohalmi, S.E. Complementation of the pha2 yeast mutant suggests functional differences for arogenate dehydratases from Arabidopsis thaliana. *Plant Physiol. Biochem.* **2011**, *49*, 882–890. [CrossRef] [PubMed]

11. SRI International Yeast Pathways Database Website Home. Available online: https://pathway.yeastgenome. org/ (accessed on 1 May 2018).

12. Chattoo, B.B.; Sherman, F.; Azubalis, D.A.; Fjellstedt, T.A.; Mehnert, D.; Ogur, M. Selection of lys2 mutants of the yeast Saccharomyces cerevisiae by the utilization of α-aminoadipate. *Genetics* **1979**, *93*, 51–65. [PubMed]

13. Whelan, W.L.; Gocke, E.; Manney, T.R. The CAN1 locus of Saccharomyces cerevisiae: Fine-structure analysis and forward mutation rates. *Genetics* **1979**, *91*, 35–51. [PubMed]

14. Halestrap, A.P.; Brand, M.D.; Denton, R.M. Inhibition of mitochondrial pyruvate transport by phenylpyruvate and α-ketoisocaproate. *Biochim. Biophys. Acta (BBA) Biomembr.* **1974**, *367*, 102–108. [CrossRef]

15. Fukuda, K.; Watanabe, M.; Asano, K.; Ouchi, K.; Takasawa, S. Isolation and genetic study of p-fluoro-dl-phenylalanine-resistant mutants overproducing β-phenethyl-alcohol in Saccharomyces cerevisiae. *Curr. Genet.* **1991**, *20*, 449–452. [CrossRef] [PubMed]

16. Lingappa, B.T.; Prasad, M.; Lingappa, Y.; Hunt, D.F.; Biemann, K. Phenethyl alcohol and tryptophol: Autoantibiotics produced by the fungus Candida albicans. *Science* **1969**, *163*, 192–194. [CrossRef] [PubMed]

17. Wishart, D.S.; Feunang, Y.D.; Marcu, A.; Guo, A.C.; Liang, K.; Vázquez-Fresno, R.; Sajed, T.; Johnson, D.; Li, C.; et al. HMDB 4.0—The Human Metabolome Database for 2018. *Nucleic Acids Res.* **2018**, 608–617. [CrossRef] [PubMed]

18. Smith, A.M.; Ammar, R.; Nislow, C.; Giaever, G. A survey of yeast genomic assays for drug and target discovery. *Pharmacol. Ther.* **2010**, *127*, 156–164. [CrossRef] [PubMed]

19. Van Andel, O.M. Fluorophenylalanine as a systemic fungicide. *Nature* **1962**, *194*. [CrossRef]

20. Ma, Y.; Sugiura, R.; Saito, M.; Koike, A.; Sio, S.O.; Fujita, Y.; Takegawa, K.; Kuno, T. Six new amino acid-auxotrophic markers for targeted gene integration and disruption in fission yeast. *Curr. Genet.* **2007**, *52*, 97–105. [CrossRef] [PubMed]

CHAPTER-7

EFFECTS OF GLUCOSE CONCENTRATION AND NACL-INDUCED OSMOTIC STRESS ON THE FORMATION OF β-D-GLUCAN IN YEAST CELL WALLS DURING ANAEROBIC FERMENTATION

INTRODUCTION: AN OVERVIEW

The synthesis of yeast cell wall β-D-glucan during anaerobic fermentation is a complex process influenced by various environmental factors. Among these factors, glucose concentration and NaCl osmotic stress play pivotal roles in shaping the dynamics of β-D-glucan formation. The concentration of glucose serves as a critical substrate, influencing the metabolic pathways that contribute to the production of cell wall components. Concurrently, NaCl-induced osmotic stress introduces an additional layer of complexity, as yeast cells respond to environmental challenges by modulating their cell wall composition. This study investigates the intricate interplay between glucose concentration and NaCl osmotic stress, providing valuable insights into the regulatory mechanisms governing yeast cell wall β-D-glucan synthesis during anaerobic fermentation. Unraveling these dynamics contributes to a deeper understanding of yeast physiology and holds significance for optimizing bioprocessing conditions in various industrial applications.

Saccharomyces cerevisiae is the model microorganism most commonly used in the study of yeast and higher eukaryotic cell physiology [1]. A major substrate for its growth is sugars, which are degraded through an aerobic or anaerobic fermentation process with the production of CO_2 or alcohol and CO_2, respectively [2]. Nowadays, the biotechnological research and industrial applications of *S. cerevisiae* have extended far beyond food and beverage production to medicine development, molecular biology and genetics, and environmental protection technologies [3].

In yeast, the plasma membrane is surrounded by the cell wall, a structure of 100–200 nm width that consists of consecutive layers. The main structural elements of the yeast cell wall are constructed of three groups of polysaccharides: (a) mannoproteins (or mannans): α-mannose polymers, which account for around 40% of the cell wall dry weight [4]; (b) α- and β-glucans: D-glucose homopolymers with α- or β-glycosidic bonds between C-1 and C-2, C-3, C-4, and C-6 of glucopyranose rings; they account for approximately 30–60% of the cell wall dry weight (β-D-glucans are the main cell wall glucans in which two different types are found: a major β-1,3-D-glucan (85%), which is the main β-glucan, representing more than 50–55% of cell wall, and a minor β-1,6-D-glucan (15%), representing 5–10% of cell wall dry weight [5–8]); and (c) chitin: β-1,4-D-*N*-acetyl-glucosamine homopolymers, equates to around 2% of the yeast cell wall dry weight [8,9]. The plasma membrane is an extension of the cell wall to the interior of the yeast cell. Its width is 7.5 nm and it consists of lipids and proteins that are found in much lower proportions compared to the cell wall components [10,11]. Yeast β-glucans (β-1,3-D-glucan and β-1,6-D-glucan) belong to biological response modifiers (BRMs) due to their ability to enhance and stimulate the human immune system presenting antitumor, anti-inflammatory, antimicrobial, wound healing, coronary heart disease prevention, hepatoprotective, weight loss, and antidiabetic properties [12–14].

Depending on the yeast growth and cultivation conditions, the cell wall dry weight may vary between 10% and 25% of the total cell dry weight [15]. The cell wall has four main functions: to (1) protect the cell from osmotic stress by maintaining its turgidity [10]; (2) protect the cell from mechanical stress;(3) maintain the cell's shape; and (4) act as a hold for the cell's surface proteins [15,16].

During yeast cell batch fermentation, the cell wall is modified. These changes are affected by the yeast strain and cell growth conditions [17–20]. The conditions of cell growth, such as pH, temperature, aeration, carbon sources, limitation of nitrogen, and the mode of cultivation, affect the cell wall polysaccharide composition, structure and morphology [21]. Also, the addition of supplements (SDS, ethylene-diaminete-tra-acetic acid (EDTA), NaCl) in the yeast culture medium and the conditions during alcoholic fermentation for beer production impact variously on the cell wall β-glucan concentration at the end of the fermentation process [22] as it also modifies the polysaccharide structure of produced spent yeast biomass [23].

The remodelling of the yeast cell wall, as a response to various environmental stresses, is controlled by: (a) the cell wall integrity (CWI) and (b) high osmotic glycerol (HOG) through mitogen-activated protein kinase (MAPK) signalling pathways that modulate the cell wall gene expression and lead to the construction of a newly modified cell wall [24,25]. In addition to the HOG pathway, the cell growth, carbon storage and stress response are regulated by the RAS-cAMP PKA pathway [26], while during rapid hyperosmotic shocks the yeast cell sensitivity is regulated by the cell wall elasticity increase through the inactivation of the Crh family of cell wall cross-linking enzymes [25].

The impact of glucose concentration and NaCl osmotic stress on yeast performance during anaerobic fermentation has been well studied [27–35], but their impact on cell wall β-glucan production has not been studied until now.

The aim of this work was the investigation of yeast cell wall β-glucan production during the anaerobic fermentation process under the influence of hyperosmotic stress of fermented media glucose concentration and NaCl pre-cultured cells used for defined media inoculation. The motivations for this research study were: (a) the fact that there are few reports concerning yeast growth conditions and cell wall polysaccharide composition; (b) the lack of studies of the impact of different environmental stresses on yeast cell wall structure during the fermentation process and the yeast cell osmo-adaptation responses against an osmotic shock; (c) the lack of knowledge of the entire physiological role of β-glucan in the yeast cell; (d) the fact that the monitoring of the yeast cell wall is a dynamic and unexplored research field that offers potential biotechnological applications and purposes; e) the increasing interest in the industrial production of β-glucan from various yeast sources (medium cultures, brewery and winery waste yeast biomass) due to its proven immune-stimulant properties and its incorporation in medicines, cosmetics and functional foods; and (f) the potential benefits from the study of the yeast cell wall in the creation of new antifungal drugs. Furthermore, the study of the yeast cell wall under various conditions can offer new perspectives on the biochemical pathways involved in the alcoholic fermentation process and the biology of the stressed yeast cell.

MATERIALS AND METHODS

Yeast Cells and Culture Growth Conditions

Yeast cells *S. cerevisiae* VIN 13 strain (commercial dry yeast) used for laboratory experiments were provided by Anchor Yeast (Cape Town, South Africa). Yeast cells were grown in a medium containing (per litre of deionised water): 100 g D-glucose, 1 g $K2HPO4$, 1 g $KH2PO4$, 0.2 g $ZnSO4$, 0.2 g $MgSO4$, 2 g yeast extract and2g $NH4SO4$. All the media components were purchased from the Sigma Chemical Company (St. Louis, MO, USA).

Yeast Cell Preconditioning and Inoculum Preparation

One gram dry weight of yeast was resuspended in 100 mL of deionised water in an Erlenmeyer flask of 250 mL volume, at 30–35°C, for 30 min with NaCl 6% *w/v*. The inoculum for experimental fermentations was prepared as follows: after 24 h of pre-culturing, 10 mL was collected and centrifuged at 5000 rpm for 15 min. Cells were resuspended in deionised water and re-centrifuged. This was repeated twice prior to the determination of total cell number and cell viability in the final washed inoculum; 2×10^5 of living cells were used to inoculate 1000 mL of medium substrate [27].

Preparation of Yeast Fermentation Medium

The medium for laboratory batch fermentations consisted of the following: 200, 300, 400 g/L glucose, 1 g/L K_2HPO_4, 1 g/L KH_2PO_4, 0.2 g/L $ZnSO_4$, 0.2 g/L $MgSO_4$, 2 g/L yeast extract and 2 g/L NH_4SO_4 (Sigma-Aldrich Co., St. Louis, MO, USA). Mineral components and glucose were sterilised separately at 120°C, and 2 atm pressure for 20 min. The pH was set to 4.0 with the use of 1 N HCl. Batch fermentations were carried out in 2000 mL volume flasks containing 1000 mL of growth medium with NaCl-preconditioned cells without shaking at 25°C [36].

Cell Wall Treatment with Lytic Enzyme and Sonication

Yeast cells were treated with lytic enzyme Glucanex®200G (β-1,3-glucanase with some β-1,6-glucanase) (Novozymes, Kobenhavn, Denmark) for the cell wall lysis (pH 4.64/50 °C/5 h/10 times higher than the normal dose of 0.015 g/L, optical density of the resuspended cells with 10 mM citrate buffer adjusted to approximately 1.0 OD at 600 nm, 15% *w/v* cell concentration) [18,37,38] and sonication for the cell disruption at 35 kHz/70 W/6 min (u = 230 V/AC, I = 1.6 A, f = 50/60 Hz) in an ice bath with a Transsonic 570/H sonicator (Elma, Singen, Germany) [39].

Determination of Cell Concentration

Yeast cell number was determined by counting using a hemocytometer (Neubauer type) [36].

Cell Viability Determination

Cell viability was determined using a haemocytometer (Neubauer type) and according to the methylene blue method: 1 mL of sample medium was taken and diluted in 9 mL of deionised water. 1 mL of this solution was dissolved with 1 mL of 10% *v/v* methylene blue solution and left for 10 min. Aliquots of 1 μL were placed on the haemocytometer using a Pasteur pipette. The haemocytometer was then microscopically observed with an optical microscope (Olympus model CHK2-F-GS microscope). Yeast cell viability was calculated and expressed as follows:

where

$$\text{Viability (\%)} = a/n \times 100,$$

a: number of metabolically active cells

n: total cell number [27].

Monitoring of Fermentation Kinetics

OD at 600 nm was determined spectrometrically (phasmatophotometer mini 1240, SHIMADZU, Beijing, China) and was used for the monitoring of fermentation kinetics and the discrimination of exponential, lag and death phase of cells growth during the fermentation process [18].

Determination of Cell Dry Weight

The sediment of yeast cells was frozen (-80°C/24 h), lyophilised (-50°C /vacuum/24 h) using a Thermo Fischer (Waltham, MA, USA) drying digital unit. The weight of each yeast cells freeze dried sample was measured gravimetrically with an analytical balance (Kern, Kern and Sohn GmbH, Balingen, Germany) with an accuracy of four decimal places.

Determination of β-Glucan Concentration (%)

The determinations of total yeast β-glucan concentration in lyophilised yeast biomass were performed with the use of Enzymatic K-EBHLG Yeast β-Glucan Assay Kit (Megazyme, Bray, Ireland).

Determination of β-Glucan Yield (%)

The yield of β-glucan was calculated as the product of yeast cells' dry weight with β-glucan purity [40].

Statistical Analysis

Each experiment was carried out in triplicate and the reproducibility was within the range of±5%. Results are displayed as means of three determinations in all methods with standard deviation. Experiments were set up in a completely randomised design while for the determination of significant differences between the different treatments and the tested parameters, all results were analysed using two-way analysis of variance (ANOVA) run on XLSTAT software (Addinsoft Co., New York, NY, USA), with significant differences indicated at $p \leq 0.05$.

RESULTS

Experimental Design

Yeast cells *S. cerevisiae* VIN 13 strain were inoculated in a medium containing three different glucose concentrations: 200, 300, or 400 g/L. The inoculation of each sample of the fermentation batch was performed according to the instructions of the yeast company (Anchor Yeast, Cape Town, South Africa) for wine production with 2×10^5 living cells of 0.1 g dry yeast preconditioned in 0% and 6% NaCl for 24 h before the inoculation according to the procedure described by Logothetis et al. (2013) [41]. The volume of the inoculated substrate was 1000 mL for each sample. During alcoholic fermentation, for the monitoring of fermentation kinetics, samples of 50 mL from each fermented medium were collected every 8 h after homogenisation of the fermented medium for the direct determination of cell viability and optical absorption (OD). In a next step, these 50 mL samples were centrifuged (5000 rpm/10 min) and the sediment (yeast biomass) was freeze dried for the determination of cell dry weight and the yeast cell β-glucan concentration. Subsequently, the freeze-dried yeast cells were treated with lytic enzyme and sonication for the estimation of cell wall resistance and its correlation with cell wall β-glucan amount. After the centrifugation of the samples and before their freeze drying, they were washed three times with distilled water for the removal of the possible glucose that may have remained in the sample. This was applied in order to avoid glucose weight being taken into account in the determination of the dry weight of the sample, which might cause subsequent data errors to arise. After 192 h for the completion of the alcoholic fermentation, the suspension was separated from the yeast biomass sediment (centrifugal/5000 rpm/10 min) for the determination of various oenological analytical parameters, while the sediment was freeze-dried for β-glucan determination in yeast cell biomass.

Cell Viability

For all the samples, cell viability rose during the early exponential phase of fermentation and reached a first peak after 16 h. At this point, cells of samples (200 g/L glucose, 0% NaCl), (200 g/L glucose, 6% NaCl) and (300 g/L, 6% NaCl) showed the highest viability (%) (97.78 ± 2.04%, 97.36 ± 1.59% and97.28 ± 1.02% respectively) while NaCl-stressed cells of samples (300 g/L glucose, 6% NaCl) and (400 g/L glucose, 6% NaCl) had higher viability portions (97.28 ± 1.02% and 93.43 ± 1.44% respectively) compared to the corresponding (300 g/L, 0% NaCl) and (400 g/L glucose, 0% NaCl) ones (93.94 ± 1.42% and89.89 ± 1.13% respectively) (Figure 1). The cell viability reached a second, lower peak during the end of the exponential and the beginning of the stationary phase for all samples. The more viable cells were counted for samples (200 g/L glucose, 0% NaCl) and (200 g/L glucose, 6% NaCl) (97.84 ± 2.03% and93.54 ± 1.88% respectively) (Figure 1). At the death phase (48–120 h), the cell viability decreased for all the samples, with a higher decrease for samples (300 g/L glucose, 6% NaCl) and (400 g/L glucose, 6% NaCl) (Figure 1). From 120 to 192 h, an increase in cell viability occurred for all samples, but this was more abrupt for NaCl-stressed cells of samples (300 g/L glucose, 6% NaCl) and

(400 g/L glucose, 6% NaCl). Despite this increase, for all samples, after 192 h of fermentation, cell viability was lower when compared to values of the second peak. Finally, the viability of NaCl preconditioned and non-preconditioned yeast cells growing in a medium containing 200 g/L glucose was similar at the end of the fermentation (85.34 ± 1.3% and 83.44 ± 0.8% respectively) (Figure 1).

A)
B)

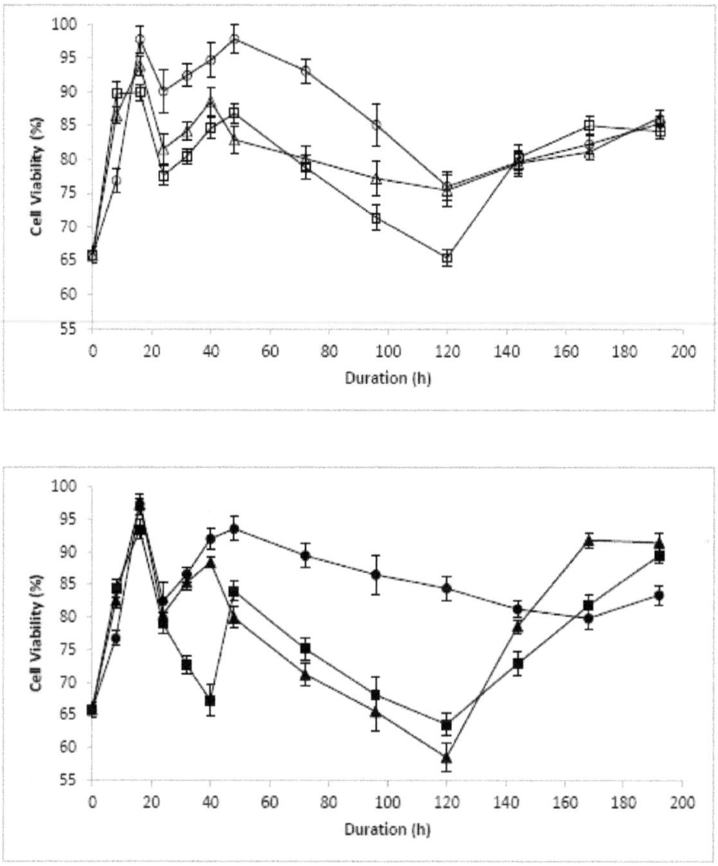

Figure 1. (A) Yeast cell viability (%) of non-NaCl-stressed cells during various fermentation times (h) with three different media containing 200, 300, or 400 g/L glucose (*n* = 3 replications for each time).(B) Yeast cell viability (%) of NaCl-stressed cells during the fermentation process (h) with three different media containing 200, 300, or 400 g/L glucose (*n* = 3 replications for each time) (0): (200 g/L glucose, 0% NaCl), (Δ): (300 g/L glucose, 0% NaCl), (D): (400 g/L glucose, 0% NaCl), (·): (200 g/L glucose, 6% NaCl), (Δ): (300 g/L glucose, 6% NaCl), (■): (400 g/L glucose, 6% NaCl).

Optical Density (OD) and Cell Count

The results of OD (600 nm) measurements showed that the lag phase was 0–8 h for all the fermented samples; the exponential phase was between 0–40 h for the samples (200 g/L glucose, 0% NaCl) and (200 g/L glucose, 6% NaCl) and 0–48 h for the others. The stationary phase for the samples (200 g/L glucose, 0% NaCl) and (200 g/L glucose, 6% NaCl) was short (40–48 h), while for the other samples it was longer (48-72 h). The death phase was determined to be between 48 and 192 h for the samples (200 g/L glucose, 0% NaCl) and (200 g/L glucose, 6% NaCl)) and 72–192 h for the samples (300 g/Lglucose, 0% NaCl), (300 g/L glucose, 6% NaCl), (400 g/L glucose, 0% NaCl), and (400 g/L glucose,6% NaCl) (Figure 2).

A)

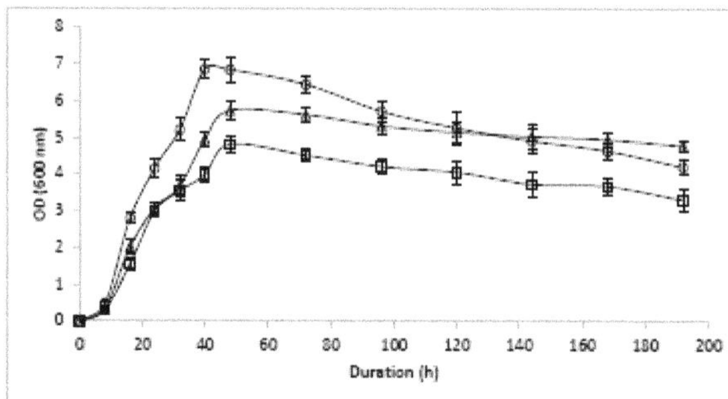

B)

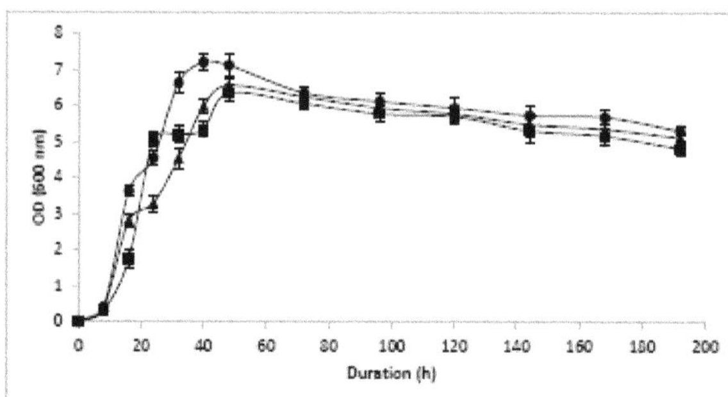

Figure 2. (A) OD (600 nm) of non-NaCl-stressed cells during various fermentation times (h) with three different media containing 200, 300, or 400 g/L glucose (*n* = 3 replications for each time); (B) OD (600 nm) of NaCl-stressed cells during the fermentation process (h) with three different media containing 200, 300, or 400 g/L glucose (*n* = 3 replications for each time) (0): (200 g/L glucose, 0% NaCl), (Δ): (300 g/L glucose, 0% NaCl), (D): (400 g/L glucose, 0% NaCl), (•): (200 g/L glucose, 6% NaCl), (Δ): (300 g/L glucose, 6% NaCl), (■): (400 g/L glucose, 6% NaCl).

The cell count was used in order to determine the total yeast cell number during the fermentation process. Our results revealed that during the death phase, the OD (600 nm) and the cell number decreased for all the samples but, surprisingly, between 120 and 192 h the number of cells grown in 400 g/L glucose (0% and 6% NaCl) increased considerably (198.75×10^6 and 186.44×10^6 cells/mL respectively) compared with the other media (200 g/L glucose, 0% NaCl), (200 g/L glucose, 6% NaCl), (300 g/L glucose, 0% NaCl)and (200 g/L, 6% NaCl) (24.15×10^6, 30.12×10^6, 52.13×10^6 and 45.54×10^6 cells/mL respectively) (Figure 3). The samples with 200 g/L glucose showed two peaks, at 24 and 48 h, with the first peak a little higher than the second. The other samples (200 g/L glucose, 0% NaCl), (200 g/L glucose, 6%

NaCl), (300 g/L glucose, 0% NaCl), (200 g/L glucose, 6% NaCl) showed one peak at 24 h, which was significant higher compared to that of samples (200 g/L glucose, 0% NaCl) and (200 g/L glucose, 6% NaCl), thus indicating a higher cell number (Figure 3).

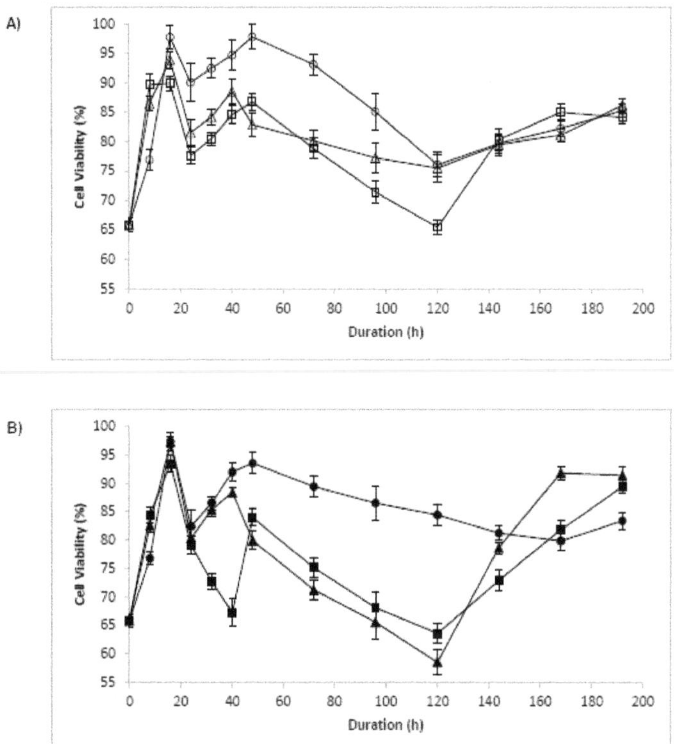

Figure 3. (A) Total cell count (cells/mL) of non-NaCl-stressed cells during various fermentation times • (h) with three different media containing 200, 300, or 400 g/L glucose after cell treatment with sonication (*n* = 3 replications for each time and error bars represent standard deviation of the average value of all replications with each range of fermentation time); (B) Total cell count (cells/mL) of NaCl-stressed cells during various fermentation times (h) with three different media containing 200, 300, or 400 g/L glucose after cell treatment with sonication (*n* = 3 replications for each time and error bars represent standard deviation of the average value of all replications with each range of fermentation time) (0): (200 g/L glucose, 0% NaCl), (Δ): (300 g/L glucose, 0% NaCl), (D): (400 g/L glucose, 0% NaCl), (•): (200 g/L glucose, 6% NaCl), (■): (300 g/L glucose, 6% NaCl), (•): (400 g/L glucose, 6% NaCl).

Dry Weight and Sample β-Glucan Yield

The freeze-dried yeast cell biomass obtained from a 50-mL sample of fermented medium at various times was determined gravimetrically. For each of the fermented media, the dry weight of yeast biomass rose constantly until the early stationary phase and then decreased constantly during the death phase only for media (200 g/L glucose, 0% NaCl), (200 g/L glucose, 6% NaCl) and (300 g/L glucose, 0% NaCl). The fermented media (300 g/L glucose, 6% NaCl), (400 g/L glucose, 6% NaCl) and (400 g/L glucose, 6% NaCl) showed a smaller decline, with a slight increase in yeast biomass for (300 g/L glucose, 6% NaCl) and (400 g/L glucose, 0% NaCl) and a more significant one for (400 g/L glucose, 6% NaCl) at the end of the fermentation (144–192 h) (Figure 4). The highest value of yeast dry weight (0.1900 ± 0.0066 g) was obtained with 200 g/L glucose, NaCl-stressed cells, with 40 h fermentation time (Figure 4).

A)

)

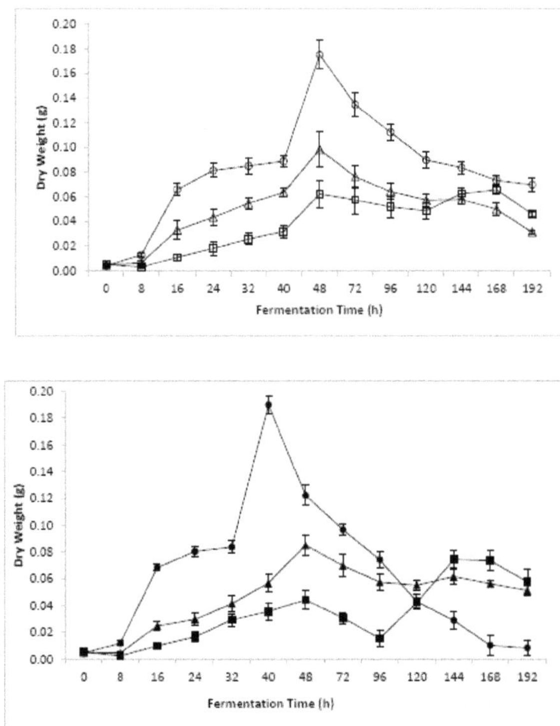

Figure 4. (A) Dry weight (g) of a 50-mL sample of non-NaCl-stressed cells during various fermentation times (h) with three different media containing 200, 300, or 400 g/L glucose (*n* = 3 replications for each time and error bars represent standard deviation of the average value of all replications with each range of fermentation time); (B) Dry weight (g) of a 50-mL sample of non-NaCl stressed cells during various fermentation times (h) with three different media containing 200, 300, or 400 g/L glucose (*n* = 3 replications for each time and error bars represent standard deviation of the average value of all replications with each range of fermentation time) (O): (200 g/L glucose, 0% NaCl), (Δ): (300 g/L glucose, 0% NaCl), (D): (400 g/L glucose, 0% NaCl), (•): (200 g/L glucose, 6% NaCl), (■): (300 g/Lglucose, 6% NaCl), (Δ): (400 g/L glucose, 6% NaCl).

The β-Glucan yield (g) was optimum for media (200 g/L glucose, 0% NaCl), (200 g/L glucose, 6% NaCl), (300 g/L glucose, 0% NaCl) and (300 g/L glucose, 6% NaCl) at the stationary phase ($p \leq 0.05$), while for medium (400 g/L glucose, 0% NaCl) it reached a peak at the stationary phase and remained constant until the end of fermentation and for medium (400 g/L glucose, 6% NaCl) it was optimum at the end of fermentation (Figure 5). The highest β-Glucan yield (0.1038 ± 0.0073 g) was obtained with fermented medium (200 g/L glucose/0% NaCl (non-stressed cells)/48 h fermentation time) at the stationary phase (Figure 5).

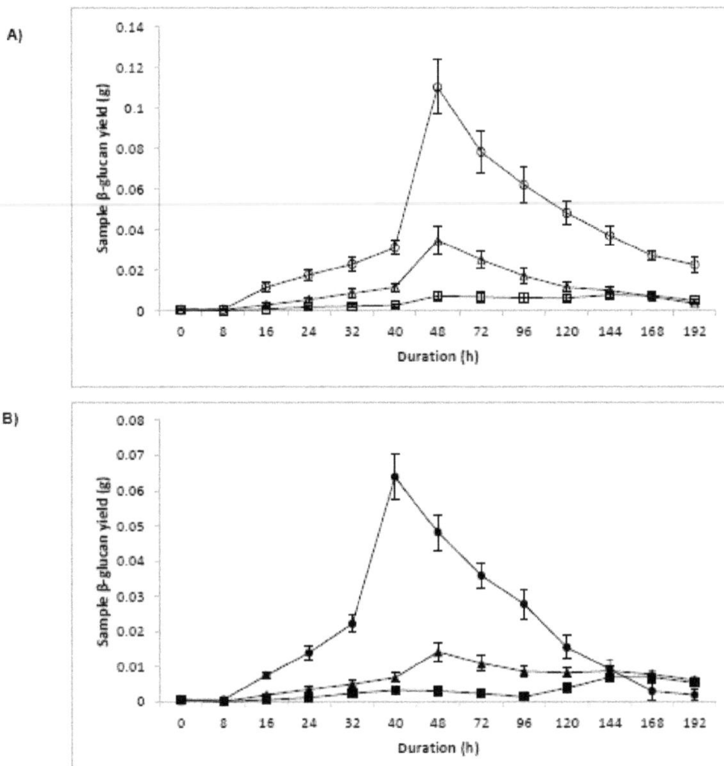

Figure 5. (A) β-Glucan yield (%) of a 50-mL sample of non-NaCl-stressed cells during various fermentation times (h) with three different media containing 200, 300, or 400 g/L glucose (*n* = 3 replications for each time and error bars represent standard deviation of the average value of all replications with each range of fermentation time). Mean values between 0–144 h were significantly different at *p* ≤ 0.05 level. (B) β-Glucan yield (%) of a 50-mL sample of non-NaCl-stressed cells during various fermentation times (h) with three different media containing 200, 300, or 400 g/L glucose (*n* = 3 replications for each time and error bars represent standard deviation of the average value of all replications with each range of fermentation time). Mean values between 0–144 h were significantly different at *p* ≤ 0.05 level. (0): (200 g/L glucose, 0% NaCl), (Δ): (300 g/L glucose, 0% NaCl), (D): (400 g/L glucose, 0% NaCl),(•): (200 g/L glucose, 6% NaCl), (Δ): (300 g/L glucose, 6% NaCl), (■): (400 g/L glucose, 6% NaCl).

β-Glucan Concentration

For the fermented media (200 g/L glucose, 0% NaCl), (200 g/L glucose, 6% NaCl), (300 g/L glucose, 0% NaCl) and (300 g/L glucose, 6% NaCl), the highest yeast cell wall β-glucan concentration was observed at the stationary phase ($p \leq 0.05$) (Figure 6). For media (400 g/L glucose, 0% NaCl) and (400 g/L glucose, 6% NaCl), the β-glucan concentration rose until the cell culture entered the stationary phase and remained stable during the death phase (Figure 6). The highest β-glucan yeast cell concentration was observed for non-NaCl-stressed and NaCl-stressed cells grown in a medium containing 200 g/L glucose (62.95 ± 3.39 % and 39.49 ± 1.78 % respectively). Also, cells grown in these media showed a β-glucan concentration higher at the end of the alcoholic fermentation compared to the cells grown with the other media (Figure 6).

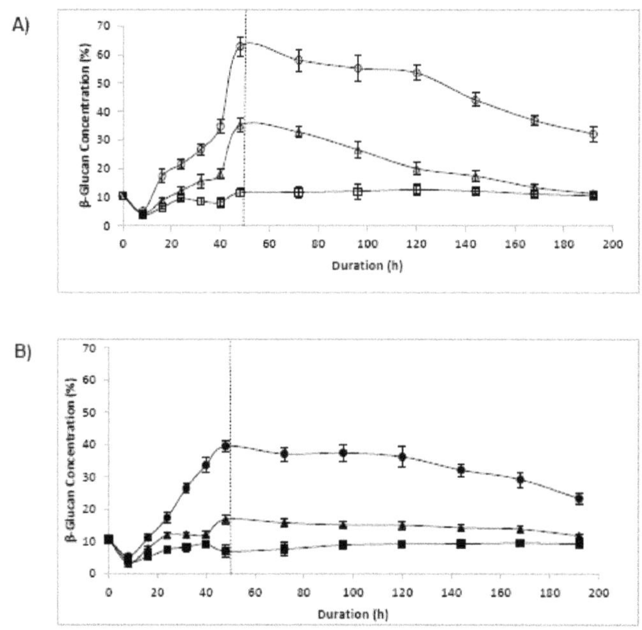

Figure 6. (A) Yeast cell wall β-glucan (%) of non-NaCl-stressed cells during various fermentation times (h) with three different media containing 200, 300, or 400 g/L glucose (n = 3 replications for each time and error bars represent standard deviation of the average value of all replications with each range of fermentation time). Mean values between 0–144 h were significantly different at $p \leq 0.05$ level.

(B) Yeast cell wall β-glucan (%) of NaCl-stressed cells during various fermentation times (h) with three different media containing 200, 300, or 400 g/L glucose (n = 3 replications for each time and error bars represent standard deviation of the average value of all replications with each range of fermentation time). Mean values between 0–144 h were significantly different at $p \leq 0.05$ level. The stationary phase is indicated with the vertical line. (0): (200 g/L glucose, 0% NaCl), (Δ): (300 g/L glucose, 0% NaCl), (■): (400 g/L glucose, 0% NaCl), (•): (200 g/L glucose, 6% NaCl), (Δ): (300 g/L glucose, 6% NaCl), (■): (400 g/L glucose, 6% NaCl).

Cell Wall Treatment with Lytic Enzyme and Sonication

The highest cell wall resistance to the end of exponential phase either to lytic enzyme or to sonication was observed for sample (200 g/L glucose, 0% NaCl), which had the highest β-glucan concentration. A decline in these results was observed for the fermented media (400 g/L glucose, 0% NaCl) and (400 g/L glucose, 6% NaCl) during the death phase (120–192 h of fermentation); despite their lowest β-glucan cell wall content, they appeared to be more resistant to the action of lytic enzyme but not to ultrasounds when compared to cells of samples (300 g/L glucose, 0% NaCl) and (300 g/L glucose, 6% NaCl) (Figures 7 and 8).

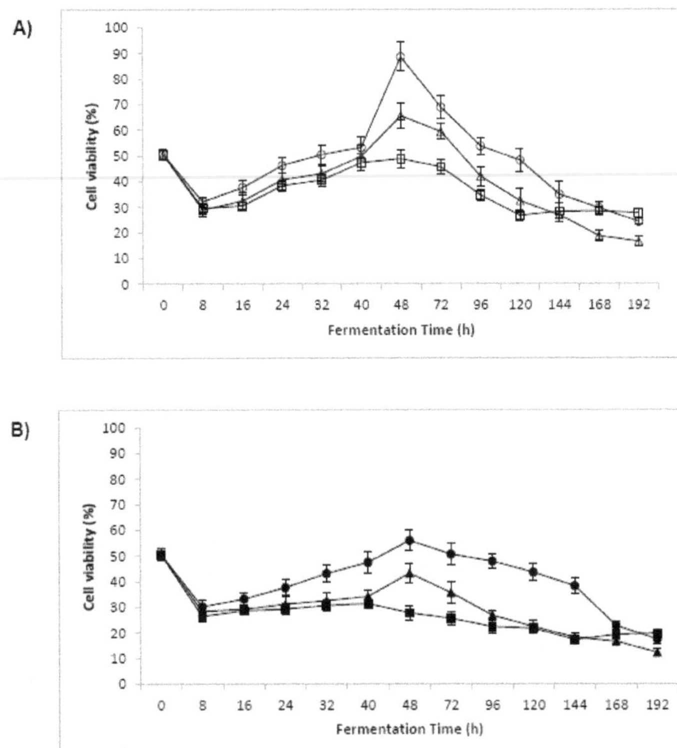

Figure 7. (A) Yeast cell viability (%) of non-NaCl-stressed cells during various fermentation times (h) with three different media containing 200, 300, or 400 g/L glucose after cell treatment with Glucanex 200G lytic enzyme (*n* = 3 replications for each time and error bars represent standard deviation of the average value of all replications with each range of fermentation time); (B) Yeast cell viability (%) of NaCl-stressed cells during various fermentation times (h) with three different media containing 200, 300, or 400 g/L glucose after cell treatment with Glucanex 200G lytic enzyme (*n* = 3 replications for each time and error bars represent standard deviation of the average value of all replications with each range of fermentation time) (0): (200 g/L glucose, 0% NaCl), (Δ): (300 g/L glucose, 0% NaCl), (D): (400 g/L glucose, 0% NaCl), (•): (200 g/L glucose, 6% NaCl), (Δ): (300 g/L glucose, 6% NaCl), (■): (400 g/L glucose, 6% NaCl).

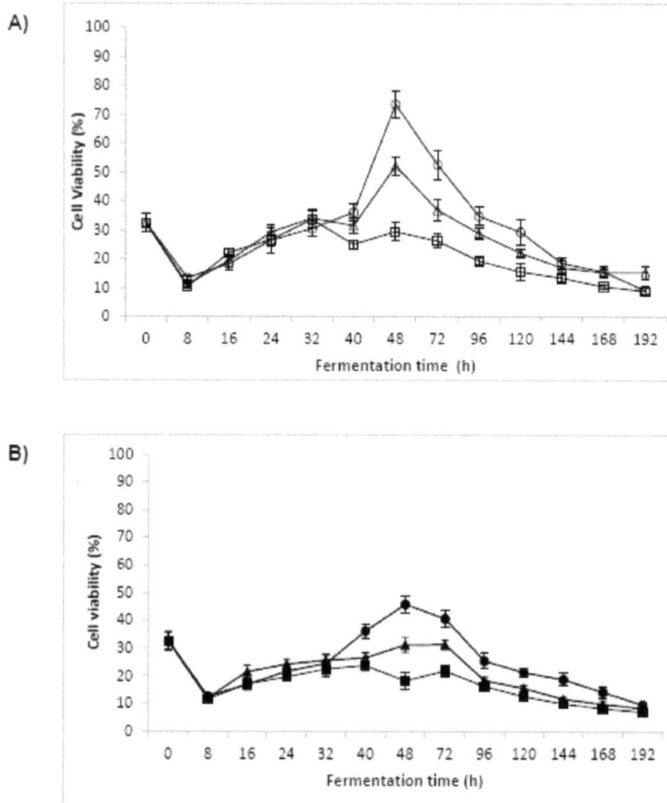

Figure 8. (A) Yeast cell viability (%) of non-NaCl-stressed cells during various fermentation times (h) with three different media containing 200, 300, or 400 g/L glucose after cell treatment with sonication (*n* = 3 replications for each time and error bars represent standard deviation of the average value of all replications with each range of fermentation time); (B) Yeast cell viability (%) of NaCl-stressed cells during various fermentation times (h) with three different media containing 200, 300, or 400 g/L glucose after cell treatment with sonication (*n* = 3 replications for each time and error bars represent standard deviation of the average value of all replications with each range of fermentation time) (0): (200 g/L glucose, 0% NaCl), (Δ): (300 g/L glucose, 0% NaCl), (■): (400 g/L glucose, 0% NaCl) (•): (200 g/L glucose, 6% NaCl), (Δ): (300 g/L glucose, 6% NaCl), (■): (400 g/L glucose, 6% NaCl).

DISCUSSION

The impact of various environmental stresses on the yeast fermentation performance has been well studied and reviewed before [27–32,42,43], but the study of the yeast cell wall composition during the fermentation process is insufficient, with only a few published works referring to the impact of fermentation growth conditions and mode of cultivation on the yeast cell wall structure [20,21,44],the mode of fermentation on yeast cell wall β-glucan content [18], and the role of additives in the culture medium in β-glucan production [22,45]. The impact of glucose and NaCl stress on yeast cell wall remodelling via the study of the fluctuation of β-glucan concentration during the fermentation process has not been studied until now.

Cell Viability

Cell viability measurement with the methylene blue method is an established technique for yeast cell staining [46], used for the estimation of clear viable yeast cells [36,38]. Cell viability measurements were conducted: (a) for the estimation of the impact of stress conditions on cell viability during the fermentation process, and (b) for the correlation of yeast β-glucan concentration with cell biomass dry weight and total cell number, as non-viable and lysed cells were expected to have a higher β-glucan concentration (expressed as % of the yeast dry weight). In our experiments, cell viability reached two peaks and then decreased for all the fermented samples (Figure 1). At this point, the glucose concentration stress seems to be critical as cell viability decreases as glucose increases but the corresponding values for NaCl-stressed and non-stressed cells were similar for the same glucose concentration for all fermented samples (Figure 1). At the death phase (48–120 h), the synergistic action of glucose and NaCl stress seems to impact negatively on the cell viability, with a larger decrease in viable cell count for fermented samples with NaCl-stressed cells grown in media containing 300 and 400 g/L glucose (Figure 1).

Our results differed from the observations of other researchers under the same glucose (200 g/L) and NaCl (6%) stress conditions with the use of the *S. cerevisiae* VIN 13 strain, which observed a constant decrease of cell viability during the fermentation process [27]. These differences might be due to the different cell NaCl pre-culturing time (24 h in our experiments, 48 h in their work), the different volume of fermented medium (1000 mL in our experiments, 250 mL in their work) and the different package containing dry yeast used for carrying out the fermentations.

Also, at the end of death phase, the NaCl-preconditioned cells grown in the denser glucose medium had the highest viable cell values. These results cannot be compared with other researchers' results that used the same yeast strain with salt pre-culturing (NaCl 6% *w/v*) but in an even denser glucose medium (550 g/L), as cell viability values are not given [41]. Our results were in accordance with the observations of other researchers who found that NaCl pre-cultured yeast cells showed an increased cell viability at the end of the fermentation process [36]. Despite this increase, for all samples, after 192 h of fermentation, cell viability was lower compared to the values of the second peak. Lower cell viability was possibly caused by cell autolysis during the death phase of the alcoholic fermentation and consequently to a rise in the cell's dry weight due to cytoplasm leaking out from the cell to the supernatant [43]. Concerning the impact of NaCl 6% (*w/v*) concentration on yeast cells, our results with NaCl pre-cultured and non-pre-cultured yeast cells growing in a medium containing 200 g/L glucose differed from results of other researchers that indicated that higher osmotic shock conditions (>5% *w/v* NaCl) resulted in a higher cell viability at the end of the fermentation compared to the non-NaCl-stressed cells, while the viability value (83.44 ± 0.8%) for NaCl-stressed cells was in accordance with the corresponding one that previous researchers reported [27,47] (Figure 1). Concerning all the fermented media with the same glucose concentration, they appeared to have almost the same cell number at the end of fermentation (Figure 1), thus indicating that NaCl stress for the same glucose concentration had almost no impact on the cell viability at the end of the fermentation (Figure 1). As a general rule, the viability results were in accordance with the results reported by Logothetis et al. (2014), which suggested that salt pre-culturing of yeast *S. cerevisiae* impacts positively on cell viability [36].

Optical Density and Cell Count

Optical density (OD) is used for the estimation of yeast growth and the determination of exponential, stationary and lag phase during batch fermentations [18,21]. For media (300 g/L glucose,0% NaCl), (300 g/L glucose, 6%), (400 g/L glucose, 0% NaCl) and (400 g/L glucose, 6% NaCl) the three fermentation phases coincided while for media (200 g/L glucose, 0% NaCl) and (200 g/L glucose, 6% NaCl) the fermentation phases also coincided but the exponential and stationary phases were shorter compared to the other media, thus indicating that NaCl stress had no impact on fermentation kinetics, while glucose concentration had a slight impact (Figure 2). Our OD results cannot be compared with those of other researchers who used the same yeast strain and stress conditions, as OD measurements are not given [27,47]. Also, our OD results differed from the report of Kim et al. (2006), but these researchers used a different *S. cerevisiae* strain with no glucose and NaCl-stressed yeast cells, and also used different media and fermentation times [18].

The cell count was used in order to determine the total yeast cell number during the fermentation process and correlate it with cell viability [47], but also with dry yeast mass weight (g) and β-glucan yield (%). During the death phase, the cell number decreased, with a simultaneous decrease in OD (600 nm) for all the fermented media; surprisingly, though, between 120 and 192 h, the number of cells grown in 400 g/L glucose increased spectacularly compared to the other samples. This cell number increase of glucose- and NaCl-stressed cells was not accompanied by a cell dry mass increase. This probably indicates a smaller yeast cell but further observations with electron microscopy must be performed in order to validate such a hypothesis (Figure 3). These results differed from other researchers' studies that used the same yeast strain, glucose and NaCl stress conditions, which suggests that NaCl-induced osmotic stress caused growth arrest in yeast cells, while an increase in osmotic stress with elevated NaCl concentration caused a decrease in yeast growth and total cell number over time [27]. During the first 40 h of fermentation, the cell count of media (200 g/L glucose, 0% NaCl) and (200 g/L glucose, 6% NaCl) was lower than that of the other media, while their dry weights appeared to be higher and their cell walls contained much more β-glucan (Figures 3 and 6), thus indicating fewer cells but larger ones, those with wider cell walls or both.

Dry Weight and Sample β-Glucan Yield

The determination of yeast dry weight at the various fermentation phases was done in order to calculate the β-glucan yield (g) at the different fermentation phases and optimise the phase at which the cells must be harvested. The highest value of dry weight (g) and β-glucan yield (g) was obtained with 200 g/L glucose/NaCl-non-stressed and 200 g/L glucose/stressed cells at 48 and 40 h, respectively, thus indicating that the stationary phase is the most appropriate time for cell harvesting, a result that is in accordance with the observations of Kim et al. (2006) [18]. From the obtained results, it is shown that glucose and NaCl stress impacts negatively on yeast dry weight and β-glucan yield (Figures 4 and 5). Also, for media (400 g/L glucose, 0% NaCl) and (400 g/L glucose, 6% NaCl), the highest value for yeast dry weight and β-glucan yield appeared at the end of the death phase and, despite the fact that β-glucan concentration was lowered, the total cell number increased. It seems that, in this way, the yeast cell maintains homeostasis in order to cope with the conditions of glucose and NaCl stress during the fermentation process, but this needs further investigation and is proposed as future research.

β-Glucan Concentration, Lytic Enzyme and Sonication

The determination of yeast β-glucan concentration was done for the estimation of polysaccharide and its accumulation in the yeast cell, while furthering the understanding of yeast cell wall β-glucan modification during the fermentation process under stress conditions, the estimation of cell wall lysis and disruption resistance against lytic enzymes and sonication was tested with cell viability measurements.

Generally, for all the fermented media, the β-glucan yeast cell wall concentration rose constantly during the exponential phase and rose to a peak value at the stationary phase and then decreased during the death phase (Figure 6). Also, the cell treatment with lytic enzyme and sonication revealed that the cell's resistance against lysis and disruption increased as the cell entered the stationary phase.

Next, as the cell entered the death phase, its resistance against lytic enzymes and sonication lowered, thus supporting the observations of a constant accumulation of β-glucan in the cell wall during the exponential phase, with its maximum quantity at the stationary phase and consequently a decrease during the death phase (Figures 7 and 8). During yeast alcoholic fermentation, the accumulation of ethanol causes an increase in yeast cell membrane permeability [42]. During the death phase, all the samples appeared to be more susceptible to lytic enzyme and sonication treatment (Figures 7 and 8) but for NaCl-stressed cells grown in a medium containing 300 g/L glucose and both NaCl-stressed and non-stressed yeast cells grown in a medium containing 400 g/L glucose, there was not a direct correlation between β-glucan concentration and cell wall resistance as these cells appeared to be more susceptible to lytic enzyme and sonication treatment, probably due to the additional impact of ethanol toxicity [48] (Figures 7 and 8).

Our results were in accordance with the observations of Kim et al. (2006), according to which, during the yeast growth in a defined medium, the cell wall resistance to the action of β-glucanase increases until the yeast cells enter the stationary phase, while their resistance to the enzyme decreases at the end of the stationary phase [18]. Also, our results support the reports of Klis et al. (2002), which indicate that as cells enter the stationary phase, they become thicker [49]. This yeast cell wall resistance is a result of the cell wall increase during the exponential phase [18].

In non-continuous fermentation, during the transition from the exponential to the stationary phase, the action of β-1,3-glucan synthetase decreases while the action of glycogen synthetase increases [50]. During the exponential phase, the autolytic action of endo β-glucanases is higher compared to that in the stationary phase of yeast cells' growth, while a significant increase in endoglucanases' activity into the soluble fraction of β-glucan during the stationary phase in non-continuous fermentation of *Saccharomyces exiguus* has also been reported [51]. The stationary phase is characterised by carbon limitation, and as the cell culture enters this, more glucan is necessary for the maintenance of the cell's viability [18]. The higher β-glucan concentration at the stationary phase can also be explained by the decreased action of cell's glucanases, with the less active cell's growth at the exponential phase, which is accompanied by a non-severe carbon limitation, similar to the one at the end of the stationary phase [18]. During aerobic fermentation, when yeast cells enter the stationary growth phase, the cells become more resistant to the action of β-1,3-glucanase and are less permeable to macromolecules with the expression of Sed1p protein [52,53].

In addition to the above, our results show that the increase of glucose osmotic stress in the growth medium, but also the NaCl hyperosmotic stress in the pre-cultured yeast cells, impacts negatively on the **β**-glucan concentration in the yeast cell wall. This negative impact on the **β**-glucan concentration was even more intense from the stationary phase until the end of the fermentation (death phase) (Figure 6).

Our results differed from those of other researchers, who reported a β-glucan increase at the end of fermentation; however, these researchers incorporated NaCl in the fermented medium and did not pre-culture yeast cells under salt hyperosmotic stress, and also used a different yeast strain [22]. Also, in their work, cell viability and total cell count were not taken into consideration and an increase in β-glucan concentration may have resulted from cell lysis, cytoplasm efflux and consequently a higher cell wall:cell ratio [14]. Our results cannot be compared with the results of Aguilar-Uscanga et al. (2003), who studied the impact of growth conditions and the mode of cultivation on the yeast cell wall structure and not the impact of stress conditions, with the collection of cell samples (50 mL volume) only in the early exponential phase [21].

Our β-glucan concentration results differed from the reports of other researchers that suggest that the cell wall integrity (CWI) pathway, in cooperation with the high osmotic response (HOG) pathway, regulates the action of zymolyase, the enzyme hydrolysing the β-1,3 glucan network, and thus result in an increase in the amount of cell wall β-glucan. These researchers used different yeast strains and different fermented medium components [24], while the data concerning osmotic stress responses were based on a previous research work in which KCl and NaCl had been diluted in the fermented substrate [54,55]. Our results are in accordance with the observations of Ene et al. (2015), who reported that the cell wall is not rigid but elastic and that sudden decreases in cell volume due to hyperosmotic conditions result in rapid increases in cell wall thickness, and thus a decrease in β-glucan content [25].

Our results are in accordance with the results of other researchers who used Glucanex 200G for the estimation of yeast cell wall resistance to lytic enzyme during batch fermentation and reported a higher cell resistance in the stationary phase [18]. Also, our results differ from other researchers' results (97 ± 0.18% breaking rate, 10–15% cell concentration), but these differences may arise from the fact that these researchers used sonication for yeast cell lysis during a β-glucan extraction protocol in dry yeast first treated with hot water for mannoprotein removal and not in yeast from various fermentation phases [38].

The yeast cells appeared to be more susceptible to sonication treatment compared to lytic enzyme treatment. The higher cell wall resistance to lytic enzymes in the various fermentation phases could be attributed to the variation in cell number during the fermentation process (Figures 3 and 7) and thus to the enzyme dispersion to a larger volume of cells [56] (e.g., the stable resistance of cells of fermented media with 400 g/L glucose to the action of the lytic enzyme at the end of the death phase where they appeared the highest cell number), while cell disruption of yeast cells treated with ultrasounds is independent of cell concentration but mainly proportional to the acoustic power [57,58] (Figures 3 and 8).

SUMMARY

Yeast β-glucan polysaccharide is a proven immune-stimulant molecule for human and animal health. In recent years, interest in β-glucan industrial production has been increasing. The yeast cell wall is modified during the fermentation process for biomass production. The impact of environmental conditions on cell wall remodelling has not been extensively investigated. The aim of this research work was to study the impact of glucose and NaCl stress on β-glucan formation in the yeast cell wall during alcoholic fermentation and the assessment of the optimum fermentation phase at which the highest β-glucan yield is obtained. VIN 13 *Saccharomyces cerevisiae* (*S. cerevisiae*) strain was pre-cultured for 24 h with 0% and 6% NaCl and inoculated in a medium consisting of 200, 300, or 400 g/L glucose. During fermentation, 50 mL of fermented medium were taken periodically for the determination of Optical Density (OD), cell count, cell viability, cell dry weight, β-glucan concentration and β-glucan yield. Next, dry yeast cell biomass was treated with lytic enzyme and sonication. At the early stationary phase, the highest β-glucan concentration and yield was observed for non-NaCl pre-cultured cells grown in a medium containing 200 g/L glucose; these cells, when treated with enzyme and sonication, appeared to be the most resistant. Stationary is the optimum phase for cell harvesting for β-glucan isolation. NaCl and glucose stress impact negatively on β-glucan formation during alcoholic fermentation. The results of this work could comprise a model study for yeast β-glucan production on an industrial scale and offer new perspectives on yeast physiology for the development of antifungal drugs.

In recent years, there has been increasing interest in industrial production of yeast β-glucan and its incorporation in functional foods and medicines due to its immunological properties in human and animal health systems. This study could comprise a quantitative indicator for the industrial production of yeast β-glucan from defined cell cultures but also from other yeast sources like breweries' [23] and wineries' spent yeast biomass [40].

The stationary growth phase appears to be optimal for β-glucan isolation from pure cell fermentation cultures. NaCl and glucose stress have a negative impact on β-glucan production. Yeast NaCl-stressed cells have a reduced β-glucan concentration compared to non-stressed for the same glucose concentrations, while this difference is more significant at the end of the fermentation for a fermented medium containing 200 g/L glucose, which is close to the wine must concentration intended for wine production. For the other two glucose concentrations (300 and 400 g/L), the differences at the end of the fermentation are not statistically significant.

In the present study, it seems that the two different stresses act synergistically, with an additive negative impact on the cell wall β-glucan concentration; as for the same glucose concentration, the preconditioned NaCl cells had a lower β-glucan concentration compared to the non-preconditioned ones. Another significant observation was that for a yeast cell grown in a medium containing 400 g/L glucose, the cell wall β-glucan concentration remained almost the same from the end of the exponential phase until the end of the fermentation for both NaCl-stressed and non-stressed cells.

The study of cell wall physiology, which is still a poorly explored research field, and yeast cell growth under various stress conditions (glucose, ethanol, temperature, SO2, etc.), could comprise a key tool for the biotechnological development of new products like functional foods and antifungal drugs for medical and agricultural applications. Further study on the impact of stress conditions on the immunological properties of the isolated β-glucan from the various fermentation phases, as well as the study of glycolysis–glycogenesis biochemical pathways in the yeast cell wall under stress conditions, is proposed as it could offer new perspectives on yeast β-glucan-based drugs, adding to a deeper understanding of yeast stress biology phenomena and a more comprehensive view of alcoholic fermentation biochemical pathways. Additionally, the study of the genes encoding β-glucan accumulation in the yeast cell wall during the fermentation process and under various environmental stresses is proposed as it could enlighten us as to the molecular mechanisms of the yeast cell.

REFERENCES

1. Reggiori, F.; Klionsky, D.J. Autophagic processes in yeast: Mechanism, machinery and regulation. *Genetics***2013**, *194*, 341–361. [CrossRef] [PubMed]
2. Ribéreau-Gayon, P.; Dubourdieu, D.; Donèche, B.; Lonvaud, A. Biochemistry of Alcoholic Fermentation and Metabolic Pathways of Wine Yeasts. In *Handbook of Enology*; John Wiley & Sons, Ltd.: Somerset, NJ, USA, 2006; pp. 53–77.
3. Branduardi, P.; Porro, D. Yeasts in Biotechnology. In *Yeast*; Wiley-VCH Verlag GmbH & Co. KGaA: Weinheim, Germany, 2012; pp. 347–370.
4. Ballou, C.E. Genetics of Yeast Mannoprotein Biosynthesis. In *Fungal Polysaccharides*; American Chemical Society: Washington, DC, USA, 1980; Volume 126, pp. 1–14.
5. François, J.M.; Walther, T.; Parrou, J.L. Genetics and Regulation of Glycogen and Trehalose Metabolism in *Saccharomyces cerevisiae*. In *Microbial Stress Tolerance for Biofuels*; Liu, Z.L., Ed.; Springer: Berlin/Heidelberg, Germany, 2012; pp. 29–55.

6. Wilson, W.A.; Hughes, W.E.; Tomamichel, W.; Roach, P.J. Increased glycogen storage in yeast results in less branched glycogen. *Biochem. Biophys. Res. Commun.* **2004**, *320*, 416–423. [CrossRef] [PubMed]

7. Arvindekar, A.U.; Patil, N.B. Glycogen—A covalently linked component of the cell wall in *Saccharomyces cerevisiae. Yeast* **2002**, *19*, 131–139. [CrossRef] [PubMed]

8. Orlean, P. Architecture and biosynthesis of the *Saccharomyces cerevisiae* cell wall. *Genetics* **2012**, *192*, 775–818. [CrossRef] [PubMed]

9. Klis, F.M. Review: Cell wall assembly in yeast. *Yeast* **1994**, *10*, 851–869. [CrossRef] [PubMed]

10. Van der Rest, M.E.; Kamminga, A.H.; Nakano, A.; Anraku, Y.; Poolman, B.; Konings, W.N. The plasma membrane of *Saccharomyces cerevisiae*: Structure, function, and biogenesis. *Microbiol. Rev.* **1995**, *59*, 304–322. [PubMed]

11. Rank, G.H.; Robertson, A.J. Protein and lipid composition of the yeast plasma membrane. In *Yeast Genetics: Fundamental and Applied Aspects*; Spencer, J.F.T., Spencer, D.M., Smith, A.R.W., Eds.; Springer: New York, NY, USA, 1983; pp. 225–241.

12. Stier, H.; Ebbeskotte, V.; Gruenwald, J. Immune-modulatory effects of dietary yeast β-1,3/1,6-D-glucan. *Nutr. J.* **2014**, *13*. [CrossRef] [PubMed]

13. Ahmad, A.; Anjum, F.M.; Zahoor, T.; Nawaz, H.; Dilshad, S.M.R. β glucan: A valuable functional ingredient in foods. *Crit. Rev. Food Sci. Nutr.* **2012**, *52*, 201–212. [CrossRef] [PubMed]

14. Varelas, V.; Liouni, M.; Calokerinos, A.C.; Nerantzis, E.T. An evaluation study of different methods for the production of β-D-glucan from yeast biomass. *Drug Test. Anal.* **2016**, *8*, 46–55. [CrossRef] [PubMed]

15. Klis, F.M.; Boorsma, A.; de Groot, P.W.J. Cell wall construction in *Saccharomyces cerevisiae. Yeast* **2006**, *23*, 185–202. [CrossRef] [PubMed]

16. Levin, D.E. Regulation of cell wall biogenesis in *Saccharomyces cerevisiae*: The cell wall integrity signaling pathway. *Genetics* **2011**, *189*, 1145–1175. [CrossRef] [PubMed]

17. Hahn-Hägerdal, B.; Karhumaa, K.; Larsson, C.U.; Gorwa-Grauslund, M.; Görgens, J.; van Zyl, W.H. Role of cultivation media in the development of yeast strains for large scale industrial use. *Microb. Cell Fact.* **2005**, *4*. [CrossRef] [PubMed]

18. Kim, K.S.; Yun, H.S. Production of soluble β-glucan from the cell wall of *Saccharomyces cerevisiae. Enzym. Microb. Technol.* **2006**, *39*, 496–500. [CrossRef]

19. Catley, B.J. Isolation and analysis of cell walls. In *Yeast, A Practical Approach*; Campbell, I., Duffus, J.H., Eds.; Oxford University Press: London, UK, 1988; pp. 163–183.

20. McMurrough, I.; Rose, A.H. Effect of growth rate and substrate limitation on the composition and structure of the cell wall of *Saccharomyces cerevisiae. Biochem. J.* **1967**, *105*, 189–203. [CrossRef] [PubMed]

21. Aguilar-Uscanga, B.; François, J.M. A study of the yeast cell wall composition and structure in response to growth conditions and mode of cultivation. *Lett. Appl. Microbiol.* **2003**, *37*, 268–274. [CrossRef] [PubMed]

22. Naruemon, M.; Romanee., S.; Cheunjit, P.; Xiao, H.; McLandsborough, L. A.; Pawadee, M. Influence of additives on *Saccharomyces cerevisiae* β-glucan production. *Int. Food Res. J.* **2013**, *20*, 1953–1959.

23. Bastos, R.; Coelho, E.; Coimbra, M.A. Modifications of *Saccharomyces pastorianus* cell wall polysaccharides with brewing process. *Carbohydr. Polym.* **2015**, *124*, 322–330. [CrossRef] [PubMed]

24. García, R.; Rodríguez-Peña, J.M.; Bermejo, C.; Nombela, C.; Arroyo, J. The high osmotic response and cell wall integrity pathways cooperate to regulate transcriptional responses to zymolyase-induced cell wall stress in *Saccharomyces cerevisiae. J. Biol. Chem.* **2009**, *284*, 10901–10911. [CrossRef] [PubMed]

25. Ene, I.V.; Walker, L.A.; Schiavone, M.; Lee, K.K.; Martin-Yken, H.; Dague, E.; Gow, N.A.R.; Munro, C.A.; Brown, A.J.P. Cell wall remodeling enzymes modulate fungal cell wall elasticity and osmotic stress resistance. *MBio* **2015**. [CrossRef] [PubMed]

26. Erasmus, D.J.; van der Merwe, G.K.; van Vuuren, H.J.J. Genome-wide expression analyses: Metabolic adaptation of *Saccharomyces cerevisiae* to high sugar stress. *FEMS Yeast Res.* **2003**, *3*, 375–399. [CrossRef]

27. Logothetis, S.; Nerantzis., E.T.; Gioulioti, A.; Kanelis, T.; Tataridis, P.; Walker, G. Influence of sodium chloride on wine yeast fermentation performance. *Int. J. Wine Res.* **2010**, *2*, 35–42. [CrossRef]

28. Ishmayana, S.; Learmonth, R.P.; Kennedy, U.J. Fermentation performance of the yeast *Saccharomyces cerevisiae* in media with high sugar concentration. In Proceedings of the 2nd International Seminar on Chemistry, Jatinangor, Indonesia, 24–25 November 2011; pp. 379–385.

29. Lei, H.; Xu, H.; Feng, L.; Yu, Z.; Zhao, H.; Zhao, M. Fermentation performance of lager yeast in high gravity beer fermentations with different sugar supplementations. *J. Biosci. Bioeng.* **2016**, *122*, 583–588. [CrossRef] [PubMed]

30. Novo, M.; Gonzalez, R.; Bertran, E.; Martínez, M.; Yuste, M.; Morales, P. Improved fermentation kinetics by wine yeast strains evolved under ethanol stress. *LWT Food Sci. Technol.* **2014**, *58*, 166–172. [CrossRef]

31. Trainotti, N.; Stambuk, B.U. NaCl stress inhibits maltose fermentation by *Saccharomyces cerevisiae*. *Biotechnol. Lett.* **2001**, *23*, 1703–1707. [CrossRef]

32. Pratt, P.L.; Bryce, J.H.; Stewart, G.G. The effects of osmotic pressure and ethanol on yeast viability and morphology. *J. Inst. Brew.* **2003**, *109*, 218–228. [CrossRef]

33. Ren, H.; Wang., X.; Liu, D.; Wang, B. A glimpse of the yeast *Saccharomyces cerevisiae* responses to NaCl stress. *Afr. J. Microbiol. Res.* ***2012***, *6*, 713–718.

34. Dhar, R.; Sägesser, R.; Weikert, C.; Yuan, J.; Wagner, A. Adaptation of *Saccharomyces cerevisiae* to saline stress through laboratory evolution. *J. Evol. Biol.* **2011**, *24*, 1135–1153. [CrossRef] [PubMed]

35. Tilloy, V.; Ortiz-Julien, A.; Dequin, S. Reduction of ethanol yield and improvement of glycerol formation by adaptive evolution of the wine yeast *Saccharomyces cerevisiae* under hyperosmotic conditions. *Appl. Environ. Microbiol.* **2014**, *80*, 2623–2632. [CrossRef] [PubMed]

36. Logothetis, S.; Nerantzis, E.T.; Tataridis, P.; Goulioti, A.; Kannelis, A.; Walker, G.M. Alleviation of stuck wine fermentations using salt-preconditioned yeast. *J. Inst. Brew.* **2014**, *120*, 174–182. [CrossRef]

37. Prieto, M.A.; Vázquez, J.A.; Murado, M.A. Comparison of several mathematical models for describing the joint effect of temperature and ph on glucanex activity. *Biotechnol. Prog.* **2012**, *28*, 372–381. [CrossRef] [PubMed]

38. Magnani, M.; Calliari, C.M.; de Macedo Jr, F.C.; Mori, M.P.; de Syllos Cólus, I.M.; Castro-Gomez, R.J.H. Optimized methodology for extraction of (1→3)(1→6)-β-D-glucan from *Saccharomyces cerevisiae* and in vitro evaluation of the cytotoxicity and genotoxicity of the corresponding carboxymethyl derivative. *Carbohydr. Polym.* **2009**, *78*, 658–665. [CrossRef]

39. Varelas, V.; Tataridis, P.; Liouni, M.; Nerantzis, E.T. Application of different methods for the extraction of yeast β-glucan. *e-J. Sci. Technol.* **2016**, *11*, 75–89.

40. Varelas, V.; Tataridis, P.; Liouni, M.; Nerantzis, E.T. Valorization of winery spent yeast waste biomass as a new source for the production of β-glucan. *Waste Biomass Valori.* **2016**, *7*, 807–817. [CrossRef]

41. Logothetis, S.; Tataridis, P.; Kanellis, A.; Nerantzis, E.T. The effect of preconditioning cells under osmotic stress on high alcohol production. *Zb. Matice Srp. za Prir. Nauke* **2013**, 405–414. [CrossRef]

42. Bauer, F.F.; Pretorius, I.S. Yeast stress response and fermentation efficiency: How to survive the making of wine—A Review. *S. Afr. J. Enol. Viticult.* **2000**, *21*, 27–51.

43. Patynowski, R.J.; Jiranek, V.; Markides, A.J. Yeast viability during fermentation and sur lie ageing of a defined medium and subsequent growth of *Oenococcus oeni*. *Aust. J. Grape Wine Res.* **2002**, *8*, 62–69. [CrossRef]

44. Morris, G.J.; Winters., L.; Coulson, G.E.; Clarke, K.J. Effect of osmotic stress on the ultrastructure and viability of the yeast *Saccharomyces cerevisiae*. *Microbiology* **1986**, *132*, 2023–2034. [CrossRef] [PubMed]

45. Naruemon, M.; Romanee., S.; Cheunjit, P.; Xiao, H.; McLandsborough, L.A.; Pawadee, M. Effect of three additives on the cell morphology and β-glucan production in *Saccharomyces cerevisiae*. *Res. J. Pharm. Biol. Chem. Sci.* **2011**, *2*, 283–295.

46. Gilliland, R.B. Determination of yeast viability. *J. Inst. Brew.* **1959**, *65*, 424–429. [CrossRef]

47. Logothetis, S.; Walker, G.; Nerantzis, E.T. Effect of salt hyperosmotic stress on yeast cell viability. *Zb. Matice Srp. za Prir. Nauke* **2007**, *113*, 271–284. [CrossRef]

48. Stanley, D.; Bandara, A.; Fraser, S.; Chambers, P.J.; Stanley, G.A. The ethanol stress response and ethanol tolerance of *Saccharomyces cerevisiae*. *J. Appl. Microbiol.* **2010**, *109*, 13–24. [CrossRef] [PubMed]

49. Klis, F.M.; Mol, P.; Hellingwerf, K.; Brul, S. Dynamics of cell wall structure in *Saccharomyces cerevisiae*. *FEMS Microbiol. Rev.* **2002**, *26*, 239–256. [CrossRef] [PubMed]

50. Fleet, G.H. Cell walls. In *The Yeasts*; Rose, A.H.H.J.D., Ed.; Academic Press: London, UK, 1991; Volume 4, pp. 199–277.

51. Inouhe, M.; Sugo, E.; Tohoyama, H.; Joho, M.; Nevins, D.J. Cell wall metabolism and autolytic activities of the yeast *Saccharomyces exiguus*. *Int. J. Biol. Macromol.* **1997**, *21*, 11–14. [CrossRef]

52. Shimoi, H.; Kitagaki, H.; Ohmori, H.; Iimura, Y.; Ito, K. Sed1p is a major cell wall protein of *Saccharomyces cerevisiae* in the stationary phase and is involved in lytic enzyme resistance. *J. Bacteriol.* **1998**, *180*, 3381–3387. [PubMed]

53. De Nobel, H.; Ruiz, C.; Martin, H.; Morris, W.; Brul, S.; Molina, M.; Klis, F.M. Cell wall perturbation in yeast results in dual phosphorylation of the slt2/mpk1 map kinase and in an slt2-mediated increase in fks2–lacz expression, glucanase resistance and thermotolerance. *Microbiology* **2000**, *146*, 2121–2132. [CrossRef] [PubMed]

54. Rep, M.; Krantz, M.; Thevelein, J.M.; Hohmann, S. The transcriptional response of *Saccharomyces cerevisiae* to osmotic shock: Hot1p and msn2p/msn4p are required for the induction of subsets of high osmolarity glycerol pathway-dependent genes. *J. Biol. Chem.* **2000**, *275*, 8290–8300. [CrossRef] [PubMed]

55. O'Rourke, S.M.; Herskowitz, I. Unique and redundant roles for HOG MAPK pathway components asrevealed by whole-genome expression analysis. *Mol. Biol. Cell* **2004**, *15*, 532–542. [CrossRef] [PubMed]

56. Kim, K.S.; Chang, J.E.; Yun, H.S. Estimation of soluble-glucan content of yeast cell wall by the sensitivity to glucanex® 200g treatment. *Enzym. Microb. Technol.* **2004**, *35*, 672–677. [CrossRef]

57. Liu, D.; Zeng, X.A.; Sun, D.W.; Han, Z. Disruption and protein release by ultrasonication of yeast cells. *Innov. Food Sci. Emerg. Technol.* **2013**, *18*, 132–137. [CrossRef]

58. Apar, D.K.; Ozmek, B. Protein releasing kinetics of bakers' yeast cells by ultrasound. *Chem. Biochem. Eng. Q.***2008**, *22*, 113–118.

CHAPTER-8

A *PICHIA ANOMALA* STRAIN (*P. ANOMALA* M1) ISOLATED FROM TRADITIONAL GREEK SAUSAGE IS AN EFFECTIVE PRODUCER OF EXTRACELLULAR LIPOLYTIC ENZYME IN SUBMERGED FERMENTATION

INTRODUCTION: AN OVERVIEW

The *Pichia anomala* strain, specifically identified as *P. anomala* M1 and originally isolated from traditional Greek sausage, emerges as a potent producer of extracellular lipolytic enzymes in submerged fermentation. This microorganism's unique origin, taken from a traditional food product, adds an intriguing dimension to its enzymatic capabilities, suggesting a potential application in the broader context of industrial processes. In this study, we explore the enzymatic prowess of *P. anomala* M1, shedding light on its ability to synthesize and secrete lipolytic enzymes during submerged fermentation. Understanding the enzymatic profile of this strain not only contributes to the growing knowledge of microbial diversity but also holds promise for enhancing industrial processes that rely on lipolytic activities.

The yeast *Pichia anomala* (syn. *Wickerhamomyces anomalus*, former *Hansenula*) has been isolated from very diverse habitats, including plants, animals, soil, water, food and hospitals [1,2]. In food and feed, *P. anomala* has either a beneficial role, as flavor enhancer, biopreservation agent, probiotic, enzyme producer, or a detrimental role, as spoilage yeast. Biopreservation is due to its unusual-for-a-yeast broad spectrum antimicrobial activity against several fungal, yeast, bacterial species and viruses.

P. anomala produces several metabolites with potential biotechnological exploitation. Examples, apart from the antimicrobial agents that make it an attractive biocontrol organism for applications in the agri-food sector, include bioethanol, isobutanol, enzymes, sophorolipids, γ-aminobutyric acid, several volatile organic compounds, and beverage starter cultures [2]. Regarding enzyme production, *P. anomala* strains have been reported as producers of phytase, amylase, β-glucosidase, peptidase and lipase [3]. *P. anomala* is therefore a biotechnologically interesting organism which deserves in-depth study in view of future commercialization, as several authors have stressed in their reports [4–6].

A preliminary study with yeast isolates from colonies on the surface of traditional Greek sausages revealed a significant number of *Pichia* spp. *Pichia* isolates were tested for lipolytic activity and upon obtaining positive results, further studies were carried out which are reported in the present work. A particular *Pichia* isolate with the highest lipolytic activity was identified as *P. anomala* and designated as M1 strain.

Lipolytic enzyme production is common among yeasts. Only a few yeasts however (mainly *Candida*, *Yarrowia* and *Geotrichum* spp.) are capable of producing lipases and esterases with interesting characteristics and in sufficient amounts that allow characterization as industrial products [7,8]. Apart from the established producers, there are several promising lipase-producing yeasts as for example, *Issatchenkia orientalis*, *Kluyveromyces marxianus* and others [9]. To the best of our knowledge, *P. anomala* is reported as a lipolytic enzyme producer in only two cases, the works of Banerjee et al. [10] and Ionita et al. [11], both being screening works. Strains of *P. anomala* were cultivated in solid media and shake flask cultures and moderate lipolytic activity, optimum at pH 6.5–7.5, was detected. In the present work, the *P. anomala* M1 strain was cultivated in stirred tank bioreactors and produced significant lipolytic activity compared to known sources. The characteristics of lipolytic enzyme production were investigated and evaluated.

Lipolytic enzymes share some important characteristics that can be application specific, e.g., substrate specificity, region specificity and chiral selectivity [12]. Novel applications have been successfully established using lipases, as for example in the synthesis of biopolymers, flavour compounds, enantiopure pharmaceuticals, agrochemicals and biosensors [12,13]. Innovations need novel biocatalysts and a useful approach for obtaining them is the isolation of microorganisms from natural sources and subsequent screening for the desired enzymatic activity. *P. anomala* exhibits many interesting and potentially exploitable characteristics. It can grow under extreme environmental stress conditions with regard to oxygen, pH, aw and sugar concentration [3,4,14,15]. As a Crabtree-negative yeast and therefore insensitive to elevated levels of glucose concentrations in its substrate, it can be grown easily in batch fermenters without the need of fed-batch cultivation. These characteristics show an organism of practical significance and therefore a potential industrial microorganism in the case of effective production of a valuable metabolite.

MATERIALS AND METHODS

Yeast Isolation

Yeast colonies were transferred from the surface of sausages on agar plates containing YP-glucose medium (YP, yeast-peptone; 1% yeast extract, 2% bacteriological peptone, 2% glucose, 2% technical agar). The plates were incubated at 28°C for 2 days. Selected colonies were sub-cultured on new plates and purified by repeated streaking. The isolated colonies were maintained on YP-glucose agar at 4°C. The isolated yeast colonies were subjected to standard tests and classification procedures according to Kreger-van Rij [16].

Molecular Characterization of an Isolated Pichia Anomala Strain

Yeast identification was done by amplification and sequence analysis of the ribosomal DNA internal transcribed space region (ITS) [17,18]. The following primers were used: ITS1 (5^t-CGGGATCCGTAGGTGAACCTGCGG-3^t) and ITS4 (5^t-CGGGATCCTCCGCTTATTGATATGC-3^t).The amplification mixture was consisted of: 20 pmol of each primer, 300 ng genomic DNA template,0.2 mM dNTP, 1.5 mM MgCl2, and 1 U Taq polymerase (final volume of 20 μL). Reactions were run for 40 cycles at 94°C for 1 min, annealing at 55°C for 1 min and extension at 72 C for 2 min, according to Tao et al. [18]. The polymerase chain reaction products were cloned into the plasmid vector pMD18T-Simple (2693 bp) and sequenced. PCR derived sequences were compared with the deposit in the GenBank ITS region (http://www.ncbi.nlm.nih.gov) and the similarity was evaluated using the BLAST program.

Yeast Cultivation

Cultures were grown in 250 mL Erlenmeyer flasks containing 50 mL of YP-glucose broth, pH 6.5, which contained in addition olive oil (1% *v*/*v*), tributyrin (1 g/L), or triolein (1 g/L), alternatively. The flasks containing sterile media were inoculated with a loopful of yeast directly from a plated culture. The inoculated flasks were incubated in a shaker incubator at 150 rpm and 28 C for 48 h. 20 h flask cultures were used to inoculate the bioreactors in volumes giving a final optical density at 600 nm of 0.1 to 0.15.

Bioreactor cultivations were carried out in two stirred tank bioreactors of different capacities: a 2 L working volume (BioFlo, New Brunswick Scientific, NJ, USA) and a 10 L working volume (BIOFLO 410, New Brunswick Scientific, NJ, USA). The agitator of the 2 L reactor had two six-bladed Rushton-type impellers

(52 mm). The reactor vessel was equipped with baffles. The agitation system of the 10 L reactor consisted of three disc turbine impellers, 8 cm in diameter, with six flat blades. The reactor vessel was also equipped with baffles. The operating stirrer speed in both reactors was 200 rpm. The airflow rate was adjusted at 1 L of air per liter of reactor working volume per minute (1 vvm). The temperature was maintained at 28°C, while the pH at 6.5 throughout fermentations by automatic addition of titrants (1 M NaOH and HCl solutions). Dissolved oxygen tension (DOT) was monitored in both reactors during fermentations.

Fermentations in the 2 L bioreactor carried out using the above-described media and lasted 72 h. In the 10 L reactor fermentations, the medium was modified to include 60 g/L glucose and 1 g/L tributyrin and fermentations lasted 96 h. Samples were taken at regular intervals, cells were removed by centrifugation at 14,000 rpm for 10 min, and the supernatants were examined for lipolytic activity. Batch fermentations were repeated at least three times and each sample was analyzed in triplicate.

Analytical Methods

The reagents used throughout the study were analytical grade by Sigma. Growth was estimated by measuring the optical density at 600 nm and by measuring the cell dry weight (CDW). Broth samples were filtered through pre-weighed glass microfiber filters (Whatman, Maidstone, UK), washed twice and dried in a microwave oven at the defrost setting for 15 min. The filters were placed in a desiccator for the next 24 h and weighed. The OD600 was correlated with the CDW and 1 unit OD600 equals 0.42 ± 0.3 mg cells/mL.

Lipolytic activity was preliminary evaluated on YP-glucose agar plates to which 1% *v*/*v* olive oil had been added. Formation of a clearer halo around yeast colonies after 5 days of incubation at 28°C was proof of lipolytic activity. The relative enzyme activity (REA) was calculated by dividing the diameter of the zone of clearance by the diameter of the respective colony (in millimeters). Experiments were carried out in triplicate and mean values are presented.

In liquid culture filtrates, lipolytic activity was assayed using the esters *p*-nitrophenyl palmitate (*p*-NPP) and *p*-nitrophenylacetate (*p*-NPA) as substrates, following the protocol of our earlier work with molds [19], which was based on the method of Winkler and Stuckmann [20]. The reaction mixture was consisted of 405 μL buffer and 45 μL of substrate. The buffer was as: 200 mg Triton X-100, 50 mg gum arabic in 50 mL of 50 mM phosphate buffer, pH 6.5. As substrate, 15 mg *p*-NPP or *p*-NPA in 10 mL 2-propanol was used. The reaction mixture was emulsified in an ultrasonic bath for 5 min. 50 μL of the sample solution containing the lipolytic enzyme was added to the mixture warmed at 40°C and after a 5 min incubation the reaction was terminated by addition of 0.5 mL of 2% Trizma base. Centrifugation was followed at 12,000 rpm/4 min and the OD of the supernatant was measured at 410 nm. One unitof lipase activity was defined as the amount of enzyme that released 1 μmol of *p*-nitrophenol/min under the specified conditions.

Neutral protease activity was determined in liquid culture filtrates using the casein assay according to Sigma's protocol [21]. Alkaline protease activity was determined using the azocasein assay according to Reichard et al. [22] with the pH of the 0.1 M Tris-HCl buffer adjusted at 9.0 instead of the suggested 8.0. One unit (U) of protease activity was defined as the absorbance variation in the assay conditions of the above methods.

Glucose concentrations were determined with the glucose oxidase/peroxidase method of Kunst et al. [23]. Ethanol, acetate and glycerol concentrations were determined calorimetrically using assay kits by Sigma-Aldrich (St. Louis, MO, USA) (MAK076, MAK086 and MAK117, respectively).

All values are given as mean ± standard deviation in triplicate for each point. Kinetic parameters of growth and production were calculated from raw data using numerical differentiation.

RESULTS AND DISCUSSION

Strain Isolation and Identification

The yeast strain *P. anomala* M1 studied in this work was isolated from colonies developed on the surface of traditional Greek sausages. Although the yeast population in the sausages is significant, there are no reports in the literature either on its biodiversity or on its role in the development of product characteristics. A preliminary study with yeast isolates from colonies on the surface of the particular type of sausages revealed a significant number of *Debaryomyces* and *Pichia* spp. (data not shown). Studies on Spanish and Italian relevant products show the dominance of *Debaryomyces* and significant numbers of *Candida*, *Pichia*, *Rhodotorula* and *Torulopsis* spp. [24,25]. The yeasts ferment sugars and have proteolytic and lipolytic activity and therefore, produce a large array of metabolites that contribute to the development of flavour and aroma [26]. In addition, some yeasts metabolize the lactic acid produced by lactic acid bacteria and cause an increase in the pH toward neutrality which is translated in the product as milder taste [25].

Colonies of the M1 strain on agar plates containing YP-glucose medium were off-white in color, flat, smooth and opaque. Pseudohyphae were not observed, while 1 to 4 ascospores were detected per ascus (Figure 1). The strain was able to grow in media containing as carbon source glucose, sucrose, galactose, and maltose. It was not able to grow on lactose. It could assimilate soluble starch, raffinose, ethanol, glycerol, D-mannitol, as well as lactic and citric acids, but not L-rhamnose.

The yeast was able to grow on substrates containing 1.5 M NaCl (osmotolerant) and in the absence of vitamins, while it produces ethanol under anaerobic conditions (negative Crabtree effect). The yeast was preliminary identified as *Pichia* while possession of traits such as high osmotolerance, fermentation of sucrose, assimilation of raffinose but not of L-rhamnose, puts it under the species classification of *P. anomala* according to Kurtzman [1].

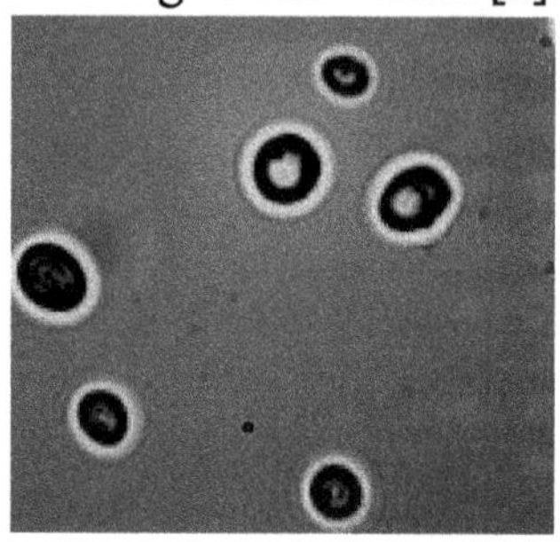

Figure 1. Pichia anomala M1 cell morphology (phase contrast microscopy, ×400).

Taxonomic identity of the M1 strain was confirmed through 18s rRNA ITS region cloning and sequencing. Following PCR amplification, a fragment of about 600 bp was obtained (Figure 2). Sequencing results of the ITS region excluding the primers were as: aaggatcattatagtattctattgagcgcttaattgcgcggcgataaacctta-
cacacattgtctagtttttttgaactttgctttgggtggtgagcctggcttactgcccaaaggtctaaacacat-
tttttaatgttaaaacctttaaccaatagtcatgaaaattttaacaaaaattaaaatcttcaaaactttcaacaacg-
gatctcttggttctcgcaacgatgaagaacgcagcgaaatgcgatacgtattgtgaattgcagattttcgtgaatc-
cgaatctttgaacgcacattgcaccctctggtattccagagggtatgcctgtttgagcgtcatttctctctcaaacct-
tcgggtttggtattgagtgatactctgtcaagggttaacttgaaatattgacttagcaagagtgtactaataag-
cagtctttctgaaataatgtattaggttcttccaactcgttatatcagctaggcaggttagaagtattttagg
ctcggcttaacaacaataaactaaaagtttgacctcacaggtaggactacccgctgaacttaa (570 bp).

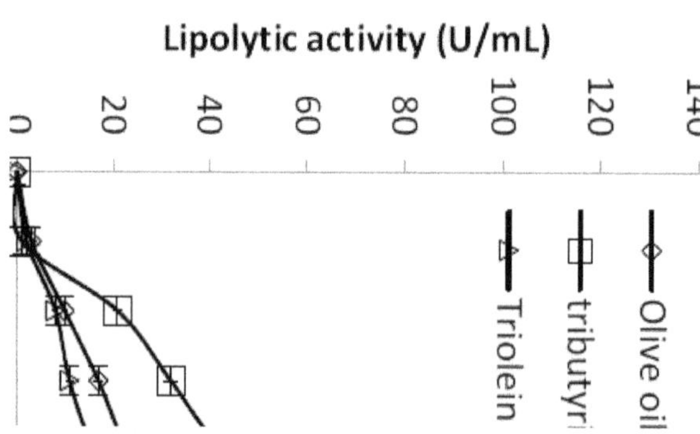

Figure 2. PCR product (pointed) of the internal transcribed spacer (ITS) region of *P. anomala* M1.

A BLAST search for the above sequence showed that it shared almost complete identity (99%) with a number of *Wickerhamomyces anomalus* strains-the name preferred by NCBI for *Pichia anomala*. The sequence was deposited to GenBank under the accession number BankIt2015357 (MF076893).

Relative Lipolytic Activity in Solid State Cultivation

Research on *P. anomala* growth conditions has shown that it is able to tolerate a relatively wide range of environmental growth conditions, e.g., temperatures ranging from 3 to 37°C; pH values from 2.0 to 12.0; and osmotic conditions as low as aw 0.92 for NaCl and 0.85 for glycerol [15]. Tao et al. [18] in their work with *P. anomala* Y1 reported that good growth was obtained in YPD medium with the pH in the range of 3.5–6.5 and temperature 25–30°C, while optimum growth conditions were reached at 30°C and pH 5.0. It would be necessary therefore to investigate for lipolytic activity by *P. anomala* M1 over a wide range of temperatures and pHs and this was feasible by evaluating first the relative lipolytic activity in solid media containing olive oil, over a temperature range at pH 6.5, since most of yeast and fungal lipases are reported to be most active and stable in neutral and alkaline pHs [19,27].

Figure 3 shows the relative lipolytic activity of *P. anomala* M1 on YP-glucose agar containing olive oil at various temperatures, ranging from 25 to 40°C. The pH of the medium was adjusted at 6.5. The highest activity was obtained at 28°C. The lipolytic activity was lower at 30°C but significantly higher compared to values obtained at 25 and 35°C. The relative lipolytic activity at 40°C was close to zero. An experiment was also carried out keeping the temperature at 28°C and changing the pH of the medium in a range from 3.0 to 7.5. As it is shown in Figure 4, the highest lipolytic activity was obtained at pH 6.5. *P. anomala* M1 was able to produce lipolytic activity within the pH range 5.0–7.0 but the relative lipolytic activity values obtained at pHs 6.5 and 6.0 were significantly higher. Lipolytic activity was not detected at pHs 3.0, 3.5, and 7.0, while it was marginally detected at pHs 4.0 and 4.5. Considering both Figures 3 and 4, it can be concluded that the optimum conditions of temperature and pH for lipolytic enzyme production by *P. anomala* M1 are 28°C and pH 6.5, and these conditions were applied in submerged fermentations.

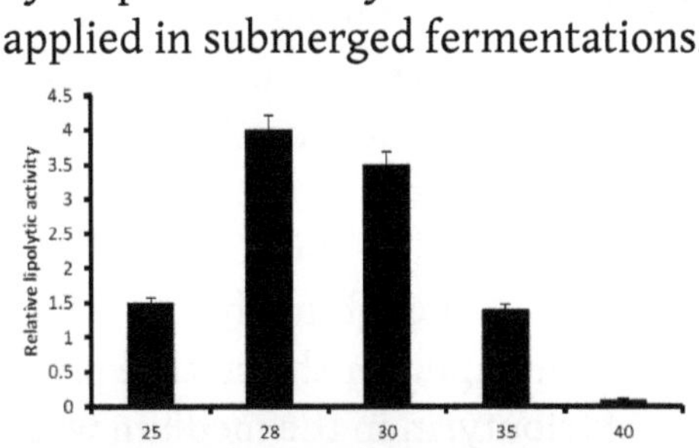

Figure 3. The effect of temperature on the relative lipolytic activity of *P. anomala* M1 grown on YP-glucose agar supplemented with olive oil at pH 6.5. Experiments were carried out in triplicate and the Y-error bars represent the standard deviation of the mean value.

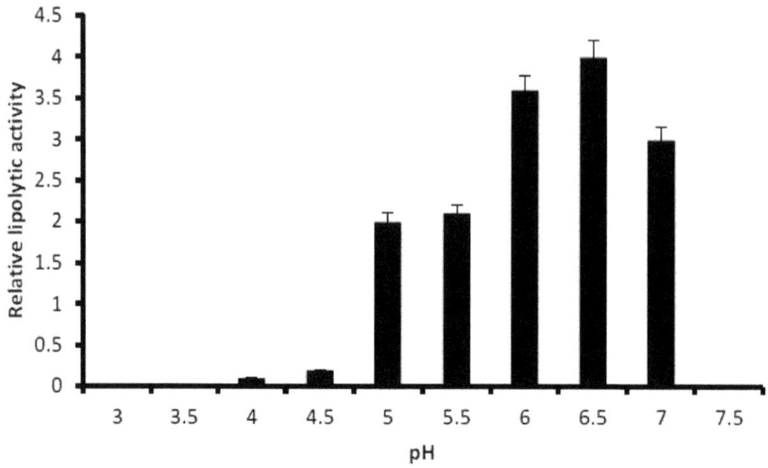

Figure 4. The effect of pH on the relative lipolytic activity of *P. anomala* M1 grown on YP-glucose agar supplemented with olive oil at 28°C. Experiments were carried out in triplicate and the Y-error bars represent the standard deviation of the mean value.

Production of Extracellular Lipase in Submerged Fermentation

Lipases can be distinguished from carboxyl esterases by their substrate spectra, using *p*-NPP (cleaved by lipases) versus *p*-NPA (cleaved by esterases). The lipolytic activity assay produced results only on the *p*-NPA substrate and therefore, the activity is due to an esterase enzyme.

Figures 5 and 6 show the time-courses of biomass and lipolytic activity of *P. anomala* grown in shaken flasks with olive oil, triburyrin (short chain triglyceride, $C_{15}H_{26}O_6$) or triolein (long chain triglyceride $C_{57}H_{104}O_6$) added in the medium as enzyme inducer. There is an obvious enhancement of both growth but mainly lipolytic activity in the case of tributyrin. Comparing results, it appears that final lipolytic activity with tributyrin was 1.91 times higher than the obtained with olive oil while 2.43 times higher than the obtained with triolein. The effect on biomass was not as sound -although results of the same trend were obtained. Biomass yield with tributyrin was 1.16 times higher compared to the obtained with olive oil and 1.47 times higher than that of triolein. In all cases, lipolytic activity increased with increasing biomass and resumed in about 40 h when cultures were well into the stationary phase of growth. Although in shaken flasks cultivation important process parameters (e.g., the air supply or the culture pH) were not under control, the effect of the inclusion of tributyrin in the medium was clear.

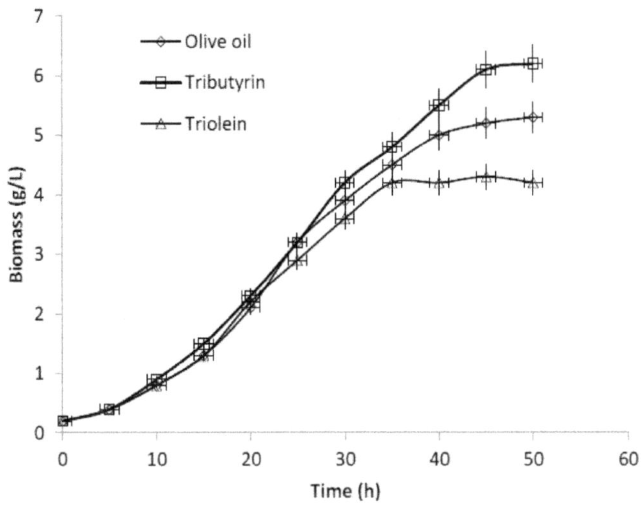

Figure 5. Time courses of biomass concentration of *P. anomala* M1 in shaken flasks cultures (200 rpm). The YP-glucose medium was supplemented alternatively with olive oil, tributyrin and triolein.

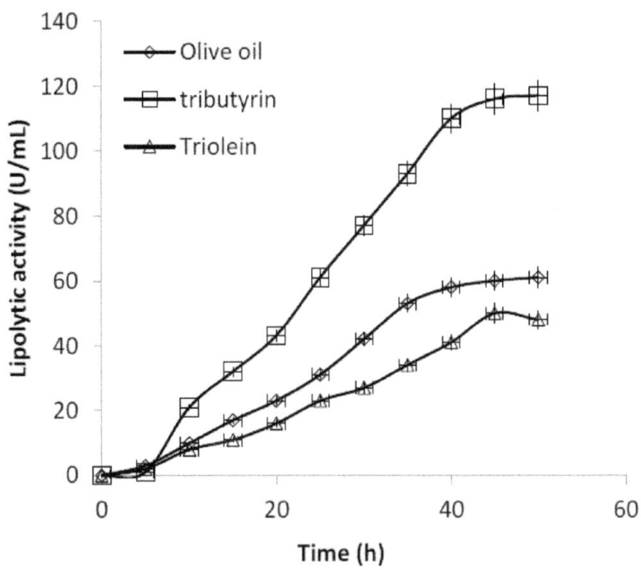

Figure 6. Time courses of extracellular lipolytic activity of *P. anomala* M1 in shaken flasks cultures (200 rpm). The YP-glucose medium was supplemented alternatively with olive oil, tributyrin and triolein.

The profile of extracellular lipolytic activity of *P. anomala* M1 was next evaluated in submerged bioreactor cultivations of two different scales under the conditions described in the previous section, using tributyrin as inducer. Concerning oxygen requirements of *P. anomala*, it is known that it is able to grow in both oxygen-replete and oxygen-limited conditions (facultative yeast). However, under aerobic conditions growth rates and biomass yields are significantly higher than those obtained under anaerobic conditions [14]. Since extracellular lipases are products of the primary metabolism with growth-associated characteristics, fermentations were carried out under aerobic conditions to ensure increased growth rates.

Figure 7 shows the time-courses of biomass, lipolytic activity and residual glucose concentrations in a run in the 2 L stirred tank bioreactor. *P. anomala* grew exponentially with a maximum specific growth rate μmax of 0.17 h^{-1} and a biomass yield Yx/s of 0.43 g of CDW g of $glucose^{-1}$ (Table 1).

Only low amounts of ethanol, glycerol and acetate were detected (results at 50 h, Table 1). Dissolved oxygen tension decreased during growth and the lowest levels monitored were 70% of saturation, which were high enough to consider oxygen limitation (Figure 7). Maximum biomass concentration (8.6 g/L) was obtained around 55 h while maximum lipolytic activity (159 U/L) around 50 h of fermentation. With depletion of glucose (50 h samples), biomass started decreasing and a sharp fall in lipolytic activity was observed. Running the process for another 24 h after glucose depletion resulted in a loss of lipolytic activity of the order of 19.2%. This was observed in several cases of lipase production in fermentation [27] and was attributed mainly to proteolytic degradation. Protease activity was determined in the present case and as shown in Figure 7, proteases appeared in the broth after 30 h of fermentation (1.2 U/mL) and proteolytic activity was increasing towards the end of the process to reach 11 U/mL. Ethanol, glycerol and acetate were not detected beyond 65 h, indicating their consumption in the absence of glucose.

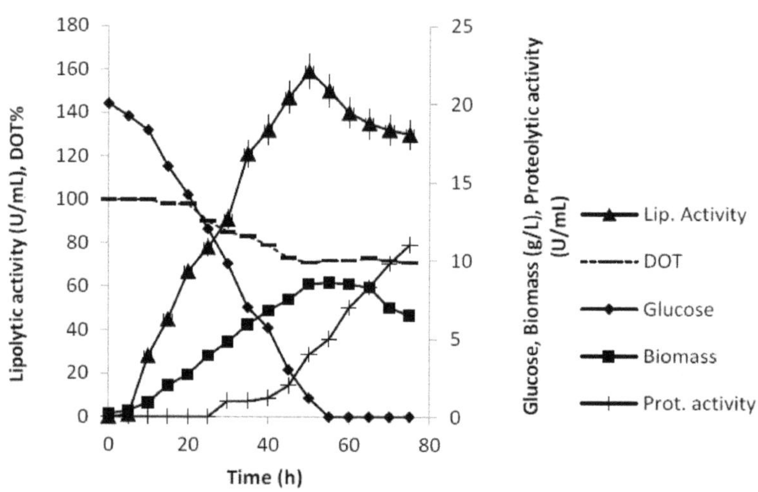

Figure 7. *P. anomala* fermentation in a 2 L working volume stirred tank reactor (200 rpm, 28°C, pH 6.5, 1 vvm aeration). The YP-20 g/L glucose medium was supplemented with tributyrin.

Table 1. Kinetic parameters of *P. anomala* M1 cultures. Values are averages ± standard deviations based on the results of three independent cultivations.

Parameter	2 L Bioreactor 20 g/L Glucose	10 L Bioreactor 60 g/L Glucose
Maximum Specific Growth Rate (h−1)	0.17 ± 0.02	0.12 ± 0.03
Maximum Specific Consumption Rate of Glucose (mmol/g/h)	1.9 ± 0.2	1.3 ± 0.3
Specific Production Rate of Lipolytic activity (U/g/h)	7 ± 0.8	51 ± 1.3
Ethanol (mmol/g/h)	0.12 ± 0.02	0.09 ± 0.03
Acetate (mmol/g/h)	0.14 ± 0.04	0.11 ± 0.01
Glycerol (mmol/g/h)	0.11 ± 0.03	0.10 ± 0.02
Yield on Glucose (g/g)		
Biomass	0.43 ± 0.14	0.24 ± 0.02
Ethanol	0.03 ± 0.002	0.05 ± 0.01
Acetate	0.05 ± 0.01	0.04 ± 0.02
Glycerol	0.02 ± 0.009	0.02 ± 0.001

Figure 7 shows the *P. anomala* M1 fermentation process in the 10 L bioreactor with 60 g/L initial glucose concentration. Both biomass production and lipolytic activity were higher compared to those obtained in the 2 L fermentation. The maximum values obtained, were 14.9 g/L biomass and 213 U/mL lipolytic activity (75 h), which is 1.57 times higher biomass, and 1.74 times higher lipolytic activity compared to those obtained from the 2 L reactor fermentation. Comparing the kinetic parameters for growth in the two reactors it appears that the values for maximum specific growth rate and the yield of biomass on glucose were lower in the 10 L fermentation: 0.12 compared to 0.17 h^{-1} and 0.24 compared to 0.43 g of CDW g of glucose $^{-1}$ g, respectively (Table 1). The lower biomass levels however produced the lipolytic enzyme (in terms of activity) at a higher rate: 0.48 U/mL/h compared to 0.06 U/mL/h and in terms of specific production rate, 51 U/g/h biomass in the 10 L process compared to 7 U/g/h biomass in the 2 L process (timing: 50 h). Glucose concentration fell to zero in 72 h and the following 24 h of the process were characterized by declining biomass and enzyme activity levels. The loss of lipolytic activity was 20.1% in the case of the 10 L fermentation. Only low amounts of ethanol, glycerol and acetate were detected in this case again (results at 50 h, Table 1). Dissolved oxygen tension decreased during fermentation but higher values were monitored compared to those of the 2 L bioreactor fermentation: about 85% of saturation at maximum biomass levels.

Since the lipolytic activity is highly dependent on the experimental conditions and the applied methodology, direct comparisons of the results with those reported for various microbial lipases become rather difficult. However, the obtained values for kinetic parameters and product titers are either comparable or even higher. Kar and co-workers [28] obtained maximum specific rates of extracellular lipase production by *Y. lipolytica* at around 50 h of fermentation as in the present study, the values however were expressed per million of cells and not per g CDW and cannot be compared since it is a different yeast species. With fungal lipases, the production period is longer as for example with *Mucor hiemalis* lipase whose maximum activity was obtained at 192 h of fermentation and reached 110 U/mL [29], or a *Penicillium nalgiovense* lipase with a maximum activity of 82 U/mL at 168 h of fermentation [19].

The applied culture conditions in the 10 L bioreactor (Figure 8) resulted in a significant increase in the specific rate of lipolytic activity while the specific rates of growth and glucose consumption were decreased compared to those of the 2 L bioreactor. The main differences between the two systems were the concentrations of glucose and the dissolved oxygen tension. *P. anomala* is a respiratory yeast that lacks a Crabtree effect and therefore, there is no induction of alcoholic fermentation by high glucose concentrations. In addition, similarly to the other respiratory yeast *K. lactis*, *P. anomala* exhibits minor fermentation despite the elevated activities of the key fermentative enzymes PDC and ADH [14,30]. This metabolic profile was apparent in the present study. Although the same stirrer speed (200 rpm) and air supply rate (1 vvm) were applied in the two reactor systems, the dissolved oxygen tension was maintained at higher levels in the 10 L bioreactor despite the increased biomass concentration. This can be explained by a more efficient mixing regime in the 10 L reactor that facilitated a smaller mixing time of the broth, however, mixing studies were not undertaken at that stage to quantify the effects. The decrease in DOT in the 2 L bioreactor may have acted as the inducing stimulus for the activation of fermentative metabolism and the increased glucose consumption rate points to an ongoing Pasteur effect. Glucose availability however, did not induce ethanol formation.

Intensive studies of *P. anomala* metabolism and physiology in response to oxygen and glucose availability carried out by Fredlund and co-workers (2004) [14] showed that with the particular yeast, the signals that induce respiratory and fermentative metabolism are completely different from those known for *Saccharomyces cerevisiae*. Respiratory growth was obtained from aerobic batch cultures with a high glucose concentration and fermentation could be induced by shifting the cultures to oxygen limitation (shift from 50% to 0% DOT). Oxygen therefore is the main regulator of the central energy metabolism while glucose does not repress respiration but possibly repress the ethanol assimilation pathway. Having this metabolic and physiological background, lipolytic enzyme production by *P. anomala* appears to be advantageous compared to other sources as it can take place under oxygen-saturated conditions using substrates with elevated glucose concentrations.

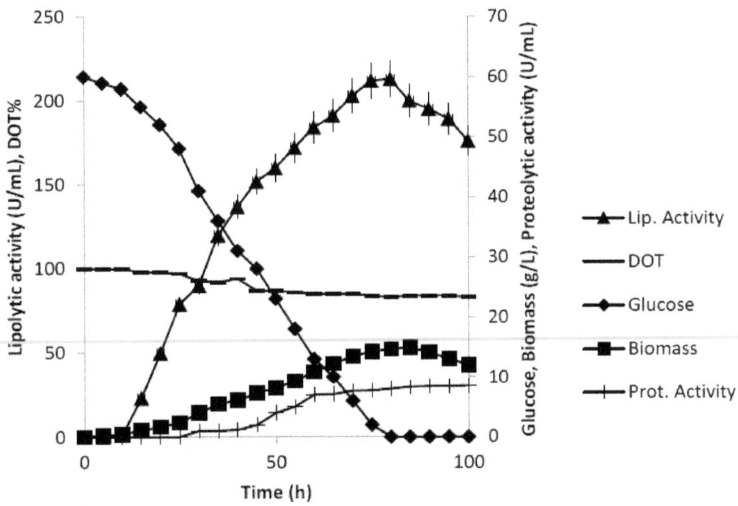

Figure 8. *P. anomala* fermentation in a 10 L working volume stirred tank reactor (200 rpm, 28°C, pH 6.5, 1 vvm aeration). The YP-60 g/L glucose medium was supplemented with tributyrin.

From the applications point of view, the non-conventional yeast *P. anomala* has been extensively studied mainly as a biopreservation agent for various post-harvest systems. Although it was reported in the past as lipase-positive yeast, its production potential in fermentation systems was not evaluated. The current study shows that a wild strain isolated from a rich in fat meat product can produce significant lipolytic activity in a scalable aerobic process. Further studies will focus on the characteristics and properties of the produced esterase as well as in the process itself. Factors for example, identified in this study as controlling process parameters, e.g., the DOT, should be studied in a scaling-up perspective.

SUMMARY

A yeast isolate, selected for its lipolytic activity from a meat product, was characterized as *Pichia anomala*. Lipolytic activity, determined on *p*-NPA as esterase, was maximum at 28°C, pH 6.5, and induced by the short chain triglyceride tributyrin. Fermentations in 2 L and 10 L stirred tank bioreactors, with 20 and 60 g/L glucose respectively, showed that in the second case lipolytic activity increased 1.74-fold, while the biomass increased 1.57-fold. Under otherwise identical aeration conditions, improved mixing in the 10 L reactor maintained higher dissolved oxygen levels which, along with the elevated glucose concentration, resulted in significant increase of specific rates of lipolytic activity (51 vs. 7 U/g/L), while specific rates of growth and glucose consumption maintained lower. The Crabtree-negative yeast (glucose insensitive growth) exhibited a Pasteur effect at lower dissolved oxygen concentrations while elevated glucose prevented ethanol formation under oxygen saturation. The particular physiological traits can be exploited to obtain significant lipolytic activity in a scalable aerobic process.

This study shows that a wild strain of *P. anomala* is capable of producing extracellular lipolytic (esterase) activity in a process with practical significance. Experiments carried out in agar plates, shaken flasks, a 2 L stirred tank bioreactor and a 10 L stirred tank bioreactor showed that by setting the appropriate conditions the physiological traits of the particular yeast can be exploited for effective enzyme production. The Crabtree-negative yeast *P. anomala* exhibits a Pasteur effect at lower dissolved oxygen concentrations but elevated glucose concentrations are not inhibitory and depress ethanol formation in an oxygen-saturated substrate. The main stimulus for the regulation of central carbon metabolism as it appears from its effects on the specific rates of growth and enzyme production is oxygen and, therefore, setting an aerobic process with a high glucose concentration in the substrate advances enzyme production.

REFERENCES

1. Kurtzman, C.P.; Fell, J.W. (Eds.) *The Yeasts, a Taxonomic Study*; Elsevier: Amsterdam, The Netherlands, 1998.
2. Passoth, V.; Fredlund, E.; Druvefors, U.Ä.; Schnürer, J. Biotechnology, physiology and genetics of the yeast *Pichia anomala*. *FEMS Yeast Res.* **2006**, *6*, 3–13. [CrossRef] [PubMed]
3. Walker, G.M. *Pichia anomala*: Cell physiology and biotechnology relative to other yeasts. *Antonie van Leeuwenhoek* **2011**, *99*, 25–34. [CrossRef] [PubMed]
4. Fletcher, E.; Feizi, A.; Kim, S.; Siewers, V.; Nielsen, J. RNA-Seq analysis of *Pichia anomala* reveals important mechanisms required for survival at low pH. *Microb. Cell Fact.* **2015**, *14*, 143. [CrossRef] [PubMed]
5. Passoth, V.; Olstorpe, M.; Schnürer, J. Past, present and future research directions with *Pichia anomala*. *Antonie Van Leeuwenhoek* **2011**, *99*, 121–125. [CrossRef] [PubMed]
6. Schneider, J.; Rupp, O.; Trost, E.; Jaenicke, S.; Passoth, V.; Goesmann, A.; Tauch, A.; Brinkrolf, K. Genome sequence of *Wickerhamomyces anomalus* DSM 6766 reveals genetic basis of biotechnologically important antimicrobial activities. *FEMS Yeast Res.* **2012**, *12*, 382–386. [CrossRef] [PubMed]

7. Goncalves, F.A.G.; Colen, G.; Takahasi, J.A. *Yarrowia lipolytica* and its multiple applications in the biotechnological industry. *Sci. World J.* **2014.** [CrossRef]
8. Satyanarayana, T.; Kunze, G. *Yeast Biotechnology: Diversity and Applications*; Springer: Berlin, Germany, 2009.
9. Stergiou, P.Y.; Foukis, A.; Sklivaniti, H.; Zacharaki, P.; Papagianni, M.; Papamichael, E.M. Experimental investigation and optimization of process variables affecting the production of extracellular lipase by *Kluyveromyces marxianus* IFO 0288. *Appl. Biochem. Biotechnol.* **2012**, *168*, 672–680. [CrossRef] [PubMed]
10. Banerjee, S.; Sengupta, I.; Majumdar, S.K. Lipase Production by *Hansenula anomala var.Schnegii*. *J. Food Sci. Technol.* **1985**, *22*, 137–139.
11. Ionita, A.; Moscovici, M.; Popa, C.; Vamanu, A.; Popa, O.; Dinu, L. Screening of yeast and fungal strains for lipolytic potential and determination of some biochemical properties of microbial lipases. *J. Mol. Catal. B. Enzym.* **1997**, *3*, 147–151. [CrossRef]
12. De Miranda, A.S.; Miranda, L.S.M.; de Suza, R.O.M.A. Lipases: Valuable catalysts for dynamic kinetic resolutions. *Biotechnol. Adv.* **2015**, *33*, 372–393. [CrossRef] [PubMed]
13. Faber, K. *Biotransformations in Organic Chemistry*, 6th ed.; Springer: Berlin, Germany, 2011.
14. Fredlund, E.; Blank, L.M.; Schnürer, J.; Sauer, U.; Passoth, V. Oxygen- and glucose-dependent regulation of central carbon metabolism in *Pichia anomala*. *Appl. Environ. Microbiol.* **2004**, *70*, 5905–5911. [CrossRef] [PubMed]
15. Fredlund, E.; Druvefors, U.; Ligsten, K.J.; Boysen, M.E.; Schnürer, J. Physiological characteristics of the biocontrol yeast *Pichia anomala* J121. *FEMS Yeast Res.* **2002**, *2*, 395–402. [PubMed]
16. Kreger-van Rij, N.J.W. *The Yeasts: A Taxonomic Study*, 3rd ed.; Elsevier: Amsterdam, The Netherlands, 1984.
17. Caggia, C.; Restuccia, C.; Pulvirenti, A.; Giudici, P. Identification of *Pichia anomala* isolated from yoghurt by RFLP of the ITS region. *Int. J. Food Microbiol.* **2001**, *71*, 71–73. [CrossRef]
18. Tao, N.; Gao, Y.; Liu, X. Isolation and characterization of a *Pichia anomala* strain: A promising candidate for bioethanol production. *Braz. J. Microbiol.* **2011**, *42*, 668–675. [CrossRef] [PubMed]
19. Papagianni, M. An evaluation of the proteolytic and lipolytic potential of *Penicillium* spp. isolated from traditional Greek sausages in submerged fermentation. *Appl. Biochem. Biotechnol.* **2014**, *172*, 767–775. [CrossRef] [PubMed]

20. Winkler, U.K.; Stuckmann, M. Glycogen, hyaluronate, and some other polysaccharides greatly enhance the formation of exolipase by *Serratia marcescens*. *J. Bacteriol.* **1979**, *138*, 663–670. [PubMed]
21. Sigma-Aldrich. Universal Protease Activity Assay: Casein as a Substrate. 2013. Available online: http://www.sigmaaldrich.com/life-science/learning-center/life-science-video/universal-protease.html. (accessed on 1 July 2017).
22. Reichard, U.; Buttner, S.; Eifferst, H.; Staib, F.; Ruchel, R. Purification and characterisation of an extracellular serine proteinase from *Aspergillus fumigatus* and its detection in tissue. *J. Med. Microbiol.* **1990**, *33*, 243–251. [CrossRef] [PubMed]
23. Kunst, A.; Draeger, B.; Ziegenhom, J. Colorimetric methods with glucose oxidase. *Methods Enzym. Anal.***1986**, *6*, 178–185.
24. Gardini, F.; Suzzi, G.; Lombardi, A.; Galgano, F.; Crudele, M.A.; Andrighetto, C.; Schirone, M.; Totalo, R. A survey of yeast in traditional sausages of southern Italy. *FEMS Yeast Res.* **2001**, *1*, 161–167. [CrossRef] [PubMed]
25. Mendoça, R.C.S.; Gouvêa, D.M.; Hungaro, H.M.; Sodré, A.F.; Querol-Simon, A. Dynamics of the yeast flora in artisanal country style and industrial dry cured sausage (yeast in fermented sausage). *Food Control* **2013**, *29*, 143–148. [CrossRef]
26. Durá, M.A.; Flores, M.; Toldrá, F. Effect of growth phase and dry-cured sausage processing conditions on *Debaryomyces* spp. generation of volatile compounds from branched-chain amino acids. *Food Chem.* **2004**, *86*, 391–399. [CrossRef]
27. Bussamara, R.; Fuentefria, A.M.; de Oliveira, E.S.; Broetto, L.; Simcikowa, M.; Valente, P.; Schrank, A.; Vainstein Henning, M. Isolation of a lipase-secreting yeast for enzyme production in a pilot-plant scale batch fermentation. *Biores. Technol.* **2010**, *101*, 268–275. [CrossRef] [PubMed]
28. Kar, T.; Delvigne, F.; Masson, M.; Destain, J.; Thonart, P. Investigation of the effect of different extracellular factors on the lipase production by *Yarrowia lipolytica* on the basis of a scale-down approach. *J. Ind. Microbiol. Biotechnol.* **2008**, *35*, 1053–1059. [CrossRef] [PubMed]
29. Hiol, A.; Jonzo, M.D.; Bruet, D.; Comeu, L. Production, purification and characterization of an extracellular lipase from *Mucor hiemalis f. hiemalis*. *Enzyme Microb. Technol.* **1999**, *25*, 80–87. [CrossRef]
30. Kiers, J.; Zeeman, A.M.; Luttic, C.; Thiele, C.; Castrillo, J.I.; Steensma, H.Y.; van Dijken, J.P.; Pronk, J.T. Regulation of alcoholic fermentation in batch and chemostat cultures of *Kluyveromyces lactis* CBS 2359. *Yeast* **1998**, *14*, 459–469. [CrossRef]

CHAPTER-9

EXPLORING THE ENHANCEMENT OF BEVERAGE FLAVOURS THROUGH THE INTEGRATION OF NON-CONVENTIONAL YEASTS

INTRODUCTION: AN OVERVIEW

In the ever-evolving landscape of beverage innovation, the quest for unique and distinctive flavours has led researchers and enthusiasts to explore unconventional sources for yeast strains. Among these, non-conventional yeasts present an exciting frontier for adding a novel twist to beverages. These yeast strains, often overlooked in traditional brewing practices, offer a spectrum of aromatic and flavour profiles that can redefine the sensory experience of beverages. In this exploration, we delve into the potential of non-conventional yeasts to impart distinct and captivating flavours to various beverages, challenging the boundaries of conventional taste expectations. From craft breweries to artisanal fermentation, the incorporation of these unconventional yeast strains opens up a world of possibilities for creating beverages that tantalize the palate and captivate the senses.

Beer is one of the most widely consumed alcoholic beverages in the world. In 2003, worldwide beer production reached around 1.82 billion hectoliters and increased to a volume of 1.93 billion hectoliters in 2013 according to the Kirin Beer University Report of 2014. Production is divided into several beer styles of which ale and lager beers are most prominent. Generally, ale is produced by top-fermenting yeasts at temperatures between 15–30°C. Ales are known for their fruity aromas which are regarded as a distinctive characteristic of top-fermenting beers. Lager beers, however, are produced by a distinct group of bottom-fermenting yeasts at fermentation temperatures between 10–15°C. The aroma of lager beers is more neutral compared to ale-type beers as they contain lower amounts of fruity flavour.

Top-fermenting yeasts generally are *S. cerevisiae* strains. At the end of fermentation, these yeasts rise to the surface of the fermenter, creating a thick cell layer. Bottom-fermenting yeasts belong to two distinct groups of lager yeasts [1]. Lager yeasts are hybrids between *S. cerevisiae* and *S. eubayanus* [2]. Lager yeasts can be divided into two groups, group I or *Saaz* type, and group II or *Frohberg* type, that are distinguished at the molecular level by ploidy differences, characteristic chromosomal rearrangements and chromosome losses [3,4].

In contrast to lager beers, lambic beers with sometimes exceptional flavor compositions are based on a larger biodiversity, including acetic and lactic acid bacteria and various yeasts, e.g., *S. cerevisiae, S. pastorianus* and *Brettanomyces bruxellensis* [5].

Beer is a complex product consisting of volatile and non-volatile components that form the final aroma. The contribution of ale yeasts to the final flavour bouquet is generally higher than that of lager yeasts. This has been attributed to the greater diversity of ale yeasts compared to the limited diversity of the two groups of lager yeast [6]. Other process parameters also influence volatile-compound formation. Specifically, sugar concentration of the wort and different aeration regimes influence the production of flavour-active esters. These esters mainly contribute to the fruitiness of the product. Dominant esters are acetate esters such as ethyl acetate (fruity), isoamyl acetate (banana) and 2-phenyl acetate (rose) and ethyl or medium-chain fatty acid esters such as ethyl hexanoate, ethyl octanoate and ethyl decanoate, which provide a fruity apple- or wine-like flavour to the beer [7]. Among the higher alcohols, n-propanol, isobutanol, 2-phenylethanol and isoamyl alcohol are most abundant. Higher alcohols such as isobutanol can contribute a rum-like aroma which gives a warm mouth feeling, while 2-phenylethanol and isoamyl alcohol are prevalent for their sweet/rose and fruity/banana-like aromas, respectively [8].

Not all flavour are desirable. Strecker aldehydes (aged flavour), aldehydes of the *Maillard* reaction, e.g., furfural, and aldehydes of fatty acid oxidation, e.g., trans-2-nonenal, are regarded as off-flavour in beer [9]. The ketone diacetyl (2,3-butanedione) is monitored during the lager beer brewing process, in particular, as it imparts an undesirable buttery flavour with a low flavour threshold [10]. Given the low concentration and volatile nature of these aroma compounds, gas chromatography coupled with mass spectrometry (GC/MS) offers an optimal technique to analyze the flavour profile of beer.

Recently, non-conventional yeasts or non-*cerevisiae* yeasts have gained importance for fermented alcoholic beverages [11,12]. They produce various mixtures of volatile compounds and so contribute to the aroma profile of beverages [6,13]. Mixed fermentations using *S. cerevisiae* in combination with a non-conventional yeast strain e.g., belonging to the genera *Lachancea*, *Pichia* or *Hanseniaspora,* could provide novel beverages with improved ester profiles [14,15]. On the other hand, the synergistic effects on aromatic compound production were observed in co-cultures with *Metschnikowia pulcherrima* and*S. cerevisiae* [16]. Non-conventional yeasts could contribute to satisfying the demand of novel and distinctive, yet natural, flavour in fermented beverages [17].

A vast number of non-conventional yeast strains has been isolated from various food sources and deposited in culture collection such as the CBS yeast collection of the *Westerdijk* Fungal Biodiversity Institute (formerly, CBS, Central bureau voor Schimmelcultures, Utrecht, Netherlands). In this study, we aimed at covering a broad spectrum of species isolated from different substrates like berries, fruits, cheese, fruit flies or even soil and spanning a broad evolutionary distance within the *Saccharomycotina* in order to identify species that could contribute with their particular flavour to lager beer fermentations.

MATERIALS AND METHODS

Strains and Media

Yeast strains used in this study are shown in Table 1, including their CBS reference number. Each strain was coded based on its coordinates in a 96-well plate. Yeast strains were sub-cultured in YPD (1% yeast extract, 2% peptone, 2% glucose) at room temperature overnight.

Table 1. Strains used in this study. Each selected strain was assigned to a code based on its coordinates on a 96-well plate.

Position	Strain Number	Taxon Name	Substrate of Isolation	Origin
B2	CBS 10151	Candida alimentaria	Cured ham	Norway
B3	CBS 12367	*Candida alimentaria*	Brie Régalou cheese	
B4	CBS 4074	*Candida diversa*	Grape must	Japan
B5	CBS 8058	*Candida kofuensis*	Berries of *Vitis coignetiae*	Japan
B6	CBS 1760	*Candida versatilis*	Pickling vat with 22% brine	USA
B7	CBS 2649	*Candida stellate*	Grape juice	France
B8	CBS 6936	*Clavispora lusitaniae*	Citrus essence	Israel
B9	CBS 4373	*Debaryomyces fabryi*	Dry white wine	South Africa
B10	CBS 767	Debaryomyces hansenii		

Position	Strain Number	Taxon Name	Substrate of Isolation	Origin
B11	CBS 2659	*Debaryomyces subglobosus*	Apple	Italy
C2	CBS 8139	*Dekkera anomala*		Netherlands
C3	CBS 615.84	*Geotrichum candidum*	Brie cheese	France
C4	CBS 95	*Hanseniaspora guilliermondii*	Fermenting bottled tomatoes	Netherlands
C5	CBS 6783	*Hanseniaspora occidentalis var. citrica*	Orange juice	Italy
C6	CBS 2585	*Hanseniaspora uvarum*	Sour dough	Portugal
C7	CBS 2568	*Hanseniaspora vineae*	*Drosophila persimilis* (fruit fly)	
C8	CBS 2352	*Hyphopichia burtonii*	Pollen, carried by wild bees	
C9	CBS 4311	*Kazachstania servazii*	Soil	Finland
C10	CBS 3019	*Kazachstania spencerorum*	Soil	South Africa
C11	CBS 2186	*Kazachstania transvaalensis*	Soil	South Africa
D2	CBS 398	*Kazachstania unispora*		
D3	CBS 7775	*Kluyveromyces aestuarii*	Neotredo reynei (shipworm)	Brazil
D4	CBS 8530	*Kluyveromyces dobzhanskii*	*Drosophila* sp.	Canada
D5	CBS 1557	*Kluyveromyces marxianus*	Stracchino cheese	Italy
D6	CBS 7005	*Lachancea fermentati*	Alpechín	Spain
D7	CBS 3082	*Lachancea kluyveri*	*Drosophila pinicola* (fruit fly)	
D8	CBS 7703	*Lachancea waltii*	Either fruit or leaf of fruit tree	
D9	CBS 5833	*Metschnikowia pulcherrima*	Berries of *Vitis labrusca* (Concord grapes)	USA

Posi-tion	Strain Number	Taxon Name	Substrate of Isolation	Origin
D10	CBS 2030	*Meyerozyma guilliermondii*	Insect frass on *Ulmus americana* (elm tree)	USA
D11	CBS 8417	*Meyerozyma guilliermondii*	Brine bath in cheese factory	Netherlands
E2	CBS 7720	*Nakaseomyces bacillisporus*	Exudate of *Quercus emoryi* (Emory oak)	USA
E3	CBS 2170	*Nakaseomyces delphensis*	Sugary deposit on dried figs	South Africa
E4	CBS 8255	*Pichia*	Kefyr	
E5	CBS 2020	Pichia farinosa	Fermenting cacao	Trinidad and Tobago
E6	CBS 2057	*Pichia fermentans*	Brewers yeast	
E7	CBS 188	*Pichia kluyveri*	Olives	
E8	CBS 5147	*Pichia kudriavzevii*	Fruit juice	
E9	CBS 191	*Pichia membranifaciens*	Wine	Italy
E10	CBS 429	*Saccharomyces cerevisiae*	Fermenting must of champagne grapes	
E11	CBS 1250	*Saccharomyces cerevisiae*	Sherry	Spain
F2	CBS 1782	*Saccharomyces cerevisiae*	Super-attenuated beer	
F3	CBS 820	*Saccharomycodes ludwigii*	Grape must	Germany
F4	CBS 2863	*Schwanniomyces occidentalis*	Soil of vineyard	Spain
F5	CBS 6741	*Schwanniomyces polymorphus var. africanus*	Soil	South Africa
F6	CBS 133	*Torulaspora delbrueckii*	Ragi	Indonesia
F7	CBS 427	*Torulaspora microellipsoides*	Apple juice	Germany
F8	CBS 248	*Wickerhamomyces anomalus*	Red currants	Netherlands

Position	Strain Number	Taxon Name	Substrate of Isolation	Origin
F9	CBS 249	*Wickerhamomyces anomalus*	Berries	
F10	CBS 261	*Wickerhamomyces anomalus*	Ragi	Indonesia
F11	CBS 262	*Wickerhamomyces anomalus*	Beer	
G2	CBS 4689	*Zygosaccharomyces bailii var. bailii*	Grape must	Italy
G3	CBS 1082	*Zygosaccharomyces bisporus*	Tea-beer fungus	Indonesia
G4	CBS 726	*Zygosaccharomyces mellis*	Wine grapes	Germany
G5	C1030	*Saccharomyces pastorianus*	Brewers' yeast	
G6	C1039	*Saccharomyces cerevisiae*	Wine yeast	
G7	C746	*Saccharomyces cerevisiae*	Brewers' yeast	
G8	CBS 1513	*Saccharomyces carlsbergensis*	Brewers' yeast	
G9	C482	*Saccharomyces cerevisiae*	Brewers' yeast	
G10	WS34/70	*Saccharomyces pastorianus*	Brewers' yeast	
G11	C598	*Saccharomyces cerevisiae*	Laboratory strain	

To evaluate growth at different temperatures or the utilization of maltose, cells were spotted on solid media plates in 10-fold dilution series and incubated for 5 days before evaluation. Growth at the different dilutions was 10-color coded and is presented as a heat map.

Fermentation Conditions

Lab-scale fermentations were carried out in 50 mL tubes filled with 40 mL nutrient-rich YPD with the glucose concentration adjusted to 16°Plato at 20°C. Each fermentation (300 rpm, using a triangular magnetic stirrer) was started with a cell density corresponding to OD600 = 0.2. For co-fermentations, the amounts of cells used equaled OD600 = 0.1. The fermentation progress was monitored for up to 14 days by daily measurement of CO_2 release. Sugar content was measured by a DMA 35 Anton Paar densitometer (medium gravity in °P). Plato measurements were taken at the beginning and at the end of each fermentation process. The fermentation was defined as finished when the CO_2 loss did not increase any further and the residual sugar concentration remained constant for 2 days. Ethanol was measured after the end of fermentation using an Anton Paar DMA 4500 M Alcolyzer. Aroma profiles were analysed by headspace gas chromatography.

Sample Preparation for Solid-Phase Micro Extraction–Gas Chromatography/ Mass Spectrometry *(SPME-GC/MS) Analysis*2.5 mL of the samples were put in 20 mL vials with the addition of sodium chloride (final concentration 40 mg/ml), 50 µL NaN_3 0.1%, 25 µL of the internal standard (2-octanol final concentration of 200 µg/L) and ascorbic acid (final concentration of 20 mg/mL).

All samples were incubated for 10 min at 40°C, then the volatile compounds were collected on a divinylbenzene/carboxen/polydimethylsiloxane fiber (DVB-CAR-PDMS) coating 50/30 µm, and 2-cm length SPME fibre purchased from Supelco (Sigma-Aldrich, Milan, Italy) for 40 min.

Analytical Methods for GC/MS

GC analysis was performed on a Trace GC Ultra gas chromatograph coupled with a TSQ Quantum Tandem mass spectrometer (Thermo Electron Corporation, Waltham, MA USA), with adaptations as described in Ravasio et al. [18]. GC separation was performed on a 30 m Solgelwax PEG capillary column with an internal diameter of 0.25 mm and a film thickness of 0.25 m (SGE Analytical Science, Melbourne, Australia). The GC oven was kept at 40°C for 4 min and then increased by 6°C/min to 250°C and kept at the final temperature for 5 min. The injector and transfer-line temperatures were kept at 250°C as well. Helium was used as the carrier gas with a flow rate of 1.2 mL/min. The time for thermal desorption of analytes was 4 min. The MS detector was operated in full scan mode at 70 eV with a scan range from 35 to 350 *m/z*. Data analysis was performed using the software ThermoXcalibur (Version 2.2 SP 1.48, Thermo scientific, Waltham, MA, USA). Identification of compounds was based on comparison with a mass spectral database (NIST version 2.0, Gaithersburg, MD, USA) and with reference standards when available. The relative amount of each volatile was expressed as µg/L of 2-octanol.

Multivariate Data Analysis

Multivariate data analysis was performed using StatSoft, Inc. STATISTICA version 8.0 (data analysis software system, 2007, StatSoft (Europe) GmbH, Berikon, Switzerland). Principal component analysis was employed to simplify data interpretation. The matrix initially contained the 60 strains considered in this study and the average of the relative 62 VOCs detected and was later reduced to the sub-selection of 19 strains studied further.

RESULTS

Strain Selection and Identification of Representative Isolates

In order to cover a wide range of the biodiversity of non-conventional yeasts, we selected 60 strains from 48 different species which were obtained from the strain collection of the Westerdijk Institute. The focus was on strains that were previously isolated from various fermentations, e.g., from fermented liquids, fruits, vegetables, or meat. These strains, therefore, may have evolved superior or desirable features, e.g., the production of high levels of ethanol or may have been recognized based on their contribution to flavour. All strains were run in lab-scale fermentation trials in nutrient-rich broth (YPD with adjusted glucose content at 16 Plato) to assess their aroma production. Flavour profile were analyzed using GC/MS and compared to a set of brewing and wine yeast strains (Supplementary Table S1).

Aroma Profile of Fermentations

In total, we identified 62 different volatiles in the samples of all strains (Table 2). Major volatile-aroma compounds that were detected included esters, alcohols, aldehydes, ketones and acids. Most species harbor the ability to produce a large variety of flavour. Thus, we did not identify single species that produced only very few compounds or species with a very high diversity, as shown in Figure 1 for a selection of species. However, there was an enormous difference in the amounts of specific volatiles produced. Esters were the most prominent group of volatiles. In total, 22 different esters could be identified and within this group ethyl-esters dominated, such as ethyl hexanoate and ethyl acetate, associated with fruity wine/apple-like and sweet pear drop flavour, respectively. Alcohols comprised the second major group of compounds. Besides ethanol, we identified 14 different alcohols. Yet, only two compounds, 2-phenylethanol, perceived as rose flavour, and isoamyl alcohol, a banana-like flavour, were produced by all strains analyzed. During anaerobic fermentations the formation of aroma alcohols is flavoured over the production of aroma acids [19]. In line with this, we identified only six acids, of which acetic acid and butyric acid were prominent.

Table 2. List of 62 volatiles that were detected by GC/MS measurement.

Number	Alcohols
1	Benzyl alcohol
2	Butanol
3	Dodecanol
4	Fenchyl alcohol
5	Furaneol
6	Isoamyl alcohol
7	Propanol
8	2-Ethyl-1-hexanol
9	2-Furanmethanol
10	2-Methyl propanol
11	2-Nonanol
12	2-Phenyl ethanol
13	3 Ethoxy-1-Propanol
14	3-(Methylthio)-1-propanol
Esters	
15	Butyl acetate
16	Ethyl (4E)-4-decenoate
17	Ethyl 2-methylbutyrate
18	Ethyl acetate
19	Ethyl butanoate
20	Ethyl decanoate
21	Ethyl dodecanoate
22	Ethyl heptanoate
23	Ethyl hexadecanoate
24	Ethyl hexanoate
25	Ethyl isobutyrate
26	Ethyl isovalerate
27	Ethyl octanoate
28	Ethyl propanoate
29	Ethyl tetradecanoate
30	Isoamyl acetate
31	Isoamyl butyrate
32	Isobutyl acetate
33	Isobutyl butanoate

Number	Alcohols
34	Phenethyl acetate
35	S-methyl thioacetate
36	2-Methyl propanoate
37	Acetic acid
38	Butyric acid
39	Decanoic acid
40	Hexanoic acid
41	Isovaleric acid
42	Octanoic acid
Aldehydes	
43	Acetaldehyde
44	Benzaldehyde
45	Furfural
46	Phenyl acetaldehyde
47	1-Decanal
48	1-Nonanal
49	3-Methyl butanal
50	4-Methyl benzaldehyde
51	5 Methyl furfural
52	5-Hydroxymethylfurfural
Ketons	
53	Acetoin
54	Diacetyl
55	Pyranone
56	2-Cyclopentene-1,4-dione
57	2-Dodecanone
58	2-Methyltetrahydrothiophen-3-one
59	2-Nonanone
60	2-Undecanone
Pyrazines	
61	2,5-Dimethyl-3-ethylpyrazine
62	2,6-Dimethylpyrazine

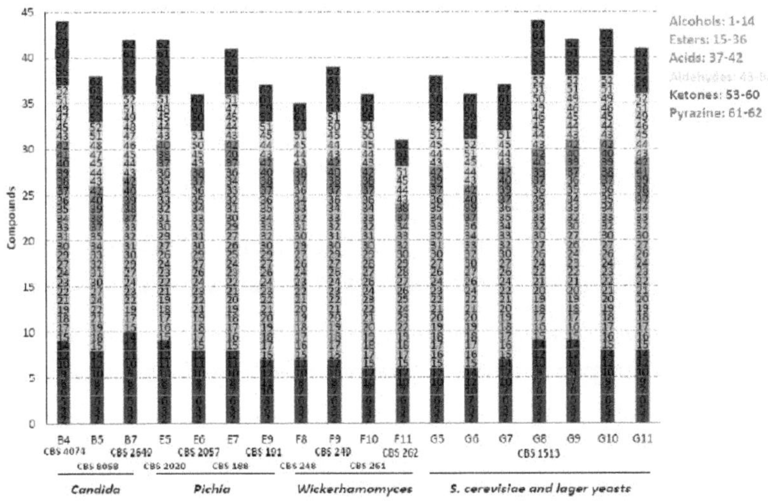

Figure 1. Number of volatile organic compounds (VOCs) produced during fermentation of selected strains. The compounds were grouped, numbered and colored according to their chemical class. The complete list of compounds is shown in Table S1 (see Table 1 for nomenclature of strains).

We were most interested in species that separated well from *S. cerevisiae* or lager yeast strains. Therefore, we narrowed down the selection of strains to those that showed clear separation in their flavour profile to the set of *S. cerevisiae* and lager yeast strains based on principal component analysis (Figure 2). This identified species belonging to the genera *Candida, Pichia* and *Wickerhamomyces.*

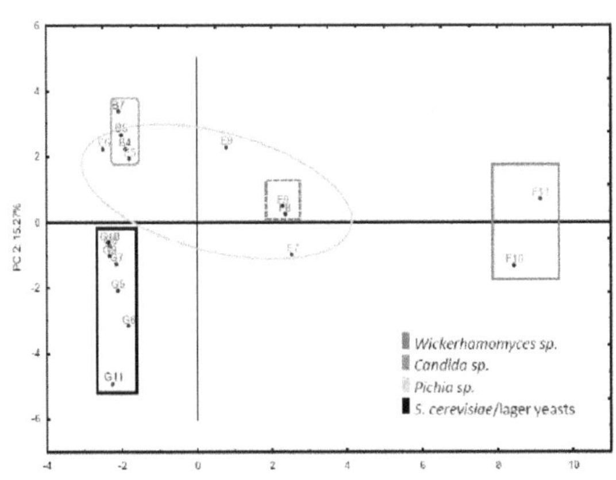

Figure 2. Principal component analysis of the set of selected strains and their chemical compounds produced. The matrix is based on the set of 19 strains and the VOCs detected. Strains are presented with their assigned coordinates (see Table 1). Species belonging to the same genus are represented by the same colour.

Based on the GC data, we generated relative quantifications of the volatile aroma compounds formed. The set of strains was then compared to the lager yeast *S. pastorianus/Weihenstephan* to identify strains that produced higher amounts of esters and alcohols (Tables 3 and 4). One strain of *Saccharomyces cerevisiae* (CBS 1250; E11) was a very good producer of alcohols, particularly of isoamyl alcohol, by contrast with lager yeast. Yet, this strain only produced moderate amounts of esters, mostly ethyl acetate. Its origin from sherry production suggests that it has been selected as the preferred strain for these fermentations.

The *Wickerhamomyces anomalus* isolates showed a remarkably high amount of both alcohol and ester production. Production of higher alcohols was strongly increased in these *W. anomalus* strains compared to lager yeast. Interestingly, here mainly isoamyl alcohol and 2-phenyl ethanol production was increased. These represent two desirable flavour associated with banana and rose flavour. The sherry isolate (CBS 1250, E11) was found to produce even more isoamyl alcohol than the *W. anomalus* strains, but far less 2-phenylethanol than these strains (Table 3). Yet, the *W. anomalus* strains produced abundant amounts of esters, particularly ethyl acetate, isoamyl acetate and 2-phenylethyl acetate. Overall, these strains produced up to 10-fold more esters than the *Weihenstephan* lager yeast strain. (Table 4). Another highly aromatic strain identified in this collection of strains was a *Pichia kluyveri* strain, CBS 188. This strain produced almost fourfold more esters than lager yeast, while its aroma alcohols profile was similar to lager yeast with the exception of fourfold higher 2-phenylethanol production.

Table 3. Strains with higher aroma alcohol production than the *Weihenstephan* lager yeast strain.

Strain Identifier	CBS1250	CBS262	CBS1082	CBS726	CBS261	CBS191	CBS188	CBS2568	CBS2649
Position	E11	F11	G3	G4	F10	E9	E7	C7	B7
Species	S. cerevisiae	W. anomalus	Z. bisporus	Z. mellis	W. anomalus	P. membrani-faciens	P. kluyveri	H. vineae	C. stellata
Butanol	4.15	3.23	4.57	2.98	1.89	1.80	1.75	1.27	1.49
Furaneol	1.81	0.00	3.49	2.07	0.00	0.00	0.00	0.00	1.82
Isoamyl alcohol	3.83	2.45	1.86	2.04	1.12	1.91	0.90	0.79	0.76
Propanol	4.28	1.18	2.19	2.37	1.68	0.00	2.04	1.87	0.00
2-Furanmethanol	0.00	0.00	3.11	2.10	0.00	0.00	0.00	0.00	9.13
2-Methyl propanol	9.10	2.92	2.78	3.62	2.31	3.61	1.24	0.59	2.08
2-Phenyl ethanol	1.42	4.26	1.98	1.94	4.81	2.19	2.97	1.53	1.45
3-(Methylthio)-1-propanol	2.38	0.00	0.00	2.98	0.00	1.83	0.00	12.58	1.98

The values reported represent the relative abundances of the specific subset of metabolites compared to the lager yeast strain (G10).

Table 4. Strains with higher aroma ester production than the *Weihenstephan* lager yeast strain.

Strain Identifier	CBS261	CBS262	CBS188	CBS249	CBS248	CBS1082	CBS133
Position	F10	F11	E7	F9	F8	G3	F6
Species	W. anomalus	W. anomalus	P. kluyveri	W. anomalus	W. anomalus	Z. bisporus	T. delbrueckii
Butyl acetate	23.64	16.84	9.18	8.76	11.73	3.84	3.56
Ethyl (4E)-4-decenoate	0.00	0.00	0.17	0.00	0.04	0.56	0.07
Ethyl 2-methylbutyrate	50.89	368.00	36.00	62.34	39.06	2.94	10.63
Ethyl acetate	31.81	29.17	15.88	13.09	10.02	8.72	5.70
Ethyl butanoate	16.59	26.92	8.00	9.26	10.75	2.58	2.71
Ethyl decanoate	0.14	0.00	2.07	0.46	0.45	2.67	0.72
Ethyl dodecanoate	0.91	0.00	2.60	0.88	1.44	2.69	0.98
Ethyl heptanoate	1.17	2.19	6.96	3.14	1.77	1.58	2.82
Ethyl hexadecanoate	12.19	13.62	16.05	9.10	8.22	11.00	2.36
Ethyl hexanoate	7.23	12.89	5.07	6.90	7.47	4.33	2.72
Ethyl isovalerate	11.84	120.99	33.97	14.43	15.00	0.00	10.79
Ethyl octanoate	0.56	0.26	1.59	0.59	0.32	0.64	0.55
Isoamyl acetate	9.36	5.84	3.21	2.92	3.10	1.27	0.99
Isobutyl acetate	152.96	59.59	24.70	40.69	26.57	6.75	9.93
Isobutyl butanoate	209.13	167.38	29.44	88.87	52.26	10.77	13.62
2-Phenylethyl acetate	5.88	4.07	2.70	2.68	2.90	0.44	0.91
methyl thioacetate	0.00	0.00	0.00	0.00	0.00	0.21	0.00
2-Methyl propanoate	0,89	0,00	1,16	1,13	0,46	0,00	0,44

The values reported represent the relative abundances of the specific subset of metabolites compared to the lager yeast strain (G10).

Fermentation Performance

We compared the fermenting capacity of the strains with very strong volatile compound formation to the lager yeast strain *S. pastorianus WS34/70* and other *S. cerevisiae* strains. All strains were adjusted to the same optical density prior to the start of fermentation to allow for comparison of the speed of fermentation between strains. Fermentation rates were followed by measuring the CO_2 release daily (Figure 3). All strains were able to ferment the liquid within one week except for the *W. anomalus* strains CBS 248; CBS 249 and CBS 261. However, in contrast to the other strains, the *W. anomalus* strains showed an almost linear fermentation curve with equal amounts of CO_2 released per day. During fermentation the *W. anomalus* strains produced oily top layers and biofilms of cells. This oily phase may have reduced the CO_2 release as we observed the lowest pH in the *W. anomalus* fermentations measured among all strains: for strain CBS 261 the pH reached 4.79 and for strain CBS 262 the pH was 4.59, while the liquids fermented by other yeast strains showed a pH of >5. This suggests that the dissolved CO_2 contributed to the acidification of the liquid in the *W. anomalus* fermentations. The most rapid fermentations as represented by fast CO_2 loss were observed with the *S. cerevisiae* and lager beer production strains. Yet, in addition to the *Wickerhamomyces* strains, several other strains showed prolonged CO_2 loss, including *P. kluyveri* CBS 188 (Figure 3).

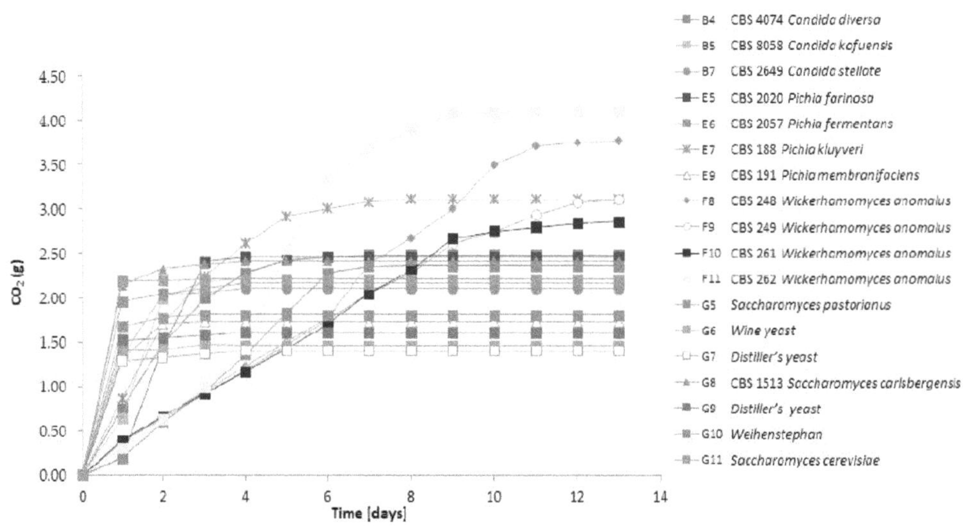

Figure 3. Fermentation profile of the selected strains representing the detected flavour diversity. The CO_2 release was measured daily for each individual strain (see Table 1 for the nomenclature of strains). Fermentations were followed for 13 days.

For most of the strains we could observe a correlation between fermentation speed, residual sugar concentration and ethanol production. The *S. cerevisiae* and lager yeast strains left 2–2.5% of sugar in the medium, while the residual sugar in *Candida* and *Pichia* strains was between 3–3.5%. *Wickerhamomyces anomalus* strains left 4.5–5% of sugars unfermented (Figure 4). Concomitantly, *S. cerevisiae* and lager yeast strains produced the highest amount of alcohol (up to 8%) and the *W. anomalus* strains only 4%.

Figure 4. Final sugar content of the selected strains at the end of fermentation. Sugar concentration was measured in °P after 13 days of fermentation.

Growth on Other Carbon Sources and Temperatures

The selected species were analysed for growth at low (10°C) and high (37°C) temperatures as well as for their ability to utilize maltose using serial-dilution plate assays. Growth at elevated temperatures was mainly restricted to *Saccharomyces* and *Zygosaccharomyces* strains, while growth at lower temperatures was often better in non-*cerevisiae* strains. Typically, all strains grew well at intermediate temperatures. Lager beer fermentation relies on maltose utilization. However, in this screening, good maltose utilization was restricted to lager yeast strains, with the exception of *Candida diversa* CBS 4074 (Figure 5), which also showed very good growth on maltose plates and *Clavispora lusitaniae* CBS 6936.

Utilization of Mixed Fermentations for Flavour Improvement

Co-fermentations are one way to improve the flavour composition of fermented beverages. We plotted volatile compound formation of several candidate strains against lager yeast (Figure 6a). This demonstrated the superior flavour-generation capacity of several strains, most notably of *W. anomalus*. Therefore, we used the same nutrient-rich high Plato fermentation broth and inoculated the *Weihenstephan* lager yeast WS34/70 in a 1:1 ratio with *W. anomalus* CBS 261. Volatiles of this mixed fermentation were determined at the end of fermentation. This clearly demonstrated an enhancement of ethyl hexadecanoate, isoamyl alcohol and 2-phenyl ethanol, improving the fruity flavour perception of the fermented liquid (Figure 6b). *W. anomalus* CBS 261 is a very strong producer of ethyl acetate, which was also pronounced in the co-fermentation.

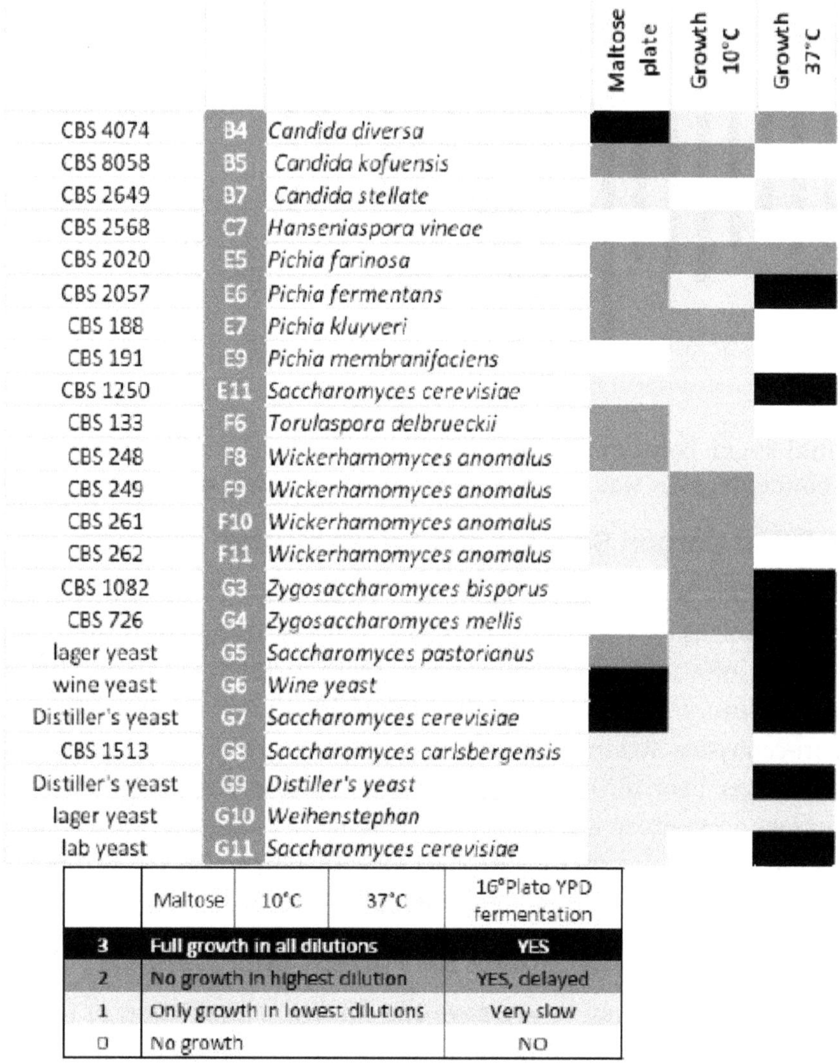

Figure 5. Heat map displaying maltose utilization and growth at the indicated temperatures of the selected strains.

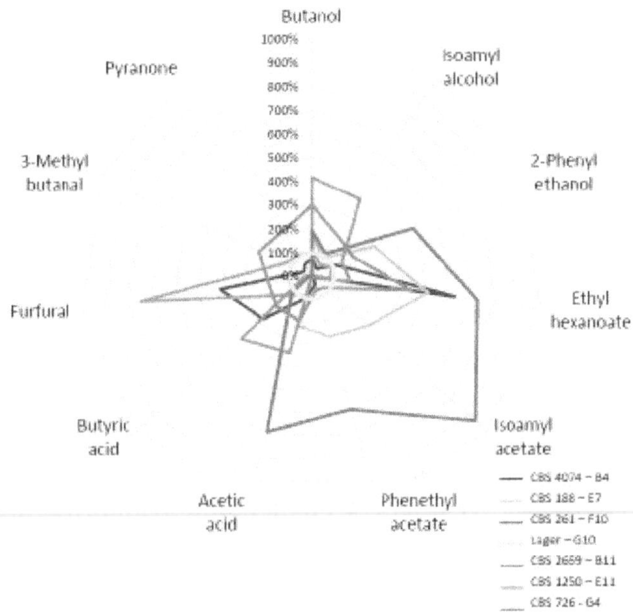

(A)

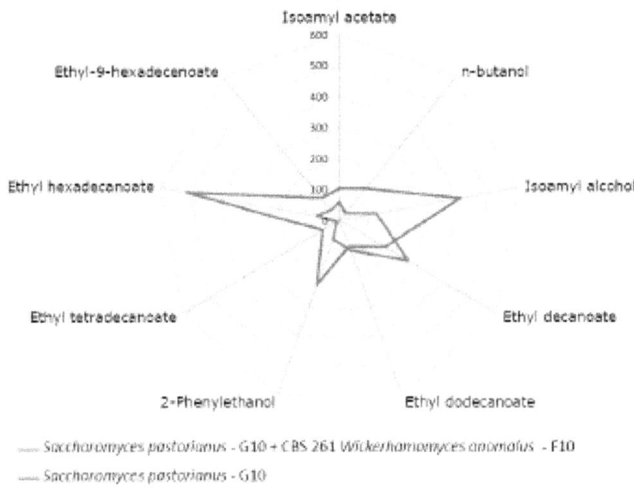

(B)

Figure 6. (A) Comparisons of selected volatiles of selected strains that produced exceptionally high concentrations of organic volatiles with the lager yeast reference. Strains used were: *Candida diversa* (CBS 4074, B4); *Debaryomyces subglobosus* (CBS 2659, B11); *Pichia kluyveri* (CBS 188, E7); *Saccharomyces cerevisiae* (CBS 1250; E11); *Wickerhamomyces anomalus* (CBS 261, F10); *Zygosaccharomyces mellis* (CBS 726, G4); *Saccharomyces pastorianus* (G10, reference). (B) Concentration of volatiles at the end of a co-fermentation using *Wickerhamomyces anomalus* (CBS 261, F10) and the *Weihenstephan* lager yeast (WS, *Saccharomyces pastorianus* G10, reference).

DISCUSSION

In our study we have screened non-*Saccharomyces* cerevisiae yeast biodiversity in order to identify strains with more pronounced volatile compound formation than present in lager yeast strains. Pronounced differences in aroma alcohol production were identified between *S. cerevisiae* and lager yeast strains, as expected from a clean pilsner beer produced by lager yeasts versus the more complex flavour produced by ale and wine yeasts. In our screening, we identified *W. anomalus* strains as the most dominant flavour producers, which also included production of substantial amounts of acetic acid. The formation of floating cell layers such as seen for the *W. anomalus* strains has been reported as a typical phenomenon in stored wines [20]. Yeast biodiversity holds a plethora of strains that show useful characteristics such as ethanol production and flavour formation. This requires a detailed evaluation of the initially identified favourable strains under different conditions, particularly in co-fermentation regimes [21]. It will be challenging to identify the most suitable co-fermentation setups, as flavour profile will certainly be influence by different ratios of non-conventional yeasts versus standard brewing strains.

In earlier studies, *W. anomalus* strains have been isolated from a range of cereal-based sources. It has been reported from sourdoughs and was found as the dominating yeast in sourdough microbial ecosystems next to *S. cerevisiae*. The prevalence of the fungus was associated with its osmotolerance and increased acid tolerance in comparison to *S. cerevisiae* [22]. Furthermore, it was shown that *W. anomalus* provides some antimicrobial activity, e.g., mycocin production, that can be used to prolong the shelf life of bread [23,24]. Other *Wickerhamomyces* species have been put to use in baking using microbread baking platforms. Specifically, bread obtained with *Kazachstania gamospora* and *Wickerhamomyces subpelliculosus* provided added broader aroma profile compared to control baker's yeast [25].

P. kluyveri CBS 188 produced a total of 41 volatiles during our fermentations. Yet, it was outstanding in its ester profile, e.g., ethyl acetate, isoamyl acetate or phenethyl acetate that was several folds higher than in the reference lager yeast strain WS34/70. *P. kluyveri* strains are found in "wild ferments" of wine but it is also commercially available to boost flavour production through sequential fermentations. *P. kluyveri* strains together with strains of *K. marxianus* were also presented as potential starter yeasts for controlled cocoa fermentation [26].

Based on increased demand for natural flavour produced from yeasts, the use of non-conventional yeasts as platform strains for the production of aroma volatile has gained attention. Hence, *K. marxianus* was suggested as a cell factory for flavour and fragrance production based on several advantageous traits, e.g., thermo-tolerance and the wide array of volatile molecules it produces [27]. We had one strain of *K. marxianus* (CBS 1557) in our collection. However, this strain was as low in aroma-compound production as our lager yeast reference. This indicates the need to obtain a larger collection of strains of one species and also to analyse volatile compound formation under different nutritional regimes and fermentation conditions [28]. It further requires the implementation of high-throughput screening tools and assays to identify suitable strains, which was actually successfully shown for *K. marxianus* strains producing ethyl acetate [29]. Additionally, it requires more effort to acquire genomic, transcriptomic and metabolomic datasets of non-conventional yeasts in order to bridge the knowledge gap about *S. cerevisiae* [12,30]. That said, our approach to screening non-conventional yeasts in non-industrial fermentation broths only provides an initial glance of the volatile production capabilities of the tested strains. These capabilities will certainly vary with fermentation and process conditions, particularly at larger scales.

The strong interest in volatile-compound formation by non-conventional yeasts may also open a new perspective on yeast ecology. Yeasts occur in diverse niches and interact with other microbes, insects and plants. These interactions may present selective forces for the production of specific volatile compounds. Interestingly, strains of *Debaryomyces hansenii* were shown to produce methyl salicylate (MeS) as a major compound under very specific conditions using pine weevil (*Hylobius abietis*) frass broth [31]. This compound essentially acts as a deterrent for pine weevil. In a follow-up case, 2-phenylethanol was identified as a strong anti-feedant compound against the pine weevil [32].

Several of the strains we analyzed produced large amounts of 2-phenylethanol, including strains from *S. cerevisiae* but most prominently the *W. anomalus* strains. *W. anomalus* strains are often isolated from tree habitats, insects and insect frass [33]. Elucidating this fascinating interplay of yeasts, insects and plants will provide substantial insight into yeast biology and ecology in the future.

The results obtained in this study indicate that yeast biodiversity harbors a large variety of strains that could enter diverse beverage production pipelines and provide additional all-natural flavour variants to improve the taste and sensory perception of lager beers and, beyond that, other fermented beverages. We used a non-industrial platform to assay strains. In future research, more detailed fermentations using specific industrial fermentation broths, e.g., wort, grape must, and other juices, should be explored with non-conventional yeasts.

SUMMARY

Fungi produce a variety of volatile organic compounds (VOCs) during their primary and secondary metabolism. In the beverage industry, these volatiles contribute to the flavour and aroma profile of the final products. We evaluated the fermentation ability and aroma profile of non-conventional yeasts that have been associated with various food sources. A total of 60 strains were analyzed with regard to their fermentation and flavour profile. Species belonging to the genera *Candida*, *Pichia* and *Wickerhamomyces* separated best from lager yeast strains according to a principal component analysis taking alcohol and ester production into account. The speed of fermentation and sugar utilization were analysed for these strains. Volatile aroma-compound formation was assayed via gas chromatography. Several strains produced substantially higher amounts of aroma alcohols and esters compared to the lager yeast strain Weihenstephan 34/70. Consequently, co-fermentation of this lager yeast strain with a Wickerhamomyces anomalus strain generated an increased fruity-flavour profile. This demonstrates that mixed fermentations utilizing non-Saccharomyces cerevisiae biodiversity can enhance the flavour profile of fermented beverages.

REFERENCES

1. Dunn, B.; Sherlock, G. Reconstruction of the genome origins and evolution of the hybrid lager yeast *Saccharomyces pastorianus*. *Genome Res.* **2008**, *18*, 1610–1623. [CrossRef] [PubMed]
2. Libkind, D.; Hittinger, C.T.; Valerio, E.; Goncalves, C.; Dover, J.; Johnston, M.; Goncalves, P.; Sampaio, J.P. Microbe domestication and the identification of the wild genetic stock of lager-brewing yeast. *Proc. Natl. Acad. Sci. USA* **2011**, *108*, 14539–14544. [CrossRef] [PubMed]
3. Nakao, Y.; Kanamori, T.; Itoh, T.; Kodama, Y.; Rainieri, S.; Nakamura, N.; Shimonaga, T.; Hattori, M.; Ashikari, T. Genome sequence of the lager brewing yeast, an interspecies hybrid. *DNA Res.* **2009**, *16*, 115–129. [CrossRef] [PubMed]
4. Walther, A.; Hesselbart, A.; Wendland, J. Genome sequence of *Saccharomyces carlsbergensis*, the world's first pure culture lager yeast. *G3 (Bethesda)* **2014**, *4*, 783–793. [CrossRef] [PubMed]
5. Spitaels, F.; Wieme, A.D.; Janssens, M.; Aerts, M.; Daniel, H.M.; Van Landschoot, A.; De Vuyst, L.; Vandamme, P. The microbial diversity of traditional spontaneously fermented lambic beer. *PLoS ONE* **2014**, *9*, e95384. [CrossRef] [PubMed]
6. Mertens, S.; Steensels, J.; Saels, V.; De Rouck, G.; Aerts, G.; Verstrepen, K.J. A large set of newly created interspecific *Saccharomyces* hybrids increases aromatic diversity in lager beers. *Appl. Environ. Microbiol.* **2015**, *81*, 8202–8214. [CrossRef] [PubMed]
7. Verstrepen, K.J.; Derdelinckx, G.; Dufour, J.P.; Winderickx, J.; Thevelein, J.M.; Pretorius, I.S.; Delvaux, F.R. Flavor-active esters: Adding fruitiness to beer. *J. Biosci. Bioeng.* **2003**, *96*, 110–118. [CrossRef]

8. Pires, E.J.; Teixeira, J.A.; Branyik, T.; Vicente, A.A. Yeast: The soul of beer's aroma—A review of flavour-active esters and higher alcohols produced by the brewing yeast. *Appl. Microbiol. Biotechnol.* **2014**, *98*, 1937–1949. [CrossRef] [PubMed]
9. Vanderhaegen, B.; Neven, H.; Coghe, S.; Verstrepen, K.J.; Verachtert, H.; Derdelinckx, G. Evolution of chemical and sensory properties during aging of top-fermented beer. *J. Agric. Food Chem.* **2003**, *51*, 6782–6790. [CrossRef] [PubMed]
10. Duong, C.T.; Strack, L.; Futschik, M.; Katou, Y.; Nakao, Y.; Fujimura, T.; Shirahige, K.; Kodama, Y.; Nevoigt, E. Identification of sc-type *ilv6* as a target to reduce diacetyl formation in lager brewers' yeast. *Metab. Eng.* **2011**, *13*, 638–647. [CrossRef] [PubMed]
11. Gamero, A.; Quintilla, R.; Groenewald, M.; Alkema, W.; Boekhout, T.; Hazelwood, L. High-throughput screening of a large collection of non-conventional yeasts reveals their potential for aroma formation in food fermentation. *Food Microbiol.* **2016**, *60*, 147–159. [CrossRef] [PubMed]
12. Varela, C. The impact of non-*saccharomyces* yeasts in the production of alcoholic beverages. *Appl. Microbiol. Biotechnol.* **2016**, *100*, 9861–9874. [CrossRef] [PubMed]
13. Steensels, J.; Verstrepen, K.J. Taming wild yeast: Potential of conventional and nonconventional yeasts in industrial fermentations. *Annu. Rev. Microbiol.* **2014**, *68*, 61–80. [CrossRef] [PubMed]
14. Viana, F.; Gil, J.V.; Genoves, S.; Valles, S.; Manzanares, P. Rational selection of non-*saccharomyces* wine yeasts for mixed starters based on ester formation and enological traits. *Food Microbiol.* **2008**, *25*, 778–785. [CrossRef] [PubMed]
15. Comitini, F.; Gobbi, M.; Domizio, P.; Romani, C.; Lencioni, L.; Mannazzu, I.; Ciani, M. Selected non-*saccharomyces* wine yeasts in controlled multistarter fermentations with *Saccharomyces cerevisiae*. *Food Microbiol.* **2011**, *28*, 873–882. [CrossRef] [PubMed]
16. Sadoudi, M.; Tourdot-Marechal, R.; Rousseaux, S.; Steyer, D.; Gallardo-Chacon, J.J.; Ballester, J.; Vichi, S.; Guerin-Schneider, R.; Caixach, J.; Alexandre, H. Yeast-yeast interactions revealed by aromatic profile analysis of sauvignon blanc wine fermented by single or co-culture of non-*saccharomyces* and *Saccharomyces* yeasts. *Food Microbiol.* **2012**, *32*, 243–253. [CrossRef] [PubMed]
17. Canonico, L.; Comitini, F.; Ciani, M. Torulaspora delbrueckii contribution in mixed brewing fermentations with different saccharomyces cerevisiae strains. *Int. J. Food Microbiol.* **2017**, *259*, 7–13. [CrossRef] [PubMed]

18. Ravasio, D.; Wendland, J.; Walther, A. Major contribution of the Ehrlich pathway for 2-phenylethanol/rose flavor production in *Ashbya gossypii*. *FEMS Yeast Res.* **2014**, *14*, 833–844. [CrossRef] [PubMed]

19. Hazelwood, L.A.; Daran, J.M.; van Maris, A.J.; Pronk, J.T.; Dickinson, J.R. The Ehrlich pathway for fusel alcohol production: A century of research on *Saccharomyces cerevisiae* metabolism. *Appl. Environ. Microbiol.* **2008**, *74*, 2259–2266. [CrossRef] [PubMed]

20. Alexandre, H. Flor yeasts of *Saccharomyces cerevisiae*—Their ecology, genetics and metabolism. *Int. J. Food Microbiol.* **2013**, *167*, 269–275. [CrossRef] [PubMed]

21. Viana, F.; Belloch, C.; Valles, S.; Manzanares, P. Monitoring a mixed starter of *Hanseniaspora vineae-Saccharomyces cerevisiae* in natural must: Impact on 2-phenylethyl acetate production. *Int. J. Food Microbiol.* **2011**, *151*, 235–240. [CrossRef] [PubMed]

22. Daniel, H.M.; Moons, M.C.; Huret, S.; Vrancken, G.; De Vuyst, L. *Wickerhamomyces anomalus* in the sourdough microbial ecosystem. *Antonie Van Leeuwenhoek* **2011**, *99*, 63–73. [CrossRef] [PubMed]

23. Coda, R.; Di Cagno, R.; Rizzello, C.G.; Nionelli, L.; Edema, M.O.; Gobbetti, M. Utilization of African grains for sourdough bread making. *J. Food Sci.* **2011**, *76*, M329–M335. [CrossRef] [PubMed]

24. Tay, S.T.; Lim, S.L.; Tan, H.W. Growth inhibition of *Candida* species by *Wickerhamomyces anomalus* mycocin and a lactone compound of *Aureobasidium pullulans*. *BMC Complement. Altern. Med.* **2014**, *14*, 439. [CrossRef] [PubMed]

25. Zhou, N.; Schifferdecker, A.J.; Gamero, A.; Compagno, C.; Boekhout, T.; Piskur, J.; Knecht, W. *Kazachstania gamospora* and *Wickerhamomyces subpelliculosus*: Two alternative baker's yeasts in the modern bakery. *Int. J. Food Microbiol.* **2017**, *250*, 45–58. [CrossRef] [PubMed]

26. Crafack, M.; Mikkelsen, M.B.; Saerens, S.; Knudsen, M.; Blennow, A.; Lowor, S.; Takrama, J.; Swiegers, J.H.; Petersen, G.B.; Heimdal, H.; et al. Influencing cocoa flavour using *Pichia kluyveri* and *Kluyveromyces marxianus* in a defined mixed starter culture for cocoa fermentation. *Int. J. Food Microbiol.* **2013**, *167*, 103–116. [CrossRef] [PubMed]

27. Morrissey, J.P.; Etschmann, M.M.; Schrader, J.; de Billerbeck, G.M. Cell factory applications of the yeast *Kluyveromyces marxianus* for the biotechnological production of natural flavour and fragrance molecules. *Yeast* **2015**, *32*, 3–16. [PubMed]

28. Urit, T.; Li, M.; Bley, T.; Loser, C. Growth of *Kluyveromyces marxianus* and formation of ethyl acetate depending on temperature. *Appl. Microbiol. Biotechnol.* **2013**, *97*, 10359–10371. [CrossRef] [PubMed]

29. Lobs, A.K.; Lin, J.L.; Cook, M.; Wheeldon, I. High throughput, colorimetric screening of microbial ester biosynthesis reveals high ethyl acetate production from *Kluyveromyces marxianus* on c5, c6, and c12 carbon sources. *Biotechnol. J.* **2016**, *11*, 1274–1281. [CrossRef] [PubMed]
30. Masneuf-Pomarede, I.; Bely, M.; Marullo, P.; Albertin, W. The genetics of non-conventional wine yeasts: Current knowledge and future challenges. *Front. Microbiol.* **2015**, *6*, 1563. [CrossRef] [PubMed]
31. Azeem, M.; Terenius, O.; Rajarao, G.K.; Nagahama, K.; Nordenhem, H.; Nordlander, G.; Borg-Karlson, A.K. Chemodiversity and biodiversity of fungi associated with the pine weevil *Hylobius abietis*. *Fungal Biol.* **2015**, *119*, 738–746. [CrossRef] [PubMed]
32. Axelsson, K.; Konstanzer, V.; Rajarao, G.K.; Terenius, O.; Seriot, L.; Nordenhem, H.; Nordlander, G.; Borg-Karlson, A.K. Antifeedants produced by bacteria associated with the gut of the pine weevil *Hylobius abietis*. *Microb. Ecol.* **2017**, *74*, 177–184. [CrossRef] [PubMed]
33. James, S.A.; Barriga, E.J.; Barahona, P.P.; Harrington, T.C.; Lee, C.F.; Bond, C.J.; Roberts, I.N. Wickerhamomyces arborarius f.A.; sp. Nov., an ascomycetous yeast species found in arboreal habitats on three different continents. *Int. J. Syst. Evol. Microbiol.* **2014**, *64*, 1057–1061. [CrossRef] [PubMed]

CHAPTER-10

EXPLORING THE POTENTIAL OF A NEWLY-DISCOVERED WINE YEAST STRAIN FOR ENHANCED PROLINE UTILIZATION IN FERMENTATION

INTRODUCTION: AN OVERVIEW

The quest for advancing winemaking processes and enhancing the quality of the final product has spurred a focused exploration into the potential of novel wine yeast strains. Among these, a particularly intriguing avenue involves yeast strains with improved capabilities for utilizing proline during fermentation. Proline, a prominent amino acid in grapes, plays a crucial role in the physiology of wine yeast, influencing fermentation kinetics and endowing the final wine with distinctive characteristics. This study delves into the utilization of a novel wine yeast strain specifically engineered or selected for its enhanced proline utilization efficiency. By scrutinizing the impact of this yeast variant on fermentation dynamics and the resulting wine profile, we aim to unravel a promising dimension in the pursuit of refining winemaking processes and elevating the sensory attributes of the end product.

Amino acids, ammonium, peptides and proteins represent the main forms of nitrogen present in grape must. The relative abundance of each class of nitrogen-containing compound has been found to vary considerably according to grape variety, vintage, region, the extent of berry development/maturation or even the clarification time [1–4]. The nitrogen-containing compounds utilised by yeast during wine fermentation are referred to as yeast assimilable nitrogen (YAN), with amino acids and ammonium contributing between 60% and 90% of YAN [2]. When available, yeast will typically utilise up to 400 mg N/L, i.e., their "maximum requirement" [5]. At lower YAN levels, which become exhausted some time during fermentation, the risk of excessive hydrogen sulfide production increases [6] and the fermentation may slow, whereas below the minimum requirement of 120 to 140 mg N/L, fermentation does not finish [2,5,7–10].

The incidence of stuck or sluggish fermentations due to nitrogen deficiencies is of great concern to winemakers around the world. To avoid such problems, winemakers often employ supplementation strategies, e.g., the addition of diammonium phosphate prior to or during fermentation or the use of YAN-rich rehydration additives. However, this practice can also influence yeast metabolism and therefore impact wine quality; for example due to the accumulation of ethyl carbamate, a known carcinogen [11,12] or changes to wine aroma profile [13].

In some grapes, proline is typically the most abundant amino acid and represents a potentially significant source of nitrogen. Proline accumulation in the berry begins just after version [14], with concentrations up to 3800 mg/L reported for grape juice from different varieties [15]. The winemaking yeast *Saccharomyces cerevisiae* has been shown to be capable of metabolizing proline when present as the sole nitrogen source [16]. However, proline is not considered to be an assimilable nitrogen source under typical winemaking conditions. In the presence of preferred nitrogen sources, the Put4p proline-specific transporter is repressed and inactivated and so proline uptake is largely prevented [17]; however, once these nitrogen sources have been exhausted, its uptake is possible, but oxygen is no longer present (i.e., fermentations have become anaerobic), so catalysis by proline oxidase cannot occur [16,18,19]. As well as being a nutrient, accumulated proline can also act as an osmo- and cryo-protectant, with an improved ethanol stress-protective effect [19,20]. Greater utilisation of proline by winemaking yeast could alleviate the need for nitrogen supplementation, enhance microbial stability by removing this potential nutrient for spoilage organisms and thus, help reduce production costs and secure wine quality. Previous attempts to enhance proline utilisation include mutagenesis of yeast or the addition of oxygen to fermentations [21–25].

This study describes the use of ethyl methane sulfonate (EMS) random mutagenesis to develop novel wine yeast with an improved capacity to utilize proline. Such a strain was sought as a means of overcoming nutrient-related fermentation problems, thereby improving fermentation reliability and overall wine quality.

MATERIALS AND METHODS

Yeast Strains and EMS Mutagenesis

A common commercial *Saccharomyces cerevisiae* strain (QA23) was obtained from Lallemand (Blagnac, France) as an agar slope stock culture and subjected to EMS mutagenesis [26]. Briefly, QA23 was cultured overnight in 100 mL of yeast extract peptone dextrose (YEPD) nutrient medium (containing 20 g/L D-glucose, 10 g/L yeast extract and 20 g/L Bacto peptone) at 28°C with shaking at 120 rpm. Aliquots of 50 mL were washed (twice) with 0.1 M sodium phosphate buffer (pH 7) and cells re-suspended in 15 mL of the same for microscopic determination of cell number. A sub-sample (1 mL) was retained as a blank and the remaining sample (14 mL) exposed to 45 µL/mL of EMS (Sigma Aldrich, St. Louis, MO, USA). Aliquots (1 mL) were collected every 10 min (for 2 h) and plated on YEPD plates (after appropriate serial dilution). The reminder of the sample from each time point (~2 mL) was kept in glycerol and stored at -80°C. Colony counts were performed to determine viability at each time point; with approximately 42% survival observed after 40 min of exposure to EMS treatment. Sporulation of strains, when required, was performed using standard protocols [27].

Selection of Isolates

EMS-treated samples (collected att= 40 min) were spread onto YEPD medium and cultured overnight at 28°C yielding approximately 10,000 colonies. The colonies were transferred to a yeast nitrogen base (YNB) agar selection medium containing 4.8 g/L methylamine (Sigma Aldrich, St. Louis, MO, USA) and 2.5 g/L proline [28] with a sterilised velvet pad, and cultured for 24 h according to methodology described previously [29]. The ammonium analogue, methylamine, elicits Nitrogen Catabolite Repression [30] but cannot serve as a nitrogen source, thus only colonies derepressed for proline utilisation will grow on the selective medium. Ninety three such colonies were observed after 2 days, checked by re-streaking individually onto YNB selection media and incubated as before. This resulted in the isolation of 8 putative mutant strains that exhibited growth. Proline utilisation by the isolates during fermentation was tested in CDGJM [2], containing 50 g/L of sugar and 500 mg/L proline. Proline consumption was determined according to the isatin method [31] and isolates found to consume significantly more proline than the wild type strain were chosen for further study.

Influence of N Source on Growth of QA23 and EMS Isolate Q7

The growth of wild type strain QA23 and EMS isolate Q7, in the presence of ammonium ions or proline was investigated. YNB medium plates (as above) with either 2.5 g/L ammonium or proline added, each with or without 4.8 g/L methylamine as a non-metabolisable repressor were used to investigate strain growth. Starter cultures of QA23 and Q7 were prepared from single colonies of each cultured in 25 mL of YEPD broth at 28°C until the desired cell number (2×10^7 cells/mL) was achieved. A dilution series (5× ten-fold dilutions) of the cultures was spotted onto each plate (10 µL), and incubated at 28°C. Plates were photographed after 3 days.

Yeast Performance during Fermentation

Small-scale fermentations (100 mL) were conducted in both CDGJM and Chenin Blanc grape juice to investigate yeast performance. The CDGJM contained 450 mg N/L, comprising 50 mg N/L of YAN (nitrogen components of CDGJM) and 3.3 g/L of proline (400 mg N/L). Chenin Blanc juice was prepared from grapes harvested in 2012 (Adelaide Hills, Adelaide, Australia). The juice initially contained 105 g/L of sugar, 40 mg/L of proline and 102 mg/L YAN, before the juice was adjusted to 200 g/L of sugar and an added 1.5 g/L proline. The media were sterilised (0.22 µm filtration) before use. Starter cultures were prepared in the appropriate medium diluted 50% with sterile water and incubated overnight at 28°C with shaking (120 rpm).

Triplicate fermentations were carried out under both aerobic and self-anaerobic conditions, at 28°C with shaking (120 rpm). Samples were collected at regular intervals to determine fermentation progress and proline utilisation. Culture density was monitored at 600 nm with a TECAN Infinite 200® PRO microplate reader (Männedorf, Switzerland). Sugar consumption was monitored with an ATAGO hand-held refractometer (Tokyo, Japan) until approx. 6° Brix before enzymatic determination [32]. Proline content was determined as above.

Extraction of Intracellular Proline from Yeast Cells

In order to investigate whether proline was metabolised or just accumulated by yeast cells, proline was extracted from the cytosolic and vacuolar compartments of cells after (i) 24 h in starter culture (i.e., prior to fermentation), (ii) 24 h of fermentation in CDGJM, (iii) mid-fermentation (i.e., sugar ≈ 100 g/L) and (iv) post-fermentation (i.e., sugar content < 2.5 g/L). Cells (approx. 3×10^8) were harvested at each time point, twice washed with sterile de-ionised water, then re-suspended in

1.5 mL of 2.5 mM potassium phosphate buffer (pH 6.0) containing 0.6 M sorbitol, 10 mM glucose and 0.2 mM $CuCl_2$, and incubated for 15 min at 30°C. Aliquots (1 mL) of each cell suspension were washed twice (12,000 rpm, 3 min) with 0.5 mL of the buffer solution (as above but without glucose). The combined 2 mL supernatants gave the cytosolic extract while the remaining cells were suspended in 3 mL of de-ionised water, boiled for 15 min, the resulting supernatant afforded the vacuolar extract after 5 min centrifugation (5000 rpm) [33, 34].

Culture samples were collected following 24 h growth in each of YEPD, starter and CDGJM (as described above), and cell numbers quantified before determination of dry cell weight. In each case, 3×10^8 cells were collected on a pre-weighed Whatman GF/C glass fibre filter, washed with an equal volume of sterile de-ionised water, dried and reweighed.

Freeze Tolerance Tests

To further confirm whether proline was accumulated by yeast cells or not, freeze tolerance tests were performed. After 24 h of fermentation, the viability of the different strains was similar (close to 100%), thus the potential benefit of strains with altered intracellular proline content was estimated via a freeze-thaw test. Approximately 1×10^8 cells were diluted in 1 mL of CDGJM. Approximately 1×10^7 cells (0.1 mL) were transferred to a 1.7 mL microfuge tube and stored at -20°C or -80°C. After 3 days, samples were thawed at room temperature, diluted with 0.9 mL 0.9% (*w*/*v*) NaCl solution, and dilutions spread onto YEPD agar plates (28 °C, 24 h) to determine residual viability [19].

Real-Time Quantitative PCR

The primers used for real-time quantitative PCR are listed in Table 1. Primers for *PUT3* and *URE2* were designed by Primer3 software (SourceForge, La Jolla, CA, USA. http://bioinfo.ut.ee/primer3-0.4.0/) and Primer-BLAST (NCBI, Bethesda, MD, USA. http://ncbi.nlm.nih.gov/tools/primerblast). Netprimer was used for analysis of secondary structures (hairpins and self-dimers) and cross-dimers (PREMIER Biosoft, Palo Alto, CA, USA. http://www.premierbiosoft.com/netprimer/). Primers were designed for annealing at around 60°C.

Table 1. Primers used in this study.

Name	Primers (5'–3')	Reference
PUT1-F	GATAAAACGGGCACTGACGA	
PUT1-R	TAACGAGCATTTGGGATTGG	[35]
PUT2-F	CCGATATGTTTGGCATATTGCA	
PUT2-R	TGGACTTGCGGATGTGTTCA	[35]
PUT3-F	GCGGTATTGAAGTCCTGTTGT	this study
PUT3-R	TGGATAGAGTGTCGCTTTGAGA	
PUT4-F	GAGCCGCACAAACTAAAACA	
PUT4-R	CGTATGAAGCGTGGATGAAG	[35]
GAP1-F	CTGTGGATGCTGCTGCTTCA	
GAP1-R	CAACACTTGGCAAACCCTTGA	[36]
URE2-F	TCCCGTATGGCTTGTAGGAGA	this study
URE2-R	CACGCAATGCCTTGATGACC	
ALG9-F	CACGGATAGTGGCTTTGGTGAACAATTAC	
ALG9-R	TATGATTATCTGGCAGCAGGAAAGAACTTGGG	[37]
TAF10-F	ATATTCCAGGATCAGGTCTTCCGTAGC	
TAF10-R	GTAGTCTTCTCATTCTGTTGATGTTGTTGTTG	[37]

RNA Extraction

Cell samples (2 mL) were collected after 24 h growth in starter media containing either: (i) 450 mg/L of YAN, without proline or (ii) 450 mg/L of nitrogen comprised of 50 mg/L of YAN (nitrogen components of CDGJM) and 400 mg/L of nitrogen (as proline, 3.3 g/L). RNA was extracted according to Chomczynski and Sacchi (2006) [38]. Yeast cultures (50 mL) were grown in YEPD at 30 °C until they reached 1×10^7 cells/mL. Samples were harvested (14,000 rpm, 5 min), washed in sterile de-ionised water and then transferred into a 2 mL screw cap microfuge tube. Cells were again pelleted and resuspended in 500 µL Trizol (Invitrogen, Carlsbad, NM, USA) and kept in -80 °C before use. Glass beads were used for cell breakage, after incubation at 65°C for 3 min, 100 µL of chloroform was added per 500 µL Trizol, followed by vigorous shaking for 15 s, then incubation at room temperature for 5 min. After centrifugation (14,000 rpm, 10 min, 4°C), the top clear phase was removed to a new tube and supplemented with 250 µL of isopropanol per 500 µL Trizol. After 10 min incubation at room temperature, tubes were centrifuged as before, the supernatant discarded and the pellet washed in 500 µL of 75% cold ethanol, vortexed and re-centrifuged. Purified RNA extracts were dissolved in 100 µL PCR water (Bioline, Melbourne, Australia) and RNA concentrations determined via 260/280 nm absorbance (TECAN Infinite 200® PRO spectrophotometer; Männedorf, Switzerland).

Reverse Transcription and Real-Time PCR

Real-time PCR was used to analyse gene expression involved in proline metabolism. Turbo DNase (Invitrogen) was used for DNA-free treatment of all the RNA extracts. Thereafter, cDNA was synthesized with iScript cDNA Synthesis Kit (Bio-Rad, Gladesville, New South Wales, Australia). Real-time PCR was performed with the SsoFast EvaGreen Supermix (Bio-Rad) in a 96-well plate on a Bio-Rad CFX96instrument. cDNA (1 µL, 200 ng) was added to a 10 µL PCR reaction volume including 0.5 µL of each primer (5 pmol/µL), 5 µL of EvaGreen Supermix and 3 µL of RNase-free water. The PCR conditions were: initial enzyme activation at 95°C for 30 s, followed by 39 cycles of denaturation at 95°C for 5 s and annealing/extension at 60°C for 5 s. A melting curve was prepared at the end of each run from 65 to 95°C. All samples were amplified in duplicate. Efficiency of amplifications was determined by running a standard curve of each run with serial dilutions of cDNA. The means of Ct values (the number of PCR cycles required for the fluorescent signal to cross the threshold detection) obtained were used for gene expression analysis conducted using qbase+ software (Biogazelle, Zwijnaarde, Belgium) and GraphPad Prism, version 6.6 (GraphPad Software, La Jolla, CA, USA).

Gene Sequencing

Sequencing of genes of interest, i.e., those related to proline metabolism, was achieved through a contract service (Australian Genome Research Facility, Waite Campus, Adelaide, Australia).

HPLC Analysis of Fermentation Metabolites

Fermentation metabolites including organic acids (malic, citric, lactic, acetic and succinic), glycerol and ethanol were determined with an Agilent Series 1100 HPLC, equipped with a vacuum degasser, a quaternary pump, a thermos tatted column oven, and a refractive index detector (RID) and diode array/multiple wavelength detector (DAD). Separation was achieved using an HPX-87H # 125-0140 column (300 mm × 7.8 mm; Aminex, Bio-Rad, Gladesville, New South Wales, Australia) at an operating temperature of 60°C. The injection volume was 20 μL and the mobile phase was 2.5 mM aqueous H_2SO_4 at a flow rate of 0.5 mL/min run in isocratic mode. Quantitation was performed using a standard curve prepared with standards of known concentration. Data were analyzed with Agilent ChemStation software (Version # 3.0.1 B, Agilent Technologies, Palo Alto, CA, USA).

RESULTS AND DISCUSSION

Comparison of Growth of QA23 and its EMS Isolate Q7

From the EMS-treated aliquots of a population of QA23, several isolates were identified that were able to grow on proline (as sole nitrogen source) in the presence of methylamine. The growth of QA23 and EMS isolate Q7 in the presence of ammonium and proline (with or without methylamine) was further compared via a dilution series on YNB containing ammonium or proline as sole nitrogen sources with or without methylamine (Figure 1). The influence of pH was also examined. When ammonium was the only nitrogen source and in the presence of methylamine, there was no difference between the growth of the wild type strain and Q7 (Figure 1A). In the case of media containing proline as the sole nitrogen source at pH 3.5, all strains grew equally well (Figure 1B). With inclusion of methylamine in the medium, growth by QA23 was reduced by two orders of magnitude. Methylamine appeared to be inducing NCR and thereby preventing utilisation of the only nitrogen source, proline, hence reducing the growth of the wild type strain, QA23 [39,40]. Instead Q7 remained unaffected with a high level of growth, implying it was depressed for proline transport and utilisation, confirming it warranted further investigation. Similar results were seen in proline-only or proline with methylamine media at pH 4.8. However, at pH 6.0, growth of QA23 was markedly reduced even in the absence of methylamine. Strain Q7 maintained high levels of growth on proline-only media at pH 6.0, suggesting that QA23 exhibited an interesting property i.e., an inability to grow on proline as sole nitrogen source at high pH values (not relevant to wine making), or possibly due to the closeness to the PI of proline (6.48). At the same pH value, the added presence of methylamine greatly restricted growth of Q7. To the best of our knowledge, methylamine shares a common transport system with ammonia in *S. cerevisiae*, and methylamine uptake is pH dependent, with approximately 25%, 60% and 100% of maximum activity being retained at pH 3.5, 4.8 and 6.0, respectively [30]. This report suggests that methylamine is more potent at high pH values, because its uptake by the cell becomes more effective. On the other hand, proline accumulation also can reduce cellular growth rates [41], which might explain the reduced cell growth observed for QA23 (Figure 1).

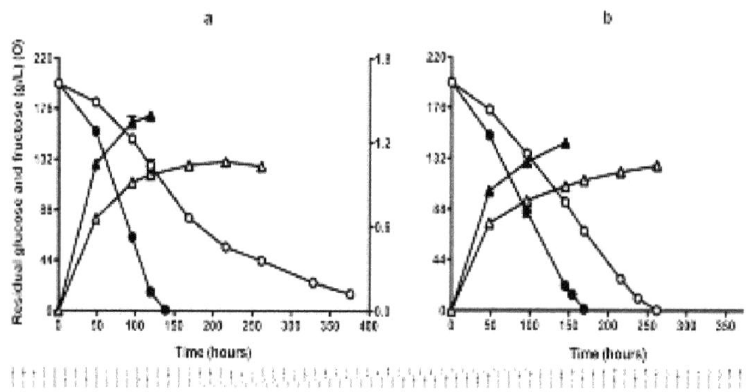

Figure 1. Growth of QA23 and EMS isolate Q7 on: (A) YNB with ammonium (±methylamine); YNB with proline (±methylamine) at pH 3.5, 4.8 and pH 6.0. Five × ten-fold serial dilutions of yeast cell suspensions (initially at OD 0.08) were applied to each plate. Plates were photographed after 3 days incubation at 28°C.

Fermentation Performance of QA23 and EMS Isolate Q7 in CDGJM.

Isolate Q7, apparently derepressed for proline utilisation, was examined in fermentations carried out in CDGJM. The trials were carried out with air-locks or loosely fitted foil to close the flasks to achieve self-anaerobic or aerobic fermentations, respectively. In the presence of 50 mg/L YAN, isolate Q7 consumed sugar at a higher rate than QA23, and completed fermentation sooner under either aerobic or self-anaerobic conditions (Figure 2). Aerobic fermentations completed about 30 h earlier than the self-anaerobic fermentations (170 h). The wild type strain QA23, however, exhibited sluggish sugar consumption under aerobic conditions, incomplete even after 370 h (Figure 2a). The reason for this is not known. It may be related to increased concentrations of some compounds, which negatively affect the fermentation [42].

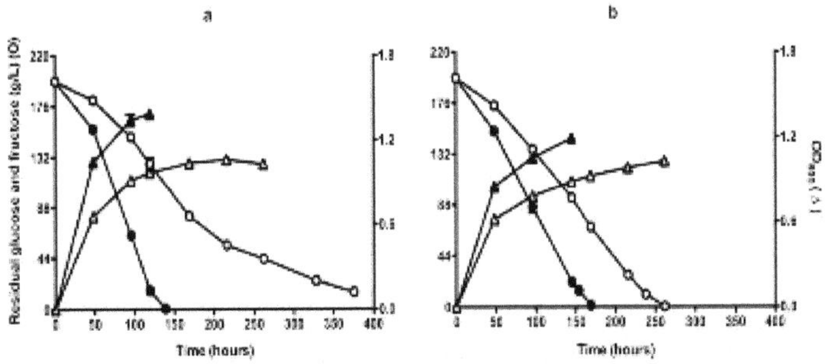

Figure 2. Sugar consumption (circles) and growth (triangles) of QA23 (open symbols) and Q7 (closed symbols) during fermentation conducted in CDGJM initially containing glucose + fructose at 200 g/L, 450 mg N/L (50 mg/L mixed amino acids and ammonia + 3.3 g/L proline). Fermentations were carried out in (a) aerobic (i.e., vessels were covered with loosely-fitted foil) or (b) self-anaerobic (i.e., vessels were fitted with air locks) conditions. Culture optical density (OD600) was measured at 600 nm. Results are the mean of triplicate cultures (±standard error).

Strain Q7 achieved about twice the culture density (estimated by OD600) of QA23 (Figure 2). These findings are supported by the observed trends in proline utilisation, namely that QA23 utilised 300 mg/L of proline, regardless of oxygen availability, whereas, Q7, which achieved higher culture densities, assimilated significantly more proline: approximately 370 mg/L (self-anaerobic) or 730 mg/L (aerobic) (Figure 3). The latter results suggest that dissolved oxygen improved fermentation rates, as well as proline utilisation, which is consistent with previous studies [21,22,43]. At this stage, it is not possible to state whether the increase in growth in the presence of oxygen, is due to enhanced sterol and unsaturated fatty acid biosynthesis, and thereby improved cell condition [22], and/or enhanced proline catabolism by proline oxidase [21]. Further work will be required to differentiate between these two mechanisms.

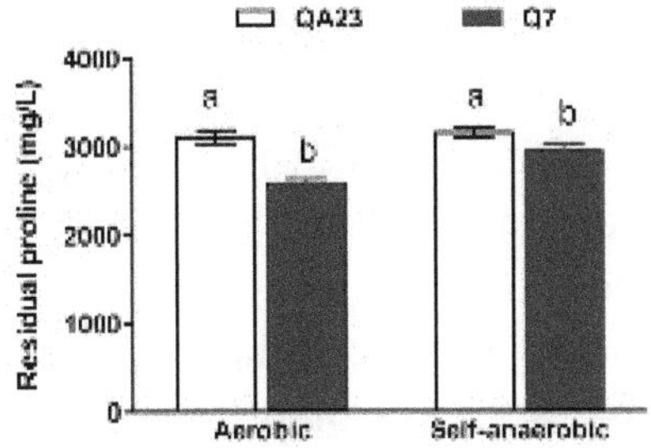

Figure 3. Proline consumption by QA23 and Q7 following fermentation in CDGJM as described in Figure 2, under aerobic and self-anaerobic conditions. The medium initially contained 3.3 g/L proline. Results are mean of three culture replicates (±standard error). Histograms identified with different letters were significantly different within a given condition.

Cell Morphology, Dry Weight and Intracellular Proline Content

Given the differences seen between QA23 and the Q7 in terms of fermentation kinetics, proline uptake and growth, we sought to determine what other features might be changed in the isolate, such as cell morphology. Prior to fermentation, yeast cells were inoculated into YEPD, followed by growth in starter CDGJM and then CDGJM (aerobic and self-anaerobic). After 24 h in YEPD, the vacuoles appeared large and well-defined in each strain. In subsequent growth conditions, vacuoles appeared more granulated. Cells of Q7 appeared to be smaller, particularly in starter CDGJM and CDGJM. This notion is supported by the fact that despite all strains producing similar dry cell weight in YEPD (approximately 5 mg/3 × 10^8 cells; Figure 4), and QA23 increasing by ~2-fold in either CDGJM, Q7 was about 40% lighter (smaller) per same number of cells.

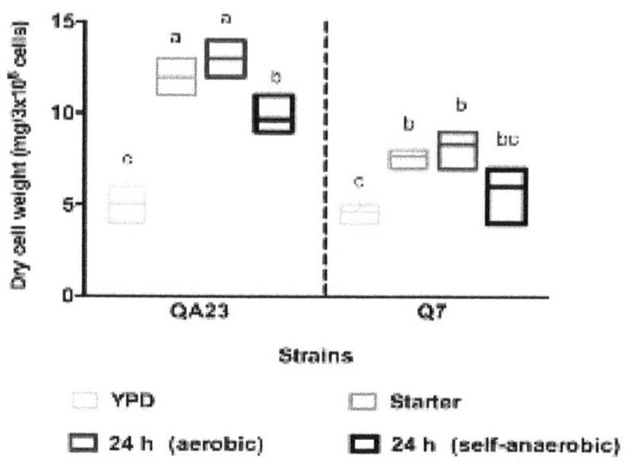

Figure 4. Comparison of dry cell weights 24 h after inoculation of QA23 and EMS isolate Q7 in yeast extract peptone dextrose (YEPD), starter, and fermentation media (as described in Figure 2). Results are the mean of three replicates. Values identified with different letters were significantly different.

To investigate whether or not proline was accumulated in the cells, culture samples were harvested at different stages of fermentation for extraction of intracellular proline. The growth medium used was starter CDGJM (100 g/L sugars) followed by experimental cultures in CDGJM (200 g/L sugars) each containing 3.3 g/L proline. Samples were taken after 24 h and at the mid and end stages of fermentation. As a reference, cultures grown in starter CDGJM without proline produced cells with no detectable intracellular proline. By comparison, QA23 cells contained readily detectable amounts of proline in the cytosolic and vacuolar fractions for all cultures and time points (Table 2). In the starter culture, most proline was localised in the vacuole rather than the cytosol (17 vs. 132 ng/10^8 cells). In subsequent QA23 cultures, highest proline contents were found at 24 h or the mid phase of fermentation, before only trace levels remained at the end of fermentation. Elevated levels of proline persisted to a greater extent in the vacuolar compared to the cytosolic fractions for QA23.

The tendency for QA23 to possess higher levels of intracellular proline compared to Q7 (Table 2), is in direct contrast to the relative amount of proline removed from the medium (Figure 3). There may be several reasons for this paradox. QA23 may take up but not metabolise proline thereby retaining higher intracellular amounts, whereas Q7 takes up higher amounts of proline, which is also catabolised or incorporated to a greater extent thereby depleting intracellular pools. The higher culture densities for Q7 (Figure 2) combined with the suggested smaller cell size (Figure 4) indicate markedly higher cell numbers for the isolates. As a consequence, the lower intracellular proline content of the isolates (expressed on a per cell basis) may be indicative of a greater dilution of accumulated proline amongst a greatly increased number of cells. Under aerobic condition, yeast cells of QA23 might respond to oxidative stress and accumulate more proline in the vacuole. Proline accumulation in the vacuole has been reported to provide a stress protective effect [20,34]. But the excess accumulation of proline in the cytosol might also be toxic to yeast cell [44], and might contribute to the slow growth of the wild type strain and slowing of the fermentation under aerobic conditions. Previous research has linked morphology and proline uptake indicating that proline accumulation can produce cytological differences and increased cell size [41], however the opposite was found for cell size in the present experiments.

Table 2. Intracellular proline content (ng/10^8 cells) of cells from QA23 and Q7 sampled at different stages of fermentation in CDGJM.

Aerobic Self-Anaerobic		Cytosolic				Vacuolar				Cytosolic	Vacuolar	
Vacuolar	24 h	mid	End	24 h	mid	End	24 h	mid	End	24 h	mid	End
132 ± 1 a	116 ± 1 a	10 ± 1 a	1 ± 0 a	136 ± 5 a	94 ± 5 a	16 ± 1 a	122 ± 4 a	13 ± 0 a	5 ± 1 a	47 ± 2 a	64 ± 2 a	4 ± 0
nd b	81 ± 3 b	nd b	nd b	nd b	1 ± 0 b	nd b	90 ± 3 b	nd b	nd b	nd b	nd b	4 ± 0

Cultures were grown in YEPD overnight before inoculation into starter CDGJM (100 g/L glucose and fructose, 3.3 g/L proline). Starter cultures were then inoculated into CDGJM (200 g/L glucose and fructose, 3.3 g/L proline) in aerobic or self-anaerobic conditions. Values followed by a different letter within columns are significantly different; 24 h: Proline was extract from cells after 24 h of fermentation; Mid: Proline was extracted from cells in the middle of fermentation. Sugar content was 100 ± 10 g/L; End: Proline was extracted from cells at the end of fermentation. Sugar content was less than 2 g/L; Values are the mean of three fermentation replicates (n = 3) ± standard error; nd = not detected.

Freeze Test

In this study, the highest intracellular proline levels were observed after 24 h of fermentation. At this time point, cell viability was also high, close to 100%, thus freeze tests were applied to gain an insight into the properties of the isolates. Cell tolerance to freezing for 3 days followed by thawing at room temperature and plating onto YEPD was used to determine residual viability (Table 3). The highest cell viability was observed for QA23 for all fermentations and under both freezing conditions. When the storage temperature was -80°C, cell viability tended to be higher. This might be because at -80°C, freezing occurs quicker than at -20°C, likely related to the rates of ice formation and changes in solute concentration [45]. Previous studies have found that proline has cryo-protective effects in *S. cerevisiae* and is equivalent to glycerol or trehalose [46]. The significantly higher cell viability of QA23 (Table 3) may be attributable to the higher intracellular proline levels of this strain (Table 2).

Table 3. Cell viability (%) after freezing for 3 days at -20°C or -80°C.

Strain	-20 °C		-80 °C	
	Aerobic	Self-Anaerobic	Aerobic	Self-Anaerobic
QA23	63 ± 6 a	57 ± 5 a	66 ± 6 a	63 ± 5 a
Q7	20 ± 2 b	29 ± 5 b	29 ± 2 b	37 ± 5 b

Values are the means of three fermentation replicates (n = 3) ± standard error. Values followed by a different letter within columns are significantly different.

Analysis of Gene Expression

An investigation into the expression of the key genes involved in proline transportation and metabolism was performed to help explain the properties of the isolates. Cells were cultured in high nitrogen media without proline and in low nitrogen media containing high levels of proline. Proline is imported into yeast cells mainly by the proline specific permease Put4p and the general amino acid permease Gap1p [40]. Proline degradation occurs in the mitochondria and is assisted by two enzymes, proline oxidase and Δ1-pyrroline-5-carboxylate dehydrogenase, encoded by *PUT1* and *PUT2*, respectively [16]. The *PUT3* protein binds to the upstream activation sequences in the promoter of *PUT1* and *PUT2*, and induces their expression, but transcription is only activated in the presence of proline [47,48]. When preferred nitrogen sources are present, the expression of PUT genes is inhibited by the *URE2* gene product [49]. These genes were targeted in a gene expression analysis.

When proline was present in the medium, *PUT1* and *PUT2* were not induced in QA23; but expression decreased to approximately one quarter and one third, respectively. Instead up-regulation was observed for *PUT1* and *PUT2* in Q7 (Figure 5). *PUT1* expression paralleled that of *PUT2*, for the two strains. *PUT3* was up-regulated in the presence of proline, consistent with previous studies, which have shown transcription is only induced in the presence of proline [47,48]. It is noteworthy that the gene that encodes the high-affinity proline transporter PUT4, was up-regulated 2-fold in Q7, but significantly down-regulated in QA23 when proline was present in the medium. The opposite trend was observed in the expression of *GAP1*. Thus when the media contained low YAN (+proline), *PUT4* was repressed in the wild type strain, but not in Q7. In contrast, *GAP1* was repressed in Q7, but not in QA23. This may indicated that proline was mainly transported into the cell via the low affinity Gap1p in QA23, but via Put4p in Q7. The reduced expression of *PUT1*, *PUT2* and *PUT4* in QA23 may be a reflection of the higher expression of *URE2* regulatory gene in the presence of proline in this strain.

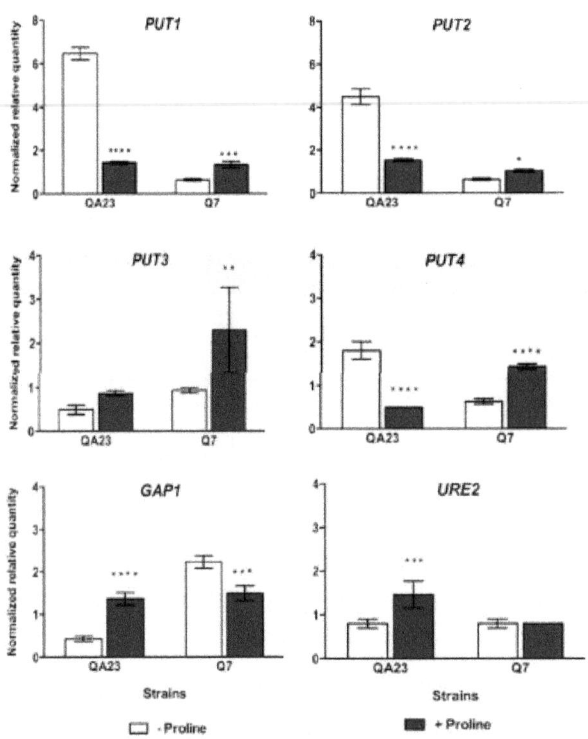

Figure 5. Normalized relative expression of key genes involved in proline transportation and metabolism in QA23 and Q7, cultivated overnight in media containing: 450 mg/L of nitrogen with mixed amino acids and ammonium, but without proline (-proline); or 50 mg/L mixed amino acids and ammonium, supplemented with 3.3 g/L proline (+proline). Initial sugar content of each medium was 100 g/L of glucose: fructose 1:1. The mRNA level of each gene was normalized against TAF10 and ALG9 levels in the same sample. Values are mean (± standard error) of duplicate determinations from duplicate cultures. Data analysis was performed using qbase+ software. Significant differences in relative expression are indicated by: *, $p < 0.05$; **, $p < 0.01$; ***, $p < 0.001$; ****, $p < 0.0001$.

Gene Sequencing

To identify the difference between the parent strain and the isolate, the key genes involved in proline transportation and metabolism *PUT1*, *PUT2* (proline metabolism), *PUT3* (inducer gene) and *PUT4* (proline transporter gene) were sequenced in these two strains (Figures S1–S4). Nucleotide sequences were aligned using Geneious 7.1.7 software (Biomatters Limited, Auckland, New Zealand). Following alignment the sequences were analysed for single nucleotide polymorphisms (SNPs) (Table 4). The sequence of *PUT1* (proline oxidase) including 81 base pairs of promoter region was examined. In Q7, *PUT1* had a SNP in the promoter at position -55. This may have led to differences in interactions with regulatory gene products, partly addressing the loss of NCR for proline metabolism. Additionally, Q7 showed substitutions at S60N (serine to asparagine) and L167V (leucine to valine). The PUT3 gene showed no differences between QA23 and Q7, whilst the *PUT4* sequence revealed a likely conservative change from a heterozygosity, Thr/Ser, to Thr at amino acid position 70.

Table 4. Summary of mutation positions in the sequence of *PUT1* and *PUT4* in QA23 and Q7.

Gene	Strain	55 (Promoter)	260 (ORF)	Amino Acid	580 (ORF)	Amino Acid
PUT1	QA23	G	G	Ser	C	Leu
	Q7	T	A	Asn	G	Val
PUT4	QA23	45 (Promoter)	465 (ORF)	Amino Acid		
		-	W	Thr/Ser		
	Q7	-	T	Thr		

Nucleotide code: W (A or T); Sequence numbers refer to nucleotide bases in the relevant gene sequences in Supplementary Materials. The position of the feature as being either in the promoter or ORF is indicated.

Protein variation effect analyser (PROVEAN: http://provean.jcvi.org/index.php, accessed on: 22 October, 2014) was applied to analyse the impact of the SNPs on the encoded proteins for PUT1. No deleterious effect was suggested for mutations in the ORF. Thus, the mutation in the promoter region (55) might be a factor influencing the strain's performance during fermentation and gene expression. However, this position is different from the reported UASNTR sequences in the PUT1 promoter region (-293 to -186) [47], thus the lack of repression could be due to modification of PUT gene regulatory regions to inhibit binding by repressor gene products, or alternatively that the repressor genes themselves are modified such that their gene products no longer bind to the regulatory regions of the PUT genes. Thus, whether the alteration could influence regulator effectiveness, e.g., URE2 or DAL80 [50], still needs to be confirmed. Further work is required to determine the genetic basis for the differences between QA23 and Q7.

Fermentation in Chenin Blanc Grape Juice

To test the performance of the isolate in real juice fermentation, Chenin Blanc (2012) containing 1540 mg/L proline and 102 mg/L YAN from the Adelaide Hills was used for the same trials as described in CDGJM. In the aerobic fermentations, QA23 became stuck after 192 h with residual sugar at about 20 g/L (Figure 6a). Isolate Q7, however, completed the fermentation in about 60 h (Figure 6a). Under self-anaerobic conditions, Q7 completed fermentations about 60 h earlier than the wild-type strain QA23 (Figure 6b). The isolate also showed significantly different growth kinetics compared to QA23, by achieving high OD600 values, thereby greater biomass yields. Proline was depleted by Q7 within 48 h in the aerobic fermentation and up to 1000 mg/L was consumed under the self-anaerobic condition at between 120–144 h (Figure 7).

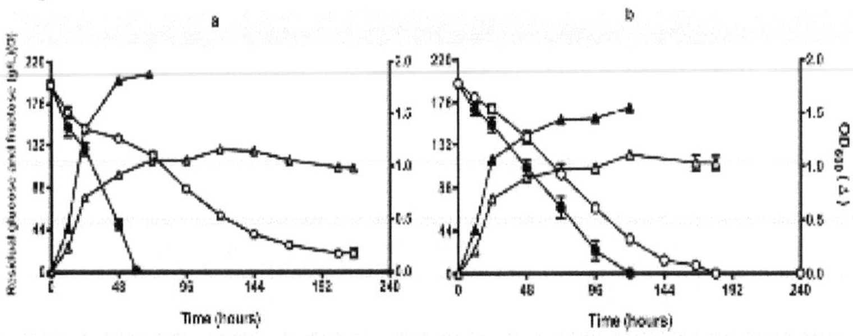

Figure 6. Sugar (circles) and cell growth (triangles) during fermentation of Chenin Blanc grape juice (2012) by QA23 (open symbols) and Q7 (closed symbols). Juice initially contained 105 g/L glucose + fructose, 102 mg N/L and 40 mg/L proline, but was modified (glucose + fructose) to give 200 g/L sugar and1.5 g/L of proline were added. Fermentations were carried out in aerobic (a) (vessels were covered with loosely-fitted foil) or self-anaerobic (b) (vessels were fitted with air locks) conditions. Culture optical density was estimated at 600nm. All results are the mean of triplicate fermentations.

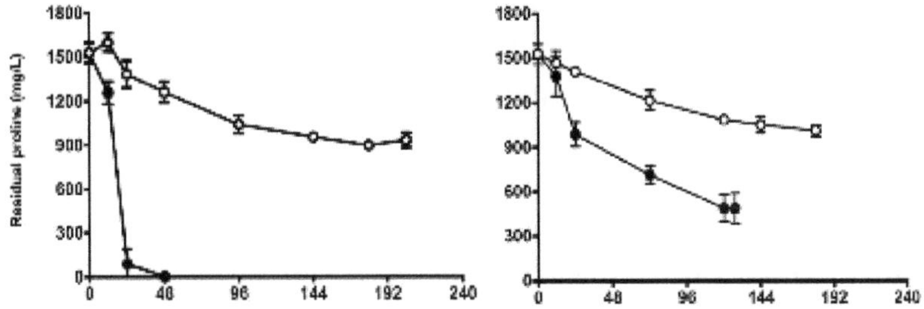

Figure 7. Proline consumption during fermentation in aerobic (a) and self-anaerobic conditions (b). Fermentation conditions as described in Figure 6. The open and closed symbols represent QA23 and Q7, respectively. Results are the mean of triplicates.

Overall Q7 fermented faster, achieved greater growth and used proline more quickly. This is not exactly the case seen in CDGJM (Figures 2 and 3). Such differences are presumably reflective of the different response of Q7 to the compositional differences between CDGJM and juice. Further trials with the isolate in a range of media would provide a greater sense of the most representative performance of the isolate.

The final self-anaerobic fermentation samples were also analyzed by HPLC to determine if there were any major differences in composition. For all strains, no significant difference in the acid, glycerol and ethanol content in the wine were observed (Table 5).

Table 5. Major fermentation metabolites derived in the final samples for Chenin Blanc fermentations by QA23 and Q7.

Metabolites (g/L)	QA23	Q7
Malic acid	3.1 ± 0.08	3.5 ± 0.22
Citric acid	3.5 ± 0.03	2.9 ± 0.36
Lactic acid	0.9 ± 0.05	0.9 ± 0.21
Acetic acid	0.2 ± 0.04	0.2 ± 0.07
Succinic acid	2.7 ± 0.04	3.2 ± 0.22
Glycerol	8.0 ± 0.08	7.7 ± 0.81
Ethanol	92.2 ± 0.41	85.7 ± 6.24

SUMMARY

Proline is the predominant amino acid in grape juice, but it is poorly assimilated by wine yeast under the anaerobic conditions typical of most fermentations. Exploiting the abundance of this naturally occurring nitrogen source to overcome the need for nitrogen supplementation and/or the risk of stuck or sluggish fermentations would be most beneficial. This study describes the isolation and evaluation of a novel wine yeast isolate, Q7, obtained through ethyl methanesulfonate (EMS) mutagenesis. The utilisation of proline by the EMS isolate was markedly higher than by the QA23 wild type strain, with approximately 700 and 300 mg/L more consumed under aerobic and self-anaerobic fermentation conditions, respectively, in the presence of preferred nitrogen sources. Higher intracellular proline contents in the wild type strain implied a lesser rate of proline catabolism or incorporation by this strain, but with higher cell viability after freezing treatment. The expression of key genes (PUT1, PUT2, PUT3, PUT4, GAP1 and URE2) involved in proline degradation, transport and repression were compared between the parent strain and the isolate, revealing key differences. The application of these strains for efficient conduct for nitrogen-limited fermentations is a possibility. The aim of this research was to develop novel wine yeasts capable of exploiting the abundance of proline present in grape juice, so as to overcome the issues of stuck/ sluggish fermentation, thereby reducing production difficulties and quality losses to benefits both the wine industry and consumers. This study reports on a novel yeast strain that utilises proline constitutively (i.e., in the presence of preferred nitrogen sources). The wild type strain also utilised proline as a nitrogen source, but at much lower rates than the EMS isolate. Further evaluations of the isolate in various juices will more fully defined its potential applicability to industry.

REFERENCES

1. Ough, C.S.; Amerine, M.A. *Methods for Analysis of Musts and Wines*; Wiley: New York, NY, USA, 1988; ISBN 0471627577.
2. Henschke, P.A.; Jiranek, V. Yeasts-metabolism of nitrogen compounds. In *Wine Microbiology and Biotechnology*; Harwood Academic Publishers GmbH: Amsterdam, the Netherlands, 1993; pp. 77–164. ISBN 978-0-41-527850-8.
3. Alexandre, H.; Charpentier, C. Biochemical aspects of stuck and sluggish fermentation in grape must.*J. Ind. Microbiol. Biotechnol.* **1998**, *20*, 20–27. [CrossRef]
4. Burin, V.M.; Gomes, T.M.; Caliari, V.; Rosier, J.P.; Bordignon Luiz, M.T. Establishment of influence the nitrogen content in musts and volatile profile of white wines associated to chemometric tools. *Microchem. J.* **2015**, *122*, 20–28. [CrossRef]
5. Jiranek, V.; Langridge, P.; Henschke, P.A. Amino acid and ammonium utilization by *Saccharomyces cerevisiae*Wine yeasts from a chemically defined medium. *Am. J. Enol. Vitic.* **1995**, *46*, 75–83.
6. Jiranek, V.; Langridge, P.; A Henschke, P. Regulation of hydrogen sulfide liberation in wine-producing*Saccharomyces cerevisiae* strains by assimilable nitrogen. *Appl. Environ. Microbiol.* **1995**, *61*, 461–467. [PubMed]
7. Ingledew, W.M.; Kunkee, R.E. Factors influencing sluggish fermentations of grape juice. *Am. J. Enol. Vitic.***1985**, *36*, 65–76.
8. Monteiro, F.F.; Bisson, L.F. Biological assay of nitrogen content of grape juice and prediction of sluggish fermentations. *Am. J. Enol. Vitic.* **1991**, *42*, 47–57.
9. Bisson, L.F. Stuck and sluggish fermentations. *Am. J. Enol. Vitic.* **1999**, *50*, 107–119.
10. Gutiérrez, A.; Chiva, R.; Sancho, M.; Beltran, G.; Arroyo-López, F.N.; Guillamon, J.M. Nitrogen requirements of commercial wine yeast strains during fermentation of a synthetic grape must. *Food Microbiol.* **2012**, *31*, 25–32. [CrossRef] [PubMed]

11. Monteiro, F.F.; Trousdale, E.K.; Bisson, L. Ethyl carbamate formation in wine: Use of radioactively labeled precursors to demonstrate the involvement of urea. *Am. J. Enol. Vitic.* **1989**, *40*, 1–8.

12. Adams, C.; Vuuren, H. Effect of timing of diammonium phosphate addition to fermenting grape must on the production of ethyl carbamate in wine. *Am. J. Enol. Vitic.* **2010**, *61*, 125–129.

13. Ugliano, M.; Henschke, P.A.; Herderich, M.; Pretorius, I.S. Nitrogen management is critical for wine flavour and style. *Aust. N. Z. Wine Ind. J.* **2007**, *22*, 24–30.

14. Stines, A.P.; Grubb, J.; Gockowiak, H.; Henschke, P.A.; Høj, P.B.; van Heeswijck, R. Proline and arginine accumulation in developing berries of *Vitis vinifera* L. in Australian vineyards: Influence of vine cultivar, berry maturity and tissue type. *Aust. J. Grape Wine Res.* **2000**, *6*, 150–158. [CrossRef]

15. Huang, Z.; Ough, C.S. Effect of vineyard locations, varieties, and rootstocks on the juice amino acid composition of severalcultivars. *Am. J. Enol. Vitic.* **1989**, *40*, 135–139.

16. Brandriss, M.C.; Magasanik, B. Genetics and physiology of proline utilization in *Saccharomyces cerevisiae:*Enzyme induction by proline. *J. Bacteriol.* **1979**, *140*, 498–503. [PubMed]

17. Wang, S.S.; Brandriss, M.C. Proline utilization in *Saccharomyces cerevisiae*: Sequence, regulation, and mitochondrial localization of the *PUT1* gene product. *Mol. Cell Biol.* **1987**, *7*, 4431–4440. [CrossRef] [PubMed]

18. Duteurtre, B.; Bourgeois, C.; Chollot, B. Study of the assimilation of proline by brewing yeast. *J. Inst. Brew.***1971**, *77*, 28–35. [CrossRef]

19. Takagi, H.; Sakai, K.; Morida, K.; Nakamori, S. Proline accumulation by mutation or disruption of the proline oxidase gene improves resistance to freezing and desiccation stresses in *Saccharomyces cerevisiae. FEMS Microbiol. Lett.* **2000**, *184*, 103–108. [CrossRef] [PubMed]

20. Takagi, H.; Taguchi, J.; Kaino, T. Proline accumulation protects *Saccharomyces cerevisiae* cells in stationary phase from ethanol stress by reducing reactive oxygen species levels. *Yeast* **2016**, *33*, 355–363. [CrossRef] [PubMed]

21. Ingledew, W.M.; Magnus, C.A.; Sosulski, F.W. Influence of oxygen on proline utilization during the wine fermentation. *Am. J. Enol. Vitic.* **1987**, *38*, 246–248.

22. Sablayrolles, J.M.; Dubois, C.; Manginot, C.; Roustan, J.-L.; Barre, P. Effectiveness of combined ammoniacal nitrogen and oxygen additions for completion of sluggish and stuck wine fermentations. *J. Ferment. Bioeng.* **1996**, *82*, 377–381. [CrossRef]

23. Gardner, J.M.; Poole, K.; Jiranek, V. Practical significance of relative assimilable nitrogen requirements of yeast: A preliminary study of fermentation performance and liberation of H2S. *Aust. J. Grape Wine Res.* **2002**, *8*, 175–179. [CrossRef]

24. Omura, F.; Fujita, A.; Miyajima, K.; Fukui, N. Engineering of yeast *PUT4* permease and its application to lager yeast for efficient proline assimilation. *Biosci. Biotechnol. Biochem.* **2005**, *69*, 1162–1171. [CrossRef] [PubMed]

25. Poole, K.; Walker, M.E.; Warren, T.; Gardner, J.; McBryde, C.; de Barros Lopes, M.; Jiranek, V. Proline transport and stress tolerance of ammonia-insensitive mutants of the *PUT4*-encoded proline-specific permease in yeast.*J. Gen. Appl. Microbiol.* **2009**, *55*, 427–439. [CrossRef] [PubMed]

26. Liccioli, T. Improving Fructose Utilization in Wine Yeast Using Adaptive Evolution. Ph.D. Thesis, The University of Adelaide, Adelaide, Australia, 2010. Available online: http: //hdl.handle.net/2440/67017 (accessed on 23 May 2013).

27. Guthrie, C.; Fink, G.R. Guide to Yeast Genetics and Molecular Biology. *Methods Enzymol.* **1991**, *194*, 1–863.

28. Poole, K. Enhancing Yeast Performance under Oenological Conditions by Enabling Proline Utilisation. Ph.D. Thesis, The University of Adelaide, Adelaide, Australia, 2002. Available online: http://hdl.handle. net/2440/58485 (accessed on 25 February 2011).

29. Salmon, J.M.; Barre, P. Improvement of nitrogen assimilation and fermentation kinetics under enological conditions by derepression of alternative nitrogen-assimilatory pathways in an industrial *Saccharomyces cerevisiae* strain. *Appl. Environ. Microbiol.* **1998**, *64*, 3831–3837. [PubMed]

30. Roon, R.J.; Even, H.L.; Dunlop, P.; Larimore, F.L. Methylamine and ammonia transport in *Saccharomyces cerevisiae*. *J. Bacteriol.* **1975**, *122*, 502–509. [PubMed]

31. Long, D.; Wilkinson, K.L.; Poole, K.; Taylor, D.K.; Warren, T.; Astorga, A.M.; Jiranek, V. Rapid method for proline determination in grape juice and wine. *J. Agric. Food Chem.* **2012**, *60*, 4259–4264. [CrossRef] [PubMed]

32. Boehringer Mannheim. D-glucose/D-fructose. In *Methods of Biochemical Analysis and Food Analysis*; Boehringer Mannheim: Mannheim, Germany, 1989; pp. 50–55.

33. Ohsumi, Y.; Kitamoto, K.; Anraku, Y. Changes induced in the permeability barrier of the yeast plasma membrane by cupric ion. *J. Bacteriol.* **1988**, *170*, 2676–2682. [CrossRef] [PubMed]

34. Matsuura, K.; Takagi, H. Vacuolar functions are involved in stress-protective effect of intracellular proline in*Saccharomyces cerevisiae*. *J. Biosci. Bioeng.* **2005**, *100*, 538–544. [CrossRef] [PubMed]

35. Kaino, T.; Takagi, H. Gene expression profiles and intracellular contents of stress protectants in Saccharomyces cerevisiae under ethanol and sorbitol stresses. *Appl. Microbiol. Biotechnol.* **2008**, *79*, 273–283. [CrossRef] [PubMed]

36. Beltran, G.; Novo, M.; Rozès, N.; Mas, A.; Guillamón, J.M. Nitrogen catabolite repression in *Saccharomyces cerevisiae* during wine fermentations. *FEMS Yeast Res.* **2004**, *4*, 625–632. [CrossRef] [PubMed]

37. Teste, M.-A.; Duquenne, M.; François, J.M.; Parrou, J.-L. Validation of reference genes for quantitative expression analysis by real-time RT-PCR in *Saccharomyces cerevisiae*. *BMC Mol. Biol.* **2009**, *10*, 99. [CrossRef] [PubMed]

38. Chomczynski, P.; Sacchi, N. The single-step method of RNA isolation by acid guanidinium thiocyanate-phenol-chloroform extraction: Twenty-something years on. *Nat. Protoc.* **2006**, *1*, 581–585. [CrossRef] [PubMed]

39. Ough, C.S.; Stashak, R.M. Further studies on proline concentration in grapes and wines. *Am. J. Enol. Vitic.***1974**, *25*, 7–12.

40. Lasko, P.F.; Brandriss, M.C. Proline transport in *Saccharomyces cerevisiae*. *J. Bacteriol.* **1981**, *148*, 241–247. [PubMed]

41. Maggio, A.; Miyazaki, S.; Veronese, P.; Fujita, T.; Ibeas, J.I.; Damsz, B.; Narasimhan, M.L.; Hasegawa, P.M.; Joly, R.J.; Bressan, R.A. Does proline accumulation play an active role in stress-induced growth reduction? *Plant J.* **2002**, *31*, 699–712. [CrossRef] [PubMed]

42. Ciani, M.; Maccarelli, F.; Fatichenti, F. Growth and fermentation behaviour of *Brettanomyces/Dekkera* yeasts under different conditions of aerobiosis. *World J. Microbiol. Biotechnol.* **2003**, *19*, 419–422. [CrossRef]

43. Aceituno, F.F.; Orellana, M.; Torres, J.; Mendoza, S.; Slater, A.W.; Melo, F.; Agosin, E. Oxygen response of the wine yeast *Saccharomyces cerevisiae* EC1118 grown under carbon-sufficient, nitrogen-limited enological conditions. *Appl. Environ. Microbiol.* **2012**, *78*, 8340–8352. [CrossRef] [PubMed]

44. Morita, Y.; Nakamori, S.; Takagi, H. Effect of proline and arginine metabolism on freezing stress of *Saccharomyces cerevisiae*. *J. Biosci. Bioeng.* **2002**, *94*, 390–394. [CrossRef]

45. Körber, C. Phenomena at the advancing ice-liquid interface: Solutes, particles and biological cells. *Q. Rev. Biophys.***1988**, *21*, 229–298. [CrossRef] [PubMed]

46. Takagi, H.; Iwamoto, F.; Nakamori, S. Isolation of freeze-tolerant laboratory strains of *Saccharomyces cerevisiae* from proline-analogue-resistant mutants. *Appl. Microbiol. Biotechnol.* **1997**, *47*, 405–411. [CrossRef] [PubMed]

47. Siddiqui, A.H.; Brandriss, M.C. The *Saccharomyces cerevisiae PUT3* activator protein associates with proline-specific upstream activation sequences. *Mol. Cell Biol.* **1989**, *9*, 4706–4712. [CrossRef] [PubMed]

48. Axelrod, J.D.; Majors, J.; Brandriss, M.C. Proline-dependent binding of *PUT3* transcriptional activator protein detected by footprinting in vivo. *Mol. Cell Biol.* **1991**, *11*, 564–567. [CrossRef] [PubMed]

49. Xu, S.; Falvey, D.A.; Brandriss, M.C. Roles of *URE2* and *GLN3* in the proline utilization pathway in *Saccharomyces cerevisiae*. *Mol. Cell Biol.* **1995**, *15*, 2321–2330. [CrossRef] [PubMed]

50. Daugherty, J.R.; Rai, R.; El Berry, H.M.; Cooper, T.G. Regulatory circuit for responses of nitrogen catabolic gene expression to the GLN3 and DAL80 proteins and nitrogen catabolite repression in *Saccharomyces cerevisiae.J. Bacteriol.* **1993**, *175*, 64–73. [CrossRef] [PubMed]

CHAPTER-11

CYTOSOLIC REDOX STATE IN *SACCHAROMYCES CEREVISIAE*, THE WINE YEAST, UNDER HYPEROSMOTIC STRESS CONDITIONS DURING THE FERMENTATION OF ICEWINE

INTRODUCTION: AN OVERVIEW

In the intricate dance of winemaking, the cytosolic redox status of wine yeast, particularly Saccharomyces cerevisiae, emerges as a pivotal player, especially when confronted with the challenges of hyperosmotic stress during icewine fermentation. As icewine production relies on the unique climatic conditions that induce the natural freezing of grapes, the yeast's ability to maintain an optimal redox balance becomes paramount. This study delves into the nuanced interplay between the cytosolic redox state and the stress of hyperosmotic conditions during the fermentation of icewine. By unraveling the molecular intricacies of S. cerevisiae's response to these environmental pressures, we aim to enhance our understanding of the physiological adaptations crucial for successful and nuanced icewine production.

Icewine is a sweet dessert wine produced from grapes naturally frozen on the vine. The low temperature during harvesting and pressing (below -8°C in Canada), traps the water crystals in grape berries, resulting in Icewine juice that is extremely concentrated in soluble solids including sugar, acids, and nitrogen compounds [1]. In Canada, the minimum concentration of soluble solids in Icewine juice for fermentation must reach 35°Brix [2], but juice between 38 and 42°Brix is commonly used for Icewine production [3]. Such a high concentration of sugar in Icewine juice places yeast cells under extreme hyperosmotic stress, resulting in prolonged fermentation times and a high concentration of volatile acidity (mainly in the form of acetic acid) [4].

Acetic acid present at high levels is a spoilage compound in wine. In Canada, the maximum allowed acetic acid concentration in Icewine is 2.1 g L^{-1}, whereas in table wines it is 1.3 g L^{-1} [2]. The average acetic acid concentration in commercial Canadian Icewines was found to be 1.30 g L^{-1}, but ranged from 0.49 to 2.29 g L^{-1} [5]. Acetic acid also contributes to the formation of another undesired aromatic compound, ethyl acetate, which is described as having a solvent-like aroma such as that found in nail polish remover [6,7]. Nurgel et al. reported that the ethyl acetate concentration in commercial Canadian Icewines ranged from 0.086 to 0.369 g L^{-1} , with an average value of 0.240 g L^{-1} [5], and some wines testing over the sensory threshold in Icewine of 0.198 g L^{-1} [6].

Acetic acid production appears linked to the osmotic stress yeast are placed under. During Icewine fermentation, the highly concentrated juice places yeast cells under extreme hyperosmotic stress. In response, *Saccharomyces cerevisiae* produces glycerol as an intracellular osmolyte to counteract the high osmolarity [8]. During glycerol formation, the oxidized cofactor NAD^+ is produced, while dihydroxyacetone phosphate is reduced to glycerol-3-phosphate, leading to a potential redox imbalance for the NAD^+/NADH cofactor system. Because of the lack of transhydrogenase in yeast to convert reducing equivalents between the NAD(H) system and the NADP(H) system, yeasts must rely on metabolite formation to maintain the redox balance [9,10]. Acetic acid has been suggested as a potential metabolite that yeast cells can produce to balance the excess NAD^+ produced during glycerol formation under hyperosmotic stress [8,11,12]. The high level of acetic acid present in Icewine may also be due to the down regulation of *ACS* genes encoding for acetyl-CoA synthetases [13]. In *S. cerevisiae*, acetic acid is an intermediate to form the central metabolite, acetyl-CoA, catalyzed by acetyl-CoA synthetases in the pyruvate dehydrogenase (PDH) pathway [14]. Heit et al. found that *ACS1* and *ACS2* were down regulated 19.0-fold and 11.2-fold, respectively, in *S. cerevisicea* K1-V1116 yeasts fermenting Icewine juice compared to diluted juice, suggesting that the lower consumption of acetic acid to form the downstream metabolites during Icewine fermentation contributes to elevated acetic acid production in Icewine [13].

The results from current literature are not in agreement concerning which Aldp isoform is responsible for acetic acid production during sugar fermentations, as different Aldps have distinct cofactor **specificities** and play different roles for acetic acid production in yeast metabolism [11–13,15].

S. cerevisiae contain five aldehyde dehydrogenases [16]. Two cytosolic isoforms are NAD^+-dependent (Ald2p and Ald3p) [16], and one is $NADP^+$-dependent (Ald6p)

[17]. Ald2p oxidizes 3-aminopropanal to β-alanine, required for pantothenic acid production [18]. Ald6p has been identified as the main cytosolic aldehyde dehydrogenase responsible for acetic acid production in *S. cerevisiae* strains during the fermentation of glucose [15,19–21], although both mitochondrial $NADP^+$-dependent Ald5p [22] and $NAD^+/NADP^+$-dependent Ald4p [23] have been reported to contribute. Ald6p was linked to acetic acid generation under osmotic stress to NADPH production to compensate for the down regulation of genes in the pentose phosphate pathway [15], thus questioning the linkage of acetic acid production under osmotic stress to glycerol formation and NADH requirements. However, Noti et al. found *ALD6* to be only slightly up regulated in some strains during the first two hours of exposure to 235 g/L sugar [21]. Contrary to these reports, additional studies have linked acetic acid to *ALD3* expression and the NAD^+/NADH cofactor system during fermentation [11,12,24] with a temporal relationship between acetaldehyde increase, *ALD3* expression, and acetic acid production when fermenting juice at 401 g/L sugar [11,12]. *ALD3* was the only isoform up regulated by acetaldehyde stress during fermentation [12]. To reconcile these differences, we hypothesize that the main Aldp responsible for acetic acid production throughout fermentation relates back to the starting sugar concentration in the juice, and the ongoing redox requirements of the cell throughout fermentation.

In *S. cerevisiae*, the NAD(H) and NADP(H) cofactor systems play important roles as electron donors/acceptors in more than 300 biochemical oxidation and reduction reactions [25]. NAD(H) is mainly involved in catabolic reactions such as glycolysis, glycerol production, and nitrogen uptake and consumption [26,27]. NADP(H) is predominantly involved in anabolic processes including the pentose phosphate pathway, which provides substrates for the synthesis of nucleic acids (RNA and DNA) and some amino acids [28], and has been indicated to be growth rate-dependent [29,30].

This study monitored daily changes in the two cofactor systems, NAD(H) and NADP(H), during the first week of fermentation when acetic acid is rapidly produced by *S. cerevisiae* K1-V1116. The fermentations were conducted in both Icewine juice and dilute Icewine juice, which mimics a table wine fermentation. The different ratios of oxidized cofactor to total cofactor for these two systems provides further evidence of which aldehyde dehydrogenases are involved in acetic acid production during Icewine versus dilute juice fermentation.

MATERIALS AND METHODS

Yeast Strain and Juices

The commercial wine yeast *S. cerevisiae* K1-V1116 (Lalvin, Montreal, QB, Canada) was used for fermentations. Vidal Icewine juice was purchased from Huebel Grapes

Estates Ltd. (Niagara-on-the-Lake, ON, Canada) and stored at -35°C until required for use. The juice was thawed at 7°C for 24 h, and racked off, followed by filtration through coarse, medium, and fine pore size pad filters using the Bueno Vino Mini Jet filter (Vineco, St. Catharines, ON, Canada). The juice was then filtered through a 0.45-μm membrane cartridge filter (Millipore, Etobicoke, ON, Canada). Sterile filtered Icewine juice (39.4°Brix, 433 ± 5g L^{-1} reducing sugars) was used for Icewine fermentation, and diluted to 20.0°Brix (201 ± 1 g L^{-1} reducing sugars) for dilute juice fermentations, which mimics table wine fermentation.

Fermentation Setup and Sampling

Dried wine yeast K1-V1116 was prepared using the stepwise acclimatization procedure described in Pigeau and Inglis [11]. A volume of 30 mL of the acclimatized starter culture was used to inoculate 2 L of Icewine juice (39.4°Brix) and dilute Icewine juice (20.0°Brix), reaching a final yeast inoculation rate of 0.5 g dry weight L^{-1}. Fermentations were incubated at 17°C in triplicate and continued until sugar consumption stopped for three consecutive days for Icewine fermentations, or until the sugar level was below 3 g L^{-1} for dilute juice fermentations. After stirring the fermentations for 5 min to ensure homogeneity, 100 mL of sample was removed daily and analyzed for cofactors in the first week of fermentation. Samples (5 mL) were also taken for metabolite analysis every day during the first week, every other day in the second week, and every third day in the following weeks until the fermentations stopped. Samples were centrifuged and the supernatants were stored at -30°C until metabolite analysis was performed.

Analysis for Metabolites

Soluble solids were determined by ABBE bench top refracto-meter (model 10450; American Optical, Buffalo, NY, USA). Acidity was determined by pH measurement using a symphony pH meter (model B10P; VWR, Mississauga, ON, Canada), and titratable acidity by titration against 0.1 mol L^{-1} NaOH to an endpoint of pH 8.2 [31]. Reducing sugars, ammonia nitrogen, amino nitrogen, glycerol, acetaldehyde, and acetic acid were determined with Megazyme assay kits (K-FRUGL, K-AMIAR, K-PANOPA, K-GCROL, K-ACHD, K-ACET; Megazyme International Ireland, Ltd., Bray, Co. Wicklow, Ireland). Ethanol was determined by gas chromatography (model 6890; Agilent Technologies Inc., Palo Alto, CA, USA) with a flame ionization detector and a capillary column (Agilent 122-7032 DB-Wax, 30 m × 250 μm diameter, 0.25 μm film thickness), using 0.1% 1-butanol as an internal standard. All metabolite measurements were performed in duplicate on each fermentation replicate. Metabolite production during the course of fermentation was calculated by the difference in the respective metabolite concentration measured between the time zero point (immediately after inoculation) and at each sampling time point during the course of fermentations. Normalized metabolite production was determined by dividing the final metabolite production by the final sugar consumed. Metabolite daily production rate was calculated as: daily production rate = (metabolite concentration on the sampling day - metabolite concentration on the previous day) / (sugar concentration on the previous day - sugar concentration on the sampling day).

Analysis for Enzyme Cofactors

To collect yeast cells, 100 mL of sample was centrifuged at 5108× *g* for 10 min at 4°C; the pellet was washed once with 2 mL sterile deionized water and re-centrifuged. The pellet was re-suspended in 2 mL g^{-1} (yeast wet weight) of Tris-dithiothreitol (DTT) buffer (100 mM Tris-H_2SO_4, pH 9.4; 10 mM DTT; sterile filtered) and incubated for 30 min at 30°C with gentle shaking at 0.059× *g*. The treated cells were then collected by centrifugation at 1464× *g* for 5 min at 4°C. The pellet was re-suspended in 2 mL spheroplast buffer (1.2 M sorbitol; 20 mM potassium phosphate, pH 7.4; sterile filtered) and centrifuged as in the previous step. Cells were then added to a mixture containing 10 mg of Zymolyase-100T g^{-1} of yeast wet weight and 4 mL spheroplast buffer. The mixture was incubated for 60 min at 35°C with shaking at 0.059× *g*. Following Zymolyase digestion, the spheroplasts were collected by centrifugation at 1464× *g* for 5 min at 4 °C, washed twice in spheroplast buffer, and pelleted as before. The spheroplasts were then re-suspended in ice-cold cell lysis buffer (0.6 M mannitol; 20 mM 4-(2-hydroxyethyl)piperazine-1-ethanesulfonic acid (HEPES)-KOH, pH 7.4; 1 mM phenyl-methyl-sulfonyl fluoride (PMSF); 0.1% *w/v* fatty acid free-bovine serum albumin (BSA); sterile filtered) and disrupted using a pre-chilled Dounce homogenizer on ice. The volume of cell lysis buffer applied was 3.5 mL for yeast samples from diluted fermentation, and 2.0 mL for samples from Icewine fermentation. The lysate was then quickly transferred to pre-chilled centrifuge tubes and fractionated at 9682× *g* for 10 min at 4°C to separate the cytosolic fraction. A small portion (200 μL) of the supernatant was immediately mixed with 200 μL of Component E from the Amplite fluorimetric NAD/NADH or NADP/NADPH Ratio Assay Kits (15263, 15264; AAT Bioquest Inc., Sunnyvale, CA, USA) and incubated for 10 min to extract NAD^+ or $NADP^+$. This sample was also diluted serially by a mixture of cell lysis buffer and Component E (at a 1:1 ratio) to ensure that the sample concentration would be within the linear range of the calibration curves of both kits. The rest of the supernatant was kept on ice for the measurement of the sum of NAD^+ and NADH or the sum of $NADP^+$ and NADPH, and serial dilution was also performed for this sample using cell lysis buffer. All samples were then tested by the abovementioned fluorimetric kits in duplicate on each fermentation replicate according to the manuals. The percentage of oxidized cofactor of either the NAD(H) or NADP(H) system was calculated as: percentage of oxidized cofactor = concentration of oxidized cofactor / (concentration of oxidized cofactor + concentration of reduced cofactor) × 100%.

Statistical Analysis

XLSTAT-Pro by Addinsoft (New York, NY, USA) was used for statistical analysis. Analysis of variance (ANOVA) with Fisher's Least Significant Difference (LSD) test ($p < 0.05$) was used to evaluate differences between variables and mean separation, respectively.

RESULTS

Fermentation Kinetics

The composition of Icewine juice in comparison to the dilute juice indicates that the starting sugar concentration is over 400 g L^{-1} in the Icewine juice and concentrated in all other juice components including acidity, yeast assimilable nitrogen, glycerol, acetaldehyde, and acetic acid (Table 1). Yeast cells exposed to a greater level of hyperosmotic stress from fermentable sugars in the Icewine fermentation consumed sugar very slowly in the first week and stopped before half of the original sugar concentration was consumed on Day 39 (Figure 1). Conversely, in the diluted condition, yeast cells rapidly consumed sugar right after inoculation with only 1 g L^{-1} of sugar left in the final wine on Day 21 (Figure 1 and Table 2).

Table 1. Vidal Icewine juice and diluted juice initial parameters (mean ± SD).

Parameter	Icewine Juice	Diluted Juice
Glucose + fructose (g L^{-1})	433 ± 5	201 ± 1
Titratable acidity (g L^{-1})	6.4 ± 0.0	3.1 ± 0.0
pH	3.76 ± 0.01	3.88 ± 0.01
Free amino nitrogen (mg N L^{-1})	238.4 ± 2.4	114.8 ± 3.0
Ammonia nitrogen (mg N L^{-1})	14.8 ± 0.1	6.6 ± 0.8
Glycerol (g L^{-1})	8.76 ± 0.06	4.04 ± 0
Acetaldehyde (mg L^{-1})	5.3 ± 0.8	2.7 ± 0.2
Acetic acid (g L^{-1})	0.04 ± 0	0.02 ± 0
Ethanol (% *v/v*)	1.3 ± 0	0.6 ± 0

Table 2. Final wine parameters in Icewine and diluted juice fermentation (mean ± SD).

Parameter	Icewine Juice Fermentation	Diluted Juice Fermentation
Glucose + fructose (g L^{-1})	252 ± 3	1 ± 1
Titratable acidity (g L^{-1})	8.3 ± 0.2	5.5 ± 0.1
pH	3.92 ± 0.01	3.63 ± 0.02

Parameter	Icewine Juice Fermentation	Diluted Juice Fermentation
Free amino nitrogen (mg N L-1)	188.5 ± 1.4	7.4 ± 0.4
Ammonia nitrogen (mg N L-1)	3.5 ± 0.1	† ND
Glycerol (g L-1)	17.44 ± 0.19	9.43 ± 0.06
Acetaldehyde (mg L-1)	48.5 ± 0.8	23.0 ± 0.8
Acetic acid (g L-1)	1.48 ± 0.10	0.12 ± 0.01
Ethanol (% *v/v*)	12.0 ± 0.2	13.9 ± 0.1

† ND indicates that no concentration was detected.

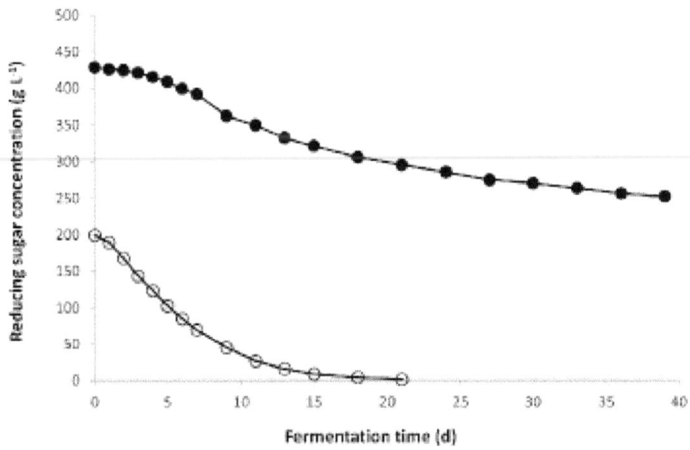

Figure 1. Sugar concentration during fermentation. Fermentations of Icewine juice ( ) and diluted juice (0) were inoculated with K1-V1116 and monitored for sugar concentration throughout fermentation. Fermentations were performed in triplicate and samples from each trial were tested in duplicate. Sugar values represent the average ± standard deviation.

Yeast Metabolite Production

Metabolite production was monitored as a function of time (Figure 2a,c,e), but because yeast consumes sugar at different rates, these metabolites were also plotted as a function of sugar consumed (Figure 2b,d,f) for direct comparison between the two conditions. Yeast produced higher levels of glycerol and acetic acid during Icewine fermentation as a function of sugar consumed, showing a 1.9-fold increase in glycerol and 16.5-fold increase in acetic acid when normalized to sugar consumed (Table 3). The peak concentration of acetaldehyde production was 4.7-fold higher in the Icewine fermentation compared to the dilute fermentation, and then declined when acetaldehyde was consumed for the formation of either ethanol or acetic acid (Figure 2c,d).

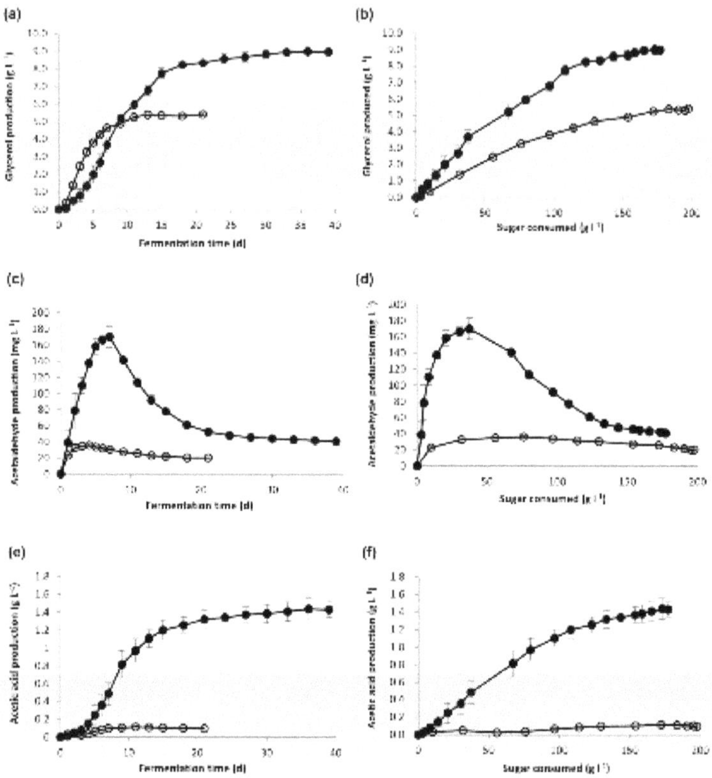

Figure 2. Metabolite production during fermentation. Fermentations of Icewine juice ( ) and diluted juice (o) were analyzed for the production of (a) glycerol, (c) acetaldehyde, and (e) acetic acid, and the generation of metabolites were also plotted out as a function of sugar consumed for (b) glycerol, (d) acetaldehyde, and (f) acetic acid. Fermentations were performed in triplicate and samples from each trial were tested in duplicate. Metabolite values represent the average ± standard deviation.

Table 3. Normalized metabolite production by yeast measured at the end of fermentation (mean ± SD).

Fermentation	Glycerol (mg g-1 of Sugar Consumed) *	Acetaldehyde (mg g-1 of Sugar Consumed) *	Ethanol (g g-1 of Sugar Consumed) *	Acetic Acid (mg g-1 of Sugar Consumed) *
Diluted juice	27.44 ± 0.30	0.23 ± 0.01	0.53 ± 0	0.49 ± 0.06
Icewine Juice	50.92 ± 0.74	0.10 ± 0	0.47 ± 0	8.10 ± 0.50

* Significantly different as determined by Fisher's Least Significant Difference test ($p < 0.05$).

Ratio of Coenzyme Concentration

The cytosolic coenzymes from both the NAD(H) and NADP(H) systems were analyzed in the first week of fermentation, when yeast cells adapted to the hyperosmotic stress under the two different sugar concentrations. The percentage of the cofactors in their oxidized form were plotted against time (Figure 3). The percentages were then linked to the daily production rate of glycerol and acetic acid (Figure 4) to understand the correlation between redox status and metabolite production under hyperosmotic stress.

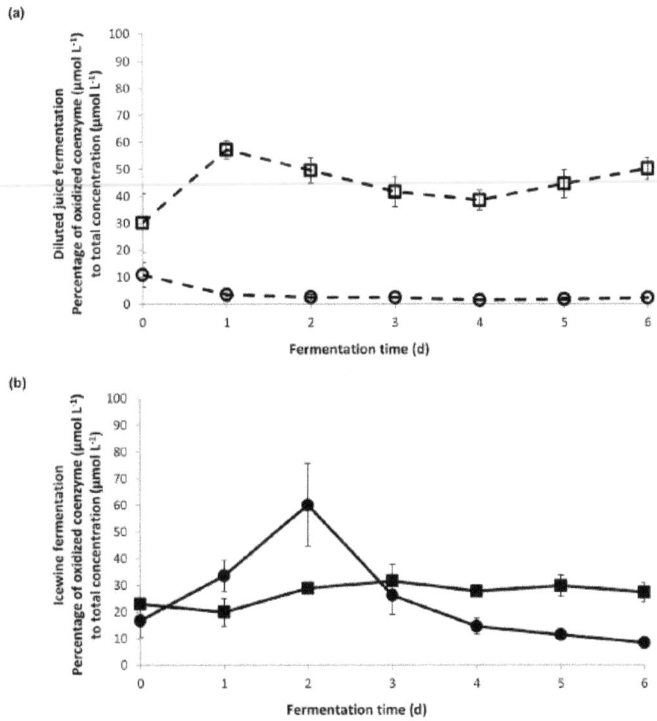

Figure 3. Percentage of NAD+ and NADP+ in the first week of fermentation. (a) The percentage of NAD+ to total NAD(H) (0) was compared to the percentage of NADP+ to total NADP(H)(D) in diluted fermentation; (b) the percentage of NAD+ to total NAD(H) ( ) was compared to the percentage of NADP+ to total NADP(H) (•) in Icewine fermentation. Fermentations were performed in triplicate and samples from each trial were tested in duplicate. Data points represent the average ± standard deviation.

During the dilute fermentation, the percent of cytosolic NADP+, which ranged from 30.2–57.2% over the course of fermentation, was consistently higher than the percent of cytosolic NAD+, which was only 3.6% from Day 1 onward (Figure 3a). The percent of the cytosolic NADP+ was consistently lower in the Icewine condition (between 19.9–31.6%, Figure 3b) compared to the dilute juice condition (between 30.2–57.2%, Figure 3a), showing a 2.9-fold higher level in the dilute juice condition on Day 1. It appears that the cytosolic NADP+-dependent Ald6p has a larger contribution to the low acetic acid production in the dilute juice condition over NAD$^+$-dependent forms of Aldp (Figure 4b).

However, a very different redox profile resulted during the Icewine fermentation (Figure 3b). During Icewine fermentation, the percent of cytosolic NAD$^+$ increased over the first two days of fermentation, peaking on Day 2 at 60.1% and remaining much higher over the remainder of the fermentation (Figure 3b) in comparison to the dilute juice fermentation (Figure 3a), showing a 24-fold increase in the Icewine condition. This spike in NAD+ corresponded to the increased production rate of glycerol in the Icewine condition (Figure 4a) and the increased production rate of acetic acid (Figure 4b). The redox profile supports the hypothesis of cytosolic NAD$^+$-dependent Aldp(s) contributing to the elevated acetic acid in Icewine fermentation along with the cytosolic NADP$^+$-dependent Ald6p.

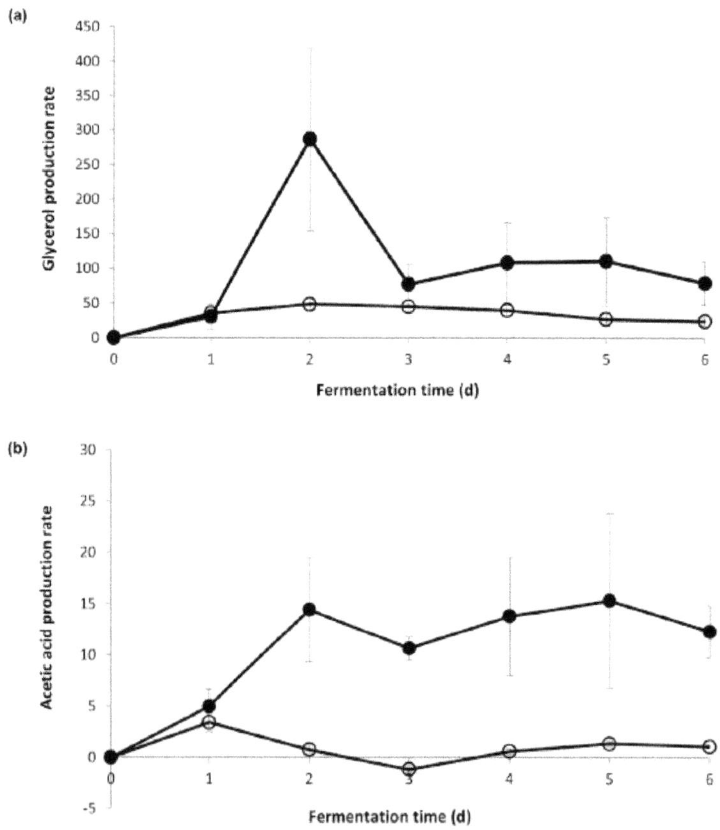

Figure 4. Metabolite production rate during fermentation. The daily metabolite production rate of • (a) glycerol and (b) acetic acid was calculated during the first week of Icewine juice fermentation ( ) and diluted juice fermentation (0). Fermentations were performed in triplicate and samples from each trial were tested in duplicate. Metabolite values represent the average ± standard deviation.

DISCUSSION

Increased acetic acid production during high sugar fermentation in *S. cerevisiae* may be due to metabolic regulation driven by the redox status of the cell, substrate availability, or product usage, or due to the transcriptional regulation of genes altering enzyme concentrations involved in the synthesis or usage of acetic acid, or a combination of all these factors.

Past studies have linked acetic acid production during fermentation to either the $NADP^+$ cofactor system [15] or the NAD^+ cofactor system [11–13] based on *ALD* gene expression patterns encoding aldehyde dehydrogenases that have different cofactor requirements. We report here, for the first time, the cytosolic redox status of both cofactor systems during fermentation at two different sugar levels.

We have previously reported increased substrate availability for the aldehyde dehydrogenases during high sugar fermentations (acetaldehyde levels quickly rise during Icewine fermentation) [11–13], which was also found in this present study, along with increased expression of *ALD3* encoding a cytosolic NAD^+-dependent isoform over that found during dilute juice fermentation [11–13]. *ALD3* was also the only *ALD* expression stimulated by acetaldehyde stress during fermentation [12]. We now report that the percent of oxidized cofactor NAD^+ on Day 2 of fermentation was 24-fold higher in the Icewine condition versus the dilute juice condition and 2.1-fold higher than the percent of oxidized cofactor $NADP^+$ on Day 2 of Icewine fermentation. Taken together, our results further confirmed a role for NAD^+-dependent Aldp isoforms along with $NADP^+$-dependent Aldp isoforms in producing elevated levels of acetic acid during the high sugar fermentations of Icewine production.

In evaluating the cytosolic redox status of fermenting yeast at two different sugar levels (201 versus 433 g L^{-1} of sugar in our present study), the discrepancy in the literature in reporting the Aldp responsible for acetic acid production during fermentation may have been due to the starting sugar concentration in the media used for fermentation. Our data supports other published work where the $NADP^+$-dependent isoform Ald6p was identified as the main aldehyde dehydrogenase responsible for acetic acid production when sugar concentrations were at or below 200 g L^{-1} sugar [15,19–21]. However, under high sugar conditions of 400 g L^{-1}, Erasmus et al. linked acetic acid generation to NADPH production via Ald6p to compensate for the down regulation of genes in the pentose phosphate pathway [15], questioning the linkage of acetic acid production under osmotic stress to glycerol formation and NADH requirements. Based on the cytosolic redox status of yeast fermenting Icewine juice at 433 g L^{-1} sugar, as well as our previous *ALD* expression analysis during Icewine fermentation [11–13], it appears that both Ald6p and Ald3p contribute to acetic acid during Icewine fermentation. Our data supports a linkage of acetic acid production under the high osmotic stress of Icewine fermentation to NADH requirements and glycerol production.

We have also previously reported that elevated acetic acid levels appear due to lack of conversion to acetyl-CoA from the down-regulation in gene expression of cytosolic and mitochondrial acetyl-CoA synthetase [13]. This was further substantiated by 2-fold lower ethyl acetate levels in the Icewines (60 mg/L) compared to that found in the dilute juice fermentations (120 mg/L), whereby acetyl-CoA is required for the esterification to ethanol by alcohol acetyltransferases [13].

We have provided further evidence to support the concept that cytosolic NAD^+-dependent aldehyde dehydrogenase(s) are responsible, in part, for increased levels of acetic acid during Icewine fermentation stimulated by substrate availability through an increase in NAD^+ and acetaldehyde, whereas the $NADP^+$-dependent isoform plays the dominate role in acetic acid production in table wine fermentation stimulated by an increase in $NADP^+$. Although substrate availability [11–13], redox status, and gene expression [11–13] have been evaluated, enzyme activities for the aldehyde dehydrogenases and downstream acetyl CoA synthetases represent missing information that could further confirmed the enzymes and cofactor system involved in the high acetic acid accumulation during high sugar fermentations such as that of Icewine.

REFERENCES

1. Bowen, A. Managing the quality of icewines. In *Managing Wine Quality: Oenology and Wine Quality*, 1st ed.; Reynolds, A.G., Ed.; Woodhead Publishing Ltd.: Cambridge, UK, 2010; pp. 523–552.
2. O. Reg. 406/00: Rules of Vintners Quality Alliance Ontario Relating to Terms for VQA Wine. Available online: https://www.ontario.ca/laws/regulation/000406 (accessed on 14 November 2017).
3. Ziraldo, D.; Kaiser, K. Science. In *Icewine: Extreme Winemaking*, 1st ed.; Key Porter Books: Toronto, ON, Canada, 2007; pp. 73–107.
4. Kontkanen, D.; Inglis, D.; Pickering, G.; Reynolds, A. Effect of yeast inoculation rate, acclimatization, and nutrient addition on Icewine fermentation. *Am. J. Enol. Vitic.* **2004**, *55*, 363–370.
5. Nurgel, C.; Pickering, G.J.; Inglis, D.L. Sensory and chemical characteristics of Canadian ice wines. *J. Sci. Food Agric.* **2004**, *84*, 1675–1684. [CrossRef]
6. Cliff, M.A.; Pickering, G.J. Determination of odour detection thresholds for acetic acid and ethyl acetate in ice wine. *J. Wine Res.* **2006**, *17*, 45–52. [CrossRef]
7. Lilly, M.; Lambrechts, M.G.; Pretorius, I.S. Effect of increased acetyltransferase activity on flavour profiles of wine and distillates. *Appl. Environ. Microbiol.* **2000**, *66*, 744–753. [CrossRef] [PubMed]

8. Blomberg, A.; Adler, L. Roles of glycerol and glycerol-3-phosphate dehydrogenase (NAD+) in acquired osmotolerance of *Saccharomyces cerevisiae. J. Bacteriol.* **1989,** *171,* 1087–1092. [CrossRef] [PubMed]
9. Lagunas, R.; Ganedo, J.M. Reduced pyridine nucleotide balance growing on *Saccharomyces cerevisiae. Eur. J. Biochem.* **1973,** *37,* 90–94. [CrossRef] [PubMed]
10. Van Dijken, J.; Scheffers, W. Redox balances in the metabolism of sugars by yeasts. *FEMS Microbiol. Rev.***1986,** *32,* 199–224. [CrossRef]
11. Pigeau, G.M.; Inglis, D.L. Upregulation of ALD3 and GPD1 in *Saccharomyces cerevisiae* during Icewine fermentation. *J. Appl. Microbiol.* **2005,** *99,* 112–125. [CrossRef] [PubMed]
12. Pigeau, G.M.; Inglis, D.L. Response of wine yeast (*Saccharomyces cerevisiae*) aldehyde dehydrogenases to acetaldehyde stress during Icewine fermentation. *J. Appl. Microbiol.* **2007,** *103,* 1576–1586. [CrossRef] [PubMed]
13. Heit, C.; Martin, S.J.; Yang, F.; Inglis, D.L. Osmoadaptation of wine yeast (*Saccharomyces cerevisiae*) to hyperosmotic stress during Icewine fermentation. *J. Appl. Microbiol.* **2017.** under review.
14. Pronk, J.T.; Steensma, H.Y.; van Dijken, J.P. Pyruvate metabolism in *Saccharomyces cerevisiae. Yeast* **1996,** *12,* 1607–1633. [CrossRef]
15. Erasmus, D.J.; van Vuuren, H.J.J. Genetic basis for osmosensitivity and genetic instability of the wine yeast *Saccharomyces cerevisiae* V1N7. *Am. J. Enol. Vitic.* ***2009,*** *60,* 145–154.
16. Navarro-Avino, J.P.; Prasad, R.; Miralles, V.J.; Benito, R.M.; Serreno, R. A proposal of nomenclature of aldehyde dehydrogenases in *Saccharomyces cerevisiae* and characterization of the stress-inducible ALD2 and ALD3 genes. *Yeast* **1999,** *15,* 829–842. [CrossRef]
17. Meaden, P.G.; Dickinson, F.M.; Mifsud, A.; Tessier, W.; Westwater, J.; Bussey, H.; Midgley, M. The *ALD6* gene of *Saccharomyces cerevisiae* encodes a cytosolic, Mg^{2+}-activated acetaldehyde dehydrogenase. *Yeast* **1997,** *13,* 1319–1327. [CrossRef]
18. White, H.W.; Skatrud, P.L.; Xue, Z.; Toyn, J.H. Specialization of function among aldehyde dehydrogenases: The *ALD2* and *ALD3* genes are required for B-alanine biosynthesis in *S. cerevisiae. Genetics* **2003,** *163,* 69–77. [PubMed]
19. Eglinton, J.M.; Heinrich, A.J.; Pollnitz, A.P.; Langridge, P.; Henschke, P.A.; de Barros Lopes, M. Decreasing acetic acid accumulation by a glycerol overproducing strain of *Saccharomyces cerevisiae* by deleting the *ALD6* aldehyde dehydrogenase gene. *Yeast* **2002,** *19,* 295–301. [CrossRef] [PubMed]

20. Luo, Z.; Walkey, C.; Madilao, L.; Measday, V.; Van Vuuren, H. Functional improvement of *Saccharomyces cerevisiae* to reduce volatile acidity in wine. *FEMS Yeast Res.* **2013**, *13*, 485–494. [CrossRef] [PubMed]
21. Noti, O.; Vaudano, E.; Pessione, E.; Garcia-Moruno, E. Short-term response of different *Saccharomyces cerevisiae* strains to hyperosmotic stress caused by inoculation in grape must: RT-qPCR study and metabolite analysis. *Food Microbiol.* **2015**, *52*, 49–58. [CrossRef] [PubMed]
22. Saint-Prix, F.; Bönquist, L.; Dequin, S. Functional analysis of the *ALD* gene family of *Saccharomyces cerevisiae* during anaerobic growth on glucose: The $NADP^+$-dependent Ald6p and Ald5p isoforms play a major role in acetate formation. *Microbiology* **2004**, *150*, 2209–2220. [CrossRef] [PubMed]
23. Remize, F.; Andrieu, E.; Dequin, S. Engineering of the pyruvate dehydrogenase bypass in *Saccharomyces cerevisiae*: Role of the cytosolic Mg^{2+} and mitochondrial K^+ acetaldehyde dehydrogenases Ald6p and Ald4p in acetate formation during alcoholic fermentation. *Appl. Environ. Microbiol.* **2000**, *66*, 3151–3159. [CrossRef] [PubMed]
24. Akamatsu, S.; Kamiya, H.; Yamashita, N.; Motoyoshi, T.; Goto-Yamamoto, N.; Ishikawa, T.; Okazaki, N.; Nishimura, A. Effects of aldehyde dehydrogenase and acetyl-CoA synthetase on acetate formation in sake mash. *J. Biosci. Bioeng.* **2000**, *90*, 555–560. [CrossRef]
25. Forster, J.; Famili, I.; Fu, P.; Palsson, B.O.; Nielsen, J. Genome-scale reconstruction of *Saccharomyces cerevisiae* metabolic network. *Genome Res.* **2003**, *13*, 244–253. [CrossRef] [PubMed]
26. Bruinenberg, P.M. The NADP(H) redox couple in yeast metabolism. *Antonie Van Leeuwenhoek* **1986**, *52*, 411–429. [CrossRef] [PubMed]
27. Quirós, M.; Martínez-Moreno, R.; Albiol, J.; Morales, P.; Vázquez-Lima, F.; Barreiro-Vázquez, A.; Ferrer, P.; Gonzalez, R. Metabolic flux analysis during the exponential growth phase of *Saccharomyces cerevisiae* in wine fermentations. *PLoS ONE* **2013**, *8*, e71909. [CrossRef]
28. Villadsen, J.; Nielsen, J.; Lidén, G. Chemicals from metabolic pathways. In *Bioreaction Engineering Principles*, 3rd ed.; Springer: New York, NY, USA, 2011; pp. 7–62.
29. Nissen, T.L.; Schulze, U.; Nielsen, J.; Villadsen, J. Flux distributions in anaerobic, glucose-limited continuous cultures of *Saccharomyces cerevisiae*. *Microbiology* **1997**, *143*, 203–218. [CrossRef] [PubMed]
30. Varela, C.; Pizarro, F.; Agosin, E. Biomass content governs fermentation rate in nitrogen-deficient wine musts.*Appl. Environ. Microbiol.* **2004**, *70*, 3392–3400. [CrossRef] [PubMed]

31. Zoecklein, B.W.; Fugelsang, K.C.; Gump, B.H. Laboratory Procedures. In *Wine Analysis and Production*, 1st ed.; Chapman and Hall, International Thomson Publishing: New York, NY, USA, 1995; pp. 374–378.

CHAPTER-12

A MICROTITER PLATE ASSAY AS A RELIABLE METHOD TO ASSURE THE IDENTIFICATION AND CLASSIFICATION OF THE VEIL-FORMING YEASTS THROUGHOUT THE AGING OF SHERRY WINES

INTRODUCTION: AN OVERVIEW

In the intricate world of Sherry wine production, the presence and activity of veil-forming yeasts play a pivotal role in shaping the distinctive characteristics of this renowned fortified wine. Veil-forming yeasts, primarily encompassing species such as Saccharomyces cerevisiae, Saccharomyces bayanus, and Saccharomyces paradoxus, contribute to the complex microbiological ecosystem during Sherry wines' aging process. The identification and classification of these yeasts are of paramount importance for winemakers, as they significantly influence the organoleptic properties and quality of the final product.

Several types of wines are characterized by growing of a yeast film on its surface, which is known as "veil or velum of flor". Two of these wines are Fino Sherry and Manzanilla, both produced from Palomino Fino (*Vitis vinifera* cv. *vinifera*) in the Jerez-Xérès-Sherry D.O., in Southern Spain [1,2]. The production of these types of wines consists of two successive processes. First, alcoholic fermentation, an anaerobic process in which sugars are converted in energy, produces as a result, ethanol and carbon dioxide gas. After the process is finished, wine is fortified to 15% (*v*/*v*) and it is put into oak barrels for storage. The process of biological ageing begins at this moment, using the "Solera" system [3].

Yeasts involved in veil formation during biological ageing are mainly *Saccharomyces cerevisiae,* and they have traditionally been divided into four races:

beticus, cheresiensis, montuliensis and rouxii. The early methods related to taxonomy involved morphological studies and biochemical analyses of differences on fermentation and assimilation tests, where the substrates used were mainly dextrose, lactose, maltose, melibiose, raffinose, galactose and sucrose. These physiological properties were traditionally tested using Durham tubes in a liquid medium, which filled up with gas if fermentation occurred [4]. Nevertheless, both morphological and physiological characteristics may be influence by culture conditions and spoilages and can provide ambiguous results. Using classical techniques can lead, in some cases, to incorrect classification of species or mis-identification of strains. Moreover, the methodology often requires evaluation of different parameters by a complex, laborious and time-consuming process that does not include the most important characteristics of the yeasts from an industrial point of view. On the other hand, sugar fermentation and assimilation tests using Durham tubes are not too reliable because some of the yeast strains do not released CO_2 immediately (these are known as slow fermentative yeasts). These yeasts were considered for a long time as non-fermentative, but they have been recently re-classified [5]. In recent years, new methods have been developed, such as rapid kits, for yeast identification (API® strips), analyses of total cell proteins and long-chain fatty acids, using gas chromatography or the determination of chemical compounds formed by the yeasts [6,7], but their cost, reproducibility and suitability to veil-forming yeast strain classification can lead to mistakes [8,9]. Therefore, it is necessary to use other identification methods to analyse them.

Recent progress in molecular biology has led to the development of several techniques for yeast identification based on similarities or dissimilarities of DNA, RNA or proteins. These include electrophoretic karyotyping [10], RFLP (Restriction Fragment Length Polymorphism) of mitochondrial DNA [11], random amplified polymorphic DNA analysis (RAPD), PCR-based techniques such as ribosomal internal transcribed spacers (ITS-PCR) [8], surveys of simple sequences repeats (SSR-Multiplex PCR) [12] or genomic DNA sequencing of the D1/D2 region of 26S rDNA. Some of these methods have been found to be of great interest in enology, due to the high level of resolution in yeast strain characterization, and for making it possible to establish a correlation between the genetic variability and the most important industrial properties of the strains.

A study carried out by [8] showed that *S. cerevisiae flor* yeasts responsible for the biological ageing of Sherry-type wines present a 24-bp deletion, located in the ITS1 region. These results have extended the analysis of this DNA region to flor yeast isolated in the Montilla-Moriles D.O. region [9]. Both studies corroborated that this deletion is fixed in *flor yeast* of the species, *S. cerevisiae*. By using molecular analysis techniques, different properties of industrial interest have been detected, such as physiologic, genetic and metabolic properties in the different identified veil-forming yeast strains. Moreover, in view of the latest yeast taxonomy, it is important to underline that there are no more races. *S. cerevisiae* (beticus) and *S. cerevisiae* (cheresiensis) are no longer considered as races or subspecies of *S. cerevisiae*, according to the last taxonomic study. Indeed, all these previously named races or subspecies are now considered as *S. cerevisiae* synonyms, based on nuclear DNA relatedness [13]. The former race, *S. cerevisiae* (rouxii), has been classified as *Zygosaccharomyces rouxii* and *S. cerevisiae* (montuliensis) is now considered as *Torulaspora delbrueckii* [14]. These latter ones are now classified as ethanol-tolerant non-*Saccharomyces* species and they also should contain the 24-bp deletion in the 131–154 ITS1 region. Other yeast species of genera different to *Saccharomyces* have been isolated and identified by molecular and/or sequencing methods from Sherry-type wines (from the initial phase of the fermentation until the ageing phase), such as *Wickerhamomyces anomalus*, *Pichia membranaefaciens*, *Zygosaccharomyces bailii* or the undesirable species *Dekkera bruxellensis* (anamorph *Brettanomyces bruxellensis*) [3,8,15]. However, the identification strategy should be used more precisely because many database sequences of these species have identical D1/D2 sequences [16]. The comparison of the D1/D2 sequences and ITS sequences between the isolated strains and those of the type strains of the species (most can be found in the CBS database). However, many authors have simply used the first hit in the blast search as the proof of taxonomic affiliation of their isolate. This strategy can be misleading because database curators do not check submissions for the correctness of taxonomic affiliation. In this case, using classical biochemical analyses, in conjunction with molecular methods, can help us to identify and classify the yeast diversity forming the velum of flor in Sherry wines. Therefore, we propose a reliable, reproducible, inexpensive and efficient biochemical test, using polystyrene 96-well plates that could reinforce the identification of the different veil-forming yeast strains.

MATERIALS AND METHODS

Yeast Strains

Six hundred yeast strains were isolated previously from the veil of flor of "Manzanilla" and "Fino" wines, from the Sanlúcar de Barrameda and Jerez wineries (Cádiz, Spain), respectively [2,17]. After a first screening, using a Mitochondrial DNA restriction analysis and electrophoretic karyotyping by pulsed-field-gel electrophoresis (PFGE) and SSR-Multiplex PCR, eighteen different profiles of *Saccharomyces* strains were shown and then grouped according to their similarities—eight from "Manzanilla", denoted as MI to MVI, MX and MXVIII, and ten from "Fino", coded as FI to FX, respectively. Additionally, five potential non-*Saccharomyces*, isolated from "Fino" veils, which showed different karyotypes to *S. cerevisiae*, were subjected to identification by ITS-PCR and sequencing of the D1/

D2 region of 26S rDNA, as stated in [18]. At least two representative members from each ITS-RFLP genotype group were randomly selected for sequencing of the LSU sRNA gene D1/D2 domain.

Mitochondrial DNA (mtDNA) Restriction Analysis

Total DNA extraction and mtDNA restriction analyses of the *Saccharomyces* yeast strains were performed by the method of [11]. Yeast DNA was digested with a *Hinf* I restriction enzyme, by incubation at 37°C, for 3 h. Fragments were separated by electrophoresis on a 1% agarose gel in 1× TBE buffer (45 mM tris-borate, 1 mM EDTA) and with 0.5 µg/mL of ethidium bromide. The image of the gel was digitalized in a Molecular Imager apparatus (Gel-Doc XR, Hercules, CA, USA) and analyzed using Quantity One 1-D software (Bio-Rad, Hercules, CA, USA).

Electrophoretic Karyotype

Electrophoretic karyotypes of the *Saccharomyces* and *non-Saccharomyces* yeast strains were obtained by PFGE. The colonies were embedded in agarose plugs and treated following a procedure based on the protocol of [19], and chromosomes were separated using a CHEF-DR III apparatus (Bio-Rad). Gels were made of 1% pulsed-field certified agarose (Bio-Rad, Hercules, CA, USA) in 0.5× TBE electrophoresis buffer. The electrophoretic conditions were as follows: an initial pulse of 60 s and a final pulse of 120 s, during 22 h, at 14°C and 6 V/cm, with an angle of 120°. Gels were stained with ethidium bromide solution (0.5 µg/mL ethidium bromide, 0.5× TBE buffer) after electrophoresis, and the image was captured under UV light with a camera (Bio-Rad Imaging System, Hercules, CA, USA). The image of the gel was digitalized in a Molecular Imager apparatus (Gel-Doc) and analyzed using Quantity One-1D software (Bio-Rad). Results were analyzed by comparing the electrophoretic mobility of the chromosomes with a reference strain (*Saccharomyces cerevisiae* YNN195).

Multiplex PCR-Microsatellite (SSR) Analysis

To perform multiplex PCR of microsatellite loci, three loci were simultaneously amplified, using the primers SC8132X (Fw: CTGCTCAACTTGTGATGGGTTTTGG; Rv: CCTCGTTACTATC GTCTTCATCTTGC), YOR267C (Fw: GGTGACTCTAACGGCAGAGTGG; Rv: GGATCTACTTGCA GTATACGGG) and SCPTSY7 (Fw: AAAAGCGTAAGCAATGG TGTAGAT; Rv: AAATGATGCCAATATTGAAAAGGT) (MWG Biotech AG, Ebersberg, Germany), proposed by Vaudano and García-Moruno [12]. PCR amplifications were performed in an Applied Biosystems SimplyAmp Termal Cycler in 50 µL volumes, constituted by: 2 µL of DNA solution (30–90 ng of DNA), 2.5 mM $MgCl_2$, 200 µM dNTPs, 10 µL of 5× PCR buffer, 2 U of GoTaq® Flexi DNA Polymerase (Promega) and 10 pmol of primers for loci, SCYOR267C and SC8132X, and 40 pmol of primers for locus SCPTSY7. By using the protocol—4 min at 94°C, 28 cycles of 30 s at 94°C, 45 s at 56°C and 30 s at 72°C, and 10 min at 72°C—PCR products were separated on an agarose gel (2.5% *w/v*) with 5 µg/mL of ethidium bromide in 1× TBE buffer at 115 V for 45 min. DNA fragment sizes were determined by comparison with a molecular ladder marker of 50 bp (Biolab, London, UK). The gel was digitalized in a Molecular Imager apparatus (Gel-Doc XR) and analyzed using Quantity One 1-D software (Bio-Rad).

Physiological and Biochemical Tests

To perform a quantitative analysis of carbon fermentation and assimilation, nitrogen assimilation, osmo tolerance, and resistance to cycloheximide, different solutions, filtered through 0.22-µm Stericup filters (Millipore, Burlington, MA, USA) prior to their addition, were incorporated into the wells of polysterene 96-well microplates (Nunc™ 96-well polystyrene conical bottom Microwell™, Thermofisher, Denmark). The basal medium for carbohydrate fermentation and assimilation tests (BMC) was as follows: sterile deionized water 100 mL, 1 g yeast nitrogen base (YNB, Difco) was prepared with 2 g (*w/v*) solutions of fructose, glucose, maltose, lactose and sucrose and 4 g (*w/v*) of raffinose for fermentation tests, and 0.78 g of glucose, galactose, sorbose, glucosamine, ribose, xylose, arabinose, rhamnose, sucrose, maltose, α-threalose, cellobiose, salicin, arbutin, melibiose, lactose, raffinose, melezitose, inulin, starch, erythrytol, ribitol, xylitol, L-arabinitol, D-glucitol, D-mannitol, myo-inositol, D-Glucono-1,5-lactone, 5-Keto-D-gluconate, D-gluconate, D-galacturonate, DL-lactate, succinate, citrate and L-tartaric acid, and 0.78 mL of glycerol, propane 1,2-diol, butane 2,3-diol, Tween 20, Tween 80 for assimilation tests. Nitrogen sources for assimilation were prepared by supplementing 100 mL of an autoclaved basal assimilation medium yeast carbon base (YNB, DIfco) (0.17 g YNB without amino acids, 0.5 g ammonium sulphate, 2 g glucose) with filter-sterilized solutions containing 0.5 g of ammonium sulphate, 0.18 g of ammonium citrate, 0.11 g of nitrate, 0.04 g of nitrite, 0.1 g of ethylamine and creatine, 0.087 g of creatinine, 0.11 g of glucosamine, 0.1 g of urea, and D-proline. Then, 180 µL of each substrate was dropped into the wells of the trays. Also, 180 µL of 2%, 20%, 30%, 40% and 50% of a 1:1 mixture of glucose and fructose solution were incorporated into the wells as an osmotolerance test. Wells for testing the resistance to cycloheximide were prepared by adding 180 µL of a sterilized 0.1% *w/v* cycloheximide solution. The medium in each well can be dehydrated by drying the plate with the lid in a Speed-Vac vacuum oven (Eppendorf), which forms a small visible residue at the base of the well. After drying, individual trays can be sealed by heat with a polypropylene-aluminium sealing sheet and can be stored at 4°C for six months, or at -20°C for one year until subsequent use, rehydrating them with the corresponding basal medium.

The selected yeast strains were previously grown for 24–48 h in 15 mL of YPD (1% yeast extract, 2% peptone, 2% glucose) at 28°C overnight; when achieved, these were washed twice with sterile, distilled-water. Twenty microliters of inoculum (McFarland standard 2, diluted by a factor of 10) were introduced into each well using a multi-channel pipette. McFarland standards are used as a reference to adjust the turbidity of yeast suspensions, so the number of yeast will be within a given range, to standardize biochemical tests. A 0.5 McFarland standard was prepared by mixing 0.05 mL of 1.175% barium chloride dehydrate with 9.95 mL of 1% sulfuric acid. The standards can be compared by eye to a suspension of yeasts in sterile water. If the yeast suspension is too turbid, it can be diluted. If the suspension is not turbid enough, more yeasts can be added (Table 1).

Table 1. McFarland standard values used as a reference to adjust the turbidity of yeast suspensions.

McFarland Standard	0.5	1	2	3	4
1% Barium chloride (mL)	0.05	0.1	0.2	0.3	0.4
1% Sulfuric acid (mL)	9.95	9.9	9.8	9.7	9.6
Cell density ± 108 cells/mL	1.5	3	6	9	12
Absorbance (610 nn)	0.1	0.25	0.45	0.60	0.67

Microplates were sealed with Breathe-Easy membranes (Sigma, Sant Louis, MI, USA) and placed into an automatic microplate reader (Multiskan go reader, Thermo Fischer Scientific, Waltham, MA, USA). The optical density of the mix was read at 610 nm every hour (one minute of shaking just before automatic reading by the microplate reader) at 25°C over 72 h. Data were collected and processed with the software, SkanIt RE for Multiskan GO 3.2. The absorbance data were exported to MS Excel for processing. A reading of optical density (OD610) = 1, measured by the microtiter reader, corresponds to 6–7 × 10^7 cells/mL. Absorbance values at 610 nm were transformed into negative values when OD < 0.4 and positive when OD > 0.4, after 72 h.

Validation and Reproducibility of the Tests in Microtiter Plates

The yeast strains used in this study were also subjected to a standard test in Durham tubes, following the protocol proposed by [20] and under the same conditions of temperature and time. These tests were repeated with the microplate tests six times with each yeast strain. The stability of the carbon and nitrogen medium of the microtiter plates in storage was also examined. The same yeast species were analyzed for their reactions in microtitre trays that had been freshly prepared and in those that had been stored for six months at 4°C and for one year at -20°C.

RESULTS

A total of 18 strains belonging to *Saccharomyces cerevisiae* species were characterized using the pulsed-field-gel electrophoresis (PFGE) technique and mitochondrial DNA restriction analysis. Moreover, analyses of polymorphic microsatellite loci and the proposed biochemical test, using 96-well plates, were used. Chromosomal profile, obtained from the studied strains, showed fourteen different patterns. Specifically, strains FII and FIII, FV and FVI, FVIII and FIX, MX and MXVIII showed common patterns. In regard to the SSR analysis, fourteen different patterns were also obtained, which complied with the results obtained in the PFGE (Figure 1).

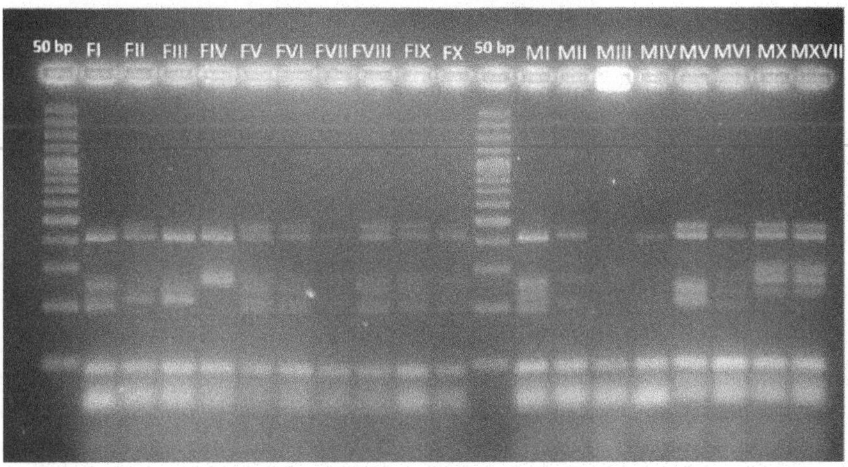

Figure 1. Patterns obtained by the simple sequences repeats (SSR)-Multiplex PCR of the *S. cerevisiae* flor yeasts, selected after the first screening. Strains FII and FIII, FV and FVI, FVIII and FIX, MX and MXVIII showed common patterns.

On the other hand, using mitochondrial DNA restriction analyses with the enzyme *Hinf* I, three different patterns were obtained. The most common pattern was H1 (corresponding to FIV, FV, FIX, MI, MII, MIII, MVIII and MX), the second was H2 (corresponding to FI, FII, MIV and MV) and the pattern H3 only corresponded to one strain (MVI).

Fifteen different strains were determined by establishing the correspondence between the three patterns obtained from the analysis of mtRFLP with *Hinf*I—coded as H1, H2 and H3—and fourteen patterns were obtained from the polymorphism study of the tested karyotypes and microsatellite loci. These fifteen different strains are FI, FII, FIII, FIV, FV, FVI, FIX, FX, MI, MII, MIII, MIV, MV, MVI and MX, according to the different combinations observed. The potential five non-*Saccharomycces* species, analyzed by the molecular method mentioned in the material and methods section (Figure 2) and after sequencing of the D1/D2 region, have been identified as *Zygosaccharomyces bailii* (NsI), *Torulaspora delbrueckii* (NsII), *Zygosaccharomyces rouxii*

(NsIII), *Rhodotorula mucilaginosa* (NsIV) and *Rhodotorula minuta* (NsV). However, the species, *Z. bailii* and *Z. rouxii*, and *R. mucilaginosa* and *R. minuta* showed very similar (less than 1%), or identical, D1/D2 sequences to those found in the databases.

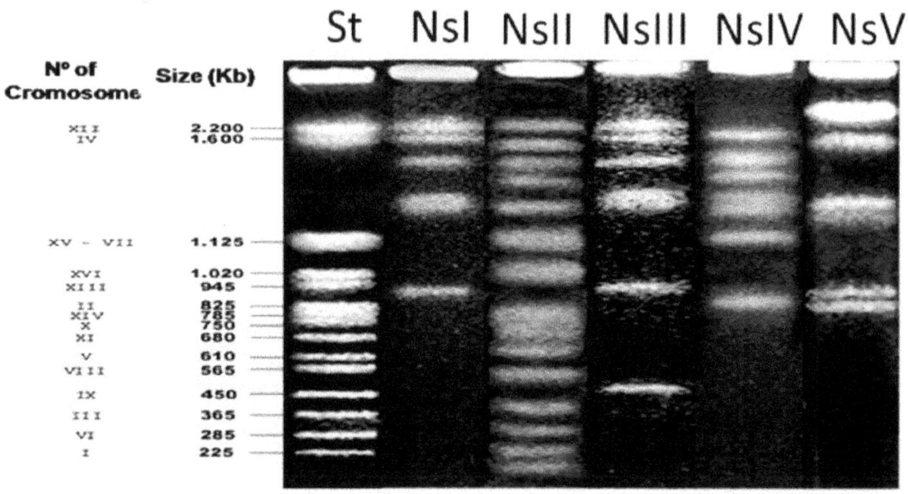

Figure 2. Karyotypes obtained by pulsed-field-gel electrophoresis (PFGE) of the *non-Saccharoymyces* veil-forming yeasts isolated in "Fino" Sherry wines. The species were identified as *Zygosaccharomyces bailii* (NsI), *Torulaspora delbrueckii* (NsII), *Zygosaccharomyces rouxii* (NsIII), *Rhodotorula mucilaginosa* (NsIV) and *Rhodotorula minuta* (NsV).

The selected strains were also analyzed to determine their fermentation and assimilation abilities, to evaluate if they shared physiological characteristics. Tables 2 and 3, show the results of those which reached a positive result, among all tested carbon and nitrogen sources.

Table 2. Fermentation of different carbohydrates by selected *S. cerevisiae* flor yeasts, analyzed with microtiter technology. Values are expressed as (+) if yeast strains were able to ferment the carbon/nitrogen source and (-) if yeast strains were not able to ferment them.

Sugar	Sugar Fermentation					
Designation	Fructose	Glucose	Lactose	Maltose	Rafinose	Sucrose
FI	-	+	-	-	-	-
FII	+	+	-	-	+	+
FIII	+	+	-	-	-	-
FIV	+	+	-	-	-	+
FV	+	+	-	-	-	+
FVI	+	+	-	-	+	+

Sugar Designation	Sugar Fermentation Fructose	Glucose	Lactose	Maltose	Rafinose	Sucrose
FIX	–	+	–	–	+	+
FX	–	+	+	–	–	–
MI	+	–	–	–	–	+
MII	+	+	+	–	+	+
MIII	–	–	+	–	–	–
MIV	+	+	+	–	+	–
MV	–	+	–	+	–	–
MVI	+	+	–	–	+	+
MX	+	+	–	–	+	+

Table 3. Assimilation of different carbon/nitrogen compounds by selected *S. cerevisiae* flor yeasts, analyzed with microtiter technology. Values are expressed as (+) if yeast strains were able to assimilate the carbon/nitrogen source and (-) if yeast strains were not able to assimilate them.

Strain Designation	Carbon/Nitrogen Compounds Ammonium Citrate	Glucose	Inulin	Melibiose	Nitrite	Starch	Urea
FI	+	–	–	–	–	–	–
FII	+	–	–	–	–	.	–
FIII	+	–	–	–	–	–	–
FIV	+	–	–	–	–	–	–
FV	+	–	–	–	–	–	–
FVI	–	–	–	–	–	–	–
FIX	–	+	+	–	–	–	–
FX	–	–	–	–	–	–	–
MI	+	–	–	–	–	–	–
MII	+	–	+	–	–	–	+
MIII	+	–	+	–	–	–	+
MIV	+	–	–	–	–	–	–
MV	–	+	–	+	+	–	–
MVI	+	–	–	–	–	+	–
MX	–	–	+	–	–	+	–

These tests, repeated six times, were also carried out in Durham tubes, obtaining the same results in all cases, with the exception of a couple of cases (fermentation of raffinose by yeast strain FII and fermentation of sucrose by yeast strain MVI), in which Durham tubes showed questionable results. However, these results obtained by the standard method, had positive results in the microtiter-read assay (OD > 0.4). Although molecular analyses have shown similarities between some strains, the diversity of the fermentation and assimilation abilities were significantly higher (Tables 2 and 3). In the case of the sequenced non-*Saccharomyces* yeast strains, *Z. bailii* did not grow on maltose, which is usually assimilated by *Z. rouxii* and *R. mucilaginosa* isolates grew on maltose, sucrose, lactose, which are not assimilated by *R. minuta*. These results are in accordance with those published in [14,21]. Thus, by applying the biochemical tests, we have ensured that these yeast strains have been satisfactorily identified (Table 4).

Table 4. Assimilation of different carbon/nitrogen compounds by non-*Saccharomyces* species, isolated from "Fino" Sherry wine, analyzed with microtiter technology. Values are expressed as (+) if yeast strains were able to assimilate the carbon/nitrogen source and (-) if yeast strains were not able to assimilate them.

Strain	Carbon/Nitrogen Compounds							
Designation	Raffinose	Galactose	Inulin	Melibiose	Maltose	Lactose	L-Sorbose	Ethanol
NsI	-	+	-	-	-	-	-	+
NsII	+	+	+	-	+	-	-	+
NsIII	-	+	-	-	+	-	-	+
NsIV	+	+	-	-	+	-	-	+
NsV	-	-	-	-	-	+	+	+

Regarding the results obtained for osmotolerance, only two isolates grew in the media supplemented with 50% of glucose: fructose, but neither belonged to the *Saccharomyces* species (Table 5). The most osmotolerant species was *Torulaspora delbrueckii*. The tests for cycloheximide resistance gave the same results when tested by either the standard or microtitre method (Table 5).

Table 5. Osmotolerance and cycloheximide resistance of the veil-forming yeasts used in this study and analyzed by microtiter technology. Values are expressed as (+) if yeast strains were able to growth in a particular concentration of glucose/fructose solution and (-) if yeast strains were not able to growth.

Strain	Osmotolerance: Growth on Media Supplemented with 1:1 (Glucose:Fructose) and Cycloheximide Resistance						
Designation	2%	20%	30%	40%	50%	Cycloheximide 0.01%	Cycloheximide 0.1%
FI	+	+	+	-	-	-	-
FII	+	+	-	-	-	-	-
FIII	+	+	+	-	-	-	-
FIV	+	+	+	-	-	-	-

Strain	Osmotolerance: Growth on Media Supplemented with 1:1 (Glucose:Fructose) and Cycloheximide Resistance						
Designation	2%	20%	30%	40%	50%	Cycloheximide 0.01%	Cycloheximide 0.1%
FV	+	+	+	-	-	-	-
FVI	+	+	+	-	-	-	-
FVII	+	+	-	-	-	-	-
FVIII	+	+	-	-	-	-	-
FIX	+	+	+	-	-	-	-
FX	+	+	+	-	-	-	-
MI	+	+	+	-	-	-	-
MII	+	+	+	-	-	-	-
MIII	+	+	+	-	-	-	-
MIV	+	+	+	-	-	-	-
MV	+	+	+	-	-	-	-
MVI	+	+	+	-	-	+	-
MX	+	+	+	-	-	+	-
MXVIII	+	+	+	-	-	-	-
NsI	+	+	-	-	-	-	-
NsII	+	+	+	+	+	-	-
NsIII	+	+	+	+	+	-	-
NsIV	+	+	-	-	-	+	-
NsV	+	+	-	-	-	+	-

The reproducibility of the tests in microtiter plates was assessed, using the selected *Saccharomyces* and non-*Saccharomyces* identified in this study. They were examined six times in microtiter plates. Responses to the identification tests for each species were the same in all replications. Moreover, the stored 96-well plates gave the same reaction responses, indicating that the multiplates could be stored for at least one year at -20°C without affecting identification results.

DISCUSSION

In recent years, multiplex microsatellite analysis has been implemented as one of the most decisive techniques in determining polymorphisms within the same species, but it has not often been applied to flor yeast analyses, since it could show few different patterns. However, in our study microsatellite analysis have shown fourteen different patterns, obtaining the same results as electrophoretic karyotyping. In any case, by combining multiplex microsatellite and RFLP of mtDNA analyses, when required, it is possible to obtain the same number of patterns as using electrophoretic karyotyping; these results are in agreement with those found in [9].

Pulsed-field gel electrophoresis (PFGE) has a good reproducibility and high discriminatory power, but it is an expensive and time-consuming technique. On the other hand, multiplex PCR of microsatellite loci [12] also has a high discriminatory power, sample preparation is easier and it is faster than PFGE.

Results obtained in the characterization of flor strains are highly variable, since some authors [8] have identified a higher degree of polymorphism between flor strains in mtDNA restriction than karyotyping. However, our study revealed the opposite: mtDNA restriction showed only three different patterns and PFGE showed fourteen different patterns. These results are similar to [9]. Mitochondrial DNA restriction analysis showed less polymorphisms due to the conditions of the wineries, since these DNA could be subjected to positive selection pressure based on their ethanol tolerance (15% *v/v*) [22] and other environmental factors, such as nutrient composition, growth temperature, and presence of toxic drugs, heavy metals, oxidizing agents, and osmotic/ionic stresses [23]. This might be the reason why the mtDNA restriction analysis showed only three different patterns of all the strains that we analyzed.

Biochemical tests offer supplementary information for characterizing flor strains. Traditionally, veil-forming yeast strains have been classified into four races belonging to the *S. cerevisiae* species, currently considered synonyms of the *S. cerevisiae* species [14]; another race, previously referred to as *S. cerevisiae* var. montuliensis, that has recently been reclassified as *Torulaspora delbrueckii* [14,24], and a last race named *S. cerevisiae rouxii*, is now *Zygosaccharomyces rouxii* [14]. This classification was based on morphological and metabolic studies of fermentation and assimilation of different carbon and nitrogen sources that differentiate these four main breeds [25]. The studies of assimilation and fermentation of sugars that we have carried out in this work have shown a high variability, so we were not able to differentiate races within *S. cerevisiae* strains. We can conclude that biochemical tests are necessary for identifying and classifying the different yeast strains that form the veil of flor of Sherry wine, because we believe that each yeast strain brings to the Sherry wines a different organoleptical characteristic. Thus, the application of this simplified method, using polysterene 96-well plates, is an inexpensive, low-time consuming and efficient technique, that can be used in combination with the molecular methods described above, to ensure veil-forming yeast diversity in Sherry wines.

SUMMARY

Yeasts involved in veil formation during biological ageing of Sherry wines are mainly *Saccharomyces cerevisiae*, and they have traditionally been divided into four races or varieties: beticus, cheresiensis, montuliensis and rouxii. Recent progress in molecular biology has led to the development of several techniques for yeast identification, based on similarity or dissimilarity of DNA, RNA or proteins. In view of the latest yeast taxonomy, there are no more races. However, molecular techniques are not enough to understand the real veil-forming yeast diversity and dynamics in Sherry wines. We propose a reliable method, using a microtiter reader, to evaluate the fermentation and assimilation of carbon and nitrogen sources, the osmotolerance and the antibiotic resistance, using 18 *S. cerevisiae* and 5 non-*Saccharomyces* yeast strains, to allow correct identification and classification of the yeast strains present in the velum of flor complex.

REFERENCES

1. Rodríguez, M.E.; Infante, J.J.; Mesa, J.J.; Rebordinos, L.; Cantoral, J.M. Enological behavior of biofilms formed by genetically-characterized strains of sherry flor yeast. *Open Biotechnol. J.* **2013**, *7*, 23–29. [CrossRef]
2. Mesa, J.J.; Infante, J.J.; Rebordinos, L.; Cantoral, J.M. Characterisation of yeasts involved in the biological ageing of sherry wines. *LWT-Food Sci. Technol.* **1999**, *32*, 114–120. [CrossRef]
3. Esteve-Zarzoso, B.; Peris-Torán, M.J.; García-Maiquez, E.; Uruburu, F.; Querol, A. Yeast population dynamics during the fermentation and biological aging of sherry wines. *Appl. Environ. Microbiol.* **2001**, *67*, 2056–2061. [CrossRef] [PubMed]
4. Martínez, P.; Codon, A.C.; Pérez, L.; Benítez, T. Physiological and molecular characterization of flor yeasts: Polymorphism of flor yeast population. *Yeast* **1995**, *11*, 1399–1411. [CrossRef] [PubMed]
5. Boekhout, T.; Kurtzman, C.P. Principles and methods used in yeast classification, and an overview of currently accepted yeast genera. In *Nonconventional Yeasts in Biotechnology*; Springer: Berlin/Heidelberg, Germany, 1996; pp. 1–81. ISBN 978-3-642-79856-6.
6. Vancanneyt, B.P.; Hennebert, G.; Kersters, K. Differentiation of yeast species based on electrophoretic whole-cell protein patters. *Syst. Appl. Microbiol.* **1991**, *14*, 23–32. [CrossRef]
7. Silva, M.M.D.; Malfeito-Ferreira, M.; Loureiro, V.; Aubyn, A.S. Long-chain fatty acid composition as a criterion for yeast distinction in the brewing industry. *J. Inst. Brew.* **1994**, *100*, 17–22. [CrossRef]

8. Esteve-Zarzoso, B.; Fernández-Espinar, M.T.; Querol, A. Authentication and identification of *Saccharomyces cerevisiae* "flor" yeast races involved in sherry ageing. *Antonie van Leewenhoek* **2004**, *85*, 151–158. [CrossRef] [PubMed]

9. Marin-Menguiano, M.; Romero-Sánchez, S.; Barrales, R.R.; Ibeas, J.I. Population analysis of biofilm yeasts during fino sherry wine aging in the Montilla Moriles D.O. region. *Int. J. Food Microbiol.* **2017**, *244*, 67–73. [CrossRef] [PubMed]

10. Rodríguez, M.E.; Infante, J.J.; Molina, M.; Rebordinos, L.; Cantoral, J.M. Using RFLP-mtDNA for the rapid monitoring of the dominant inoculated yeast strains in industrial wine fermentation. *Int. J. Food Microbiol.* **2011**, *145*, 331–335. [CrossRef] [PubMed]

11. Querol, A.; Barrio, E.; Huerta, T.; Ramón, D. Molecular monitoring of wine fermentatios conducted by active dry yeast strains. *Appl. Environ. Microbiol.* **1992**, *58*, 2948–2953. [PubMed]

12. Vaudano, E.; García-Moruno, E. Discrimination of *Saccharomyces cerevisiae* wine strains using microsatellite multiplex PCR and band pattern analysis. *Food Microbiol.* **2008**, *25*, 56–64. [CrossRef] [PubMed]

13. Legras, J.C.; Moreno-García, J.; Zara, S.; Zara, G.; García-Martínez, T.; Mauricio, J.C.; Mannazzu, I.; Coi, A.L.; Zeidan, M.B.; Dequin, S.; et al. Flor yeast: New perspectives beyond wine aging. *Front. Microbiol.* **2016**, *7*, 503. [CrossRef] [PubMed]

14. Kurtzman, C.; Fell, J.W.; Boekhout, T. *The Yeasts: A Taxonomic Study*; Elvesier: London, UK, 2011; p. 2363.

15. Fernández-Espinar, M.T.; Esteve-Zarzoso, B.; Querol, A.; Barrio, E. RFLP analysis of the ribosomal internal transcribed spacers and the 5.8 S rRNA gene region of the genus *Saccharomyces*: A fast method for species identification and the differentiation of flor yeasts. *Antonie van Leeuwenhoek* **2000**, *78*, 87–97. [CrossRef]

16. Lopandic, K.; Pfliegler, W.P.; Tiefenbrunner, W.; Gangl, H.; Sipiczki, M.; Sterlinger, K. Genotypic and phenotypic evolution of yeast interspecies hybrids during high-sugar fermentation. *Appl. Microbiol. Biotechnol.* **2016**, *100*, 6331–6343. [CrossRef] [PubMed]

17. Espinazo-Romeu, M.; Cantoral, J.M.; Matallana, E.; Aranda, A. Btn2p is involved in ethanol and biofilm formation in flor yeast. *FEMS Yeast Res.* **2008**, *8*, 1127–1136. [CrossRef] [PubMed]

18. Cordero-Bueso, G.; Rodríguez, M.E.; Garrido, C.; Cantoral, J.M. Rapid and not culture-dependent assay based on multiplex PCR-SSR analysis for monitoring inoculated yeast strains in industrial wine fermentations. *Arch. Microbiol.* **2017**, *199*, 135–143. [CrossRef] [PubMed]

19. Rodríguez, M.E.; Infante, J.J.; Molina, M.; Domínguez, M.; Rebordinos, L.; Cantoral, J.M. Genomic characterization and selection of wine yeast to conduct industrial fermentations of a white wine produced in a SW Spain winery. *J. Appl. Microbiol.* **2010**, *108*, 1292–1302. [CrossRef] [PubMed]

20. Van Der Walt, J.P.; Yarrow, D. Methods for the isolation, maintenance, classification and identification of yeasts. In *The Yeasts: A Taxonomic Study*, 3rd ed.; Kreger-van Rij, N.J.W., Ed.; Elvesier Science Publishers: Amsterdan, The Netherlands, 1984; pp. 45–104.

21. Sipiczki, M. Overwintering of vineyard yeasts: Survival of interacting yeast communities in grapes mummified on vines. *Front. Microbiol.* **2016**, *7*, 212. [CrossRef] [PubMed]

22. Ibeas, J.I.; Jiménez, J. Mitochondrial DNA loss caused by etanol in *Saccharomyces* flor yeasts. *Appl. Environ. Microbiol.* **1997**, *63*, 7–12. [PubMed]

23. Kvitek, D.J.; Will, J.L.; Gasch, A.P. Variations in Stress Sensitivity and Genomic Expression in DiverseS. *cerevisiae* Isolates. *PLoS Genet.* **2008**, 4, e1000223. [CrossRef] [PubMed]

24. Alexandre, H. Flor yeasts of *Saccharomyces cerevisiae*—Their ecology, genetics and metabolism. *Int. J. Food Microbiol.* **2013**, *167*, 269–275. [CrossRef] [PubMed]

25. Barnett, J.A.; Payne, R.W.; Yarrow, D. *Yeasts: Characterisation and Identification*, 3rd ed.; Cambridge University Press: Cambridge, UK, 2000; p. 1139.

CHAPTER-13

TWO NOVEL STRAINS OF *TORULASPORA DELBRUECKII* ISOLATED FROM THE HONEY BEE MICROBIOME AND THEIR USE IN HONEY FERMENTATION

INTRODUCTION: AN OVERVIEW

In the intricate tapestry of the natural world, the microbiome of the honey bee has recently revealed two novel strains of Torulaspora delbrueckii, a yeast species known for its significance in various fermentation processes. These newly isolated strains have captured the attention of researchers and fermentologists alike due to their unique characteristics. This study delves into the exploration of these two distinct Torulaspora delbrueckii strains identified within the honey bee microbiome, shedding light on their genetic makeup and metabolic traits. More notably, the research focuses on the application of these strains in the fascinating realm of honey fermentation, unraveling the potential implications for the flavour profiles and overall quality of fermented honey products. As we embark on this scientific journey, the aim is to not only broaden our understanding of microbial biodiversity within the honey bee community but also to unlock new possibilities for enhancing the art and science of honey fermentation using these exceptional Torulaspora delbrueckii strains.

Mankind has been fermenting foods and beverages for thousands of years [1-4], which certainly predates our knowledge that yeasts were the microbial agents responsible for metabolizing sugar into alcohol. Honey, which is produced by honey bees such as *Apis mellifera*, is a natural source of abundant, readily fermentable sugar and has been found as an ingredient in some of the earliest known fermented beverages [2]. When a dilute honey solution is fermented on its own without the addition of fruit/fruit juice or other additives, this creates a traditionally strong (8–18% alcohol by volume, ABV) beverage called mead (reviewed in [5]).

Much like craft brewing, mead making is currently experiencing a renewed interest at the amateur and professional levels worldwide. Indeed, one average, a new meadery opens in the United States every 3 days, and one opens every 7 days in the rest of the world [6]. With this proliferation of mead making has come experimentation with the traditional methods, including aging mead in oak or previously used spirit barrels and fermentation with non-*Saccharomyces* microbes to generate sour and "funky" meads [6]. In many ways, this again mirrors the craft brewing scene and current popularity of sour beer [7–9].

Another factor driving innovation in fermented beverage production is the "local movement" [10,11], i.e., the push to use locally sourced ingredients. In the case of mead, it is often relatively simple to utilize locally produced honey, but few meaderies use indigenous microbes unless they rely on those naturally found in the honey itself. Such microbes are naturally resistant to high osmotic stress from concentrated sugar in honey and typically come from pollen/flower, bee gut microbiome, and dirt/dust contamination of the honey (reviewed in [12,13]). These organisms include multiple species from dozens of different genera of bacteria (e.g., *Bacillus*, *Citrobacter*, and *Lactobacillus*), yeast (e.g., *Saccharomyces*, *Schizosaccharomyces*, and *Pichia*), and molds.

Here, we describe the bio-prospecting for novel yeasts to use in honey fermentation. We concentrated on ultra-local strains by attempting to enrich for ethanol-tolerant yeasts in honey, the hive, and honey bees themselves. Ultimately, we recovered two strains of the yeast *Torulaspora delbrueckii* from the honey bee microbiome and characterized them for their ability to ferment honey. Many *T. delbrueckii* strains have previously been examined for their fermentative abilities and were found to improve wine quality (see [14] and references therein) and create distinct beer flavour in beer [15–17]. Compared to *S. cerevisiae*, *T. delbrueckii* is generally found to generate lower levels of ethanol [16] but a wider range of fruity aromas [18,19]. However, in many of these cases,*T. delbreuckii* is used in combination with *S. cerevisiae* for sequential fermentations [14] or mixed culture fermentations [15].

Currently, there are no reports of *T. delbrueckii* being used to make mead and how this species may impact the sensory qualities of the final product. Here, we found that, compared to a traditional *Saccharomyces cerevisiae* control strain (WLP715), the mead produced by the *T. delbrueckii* strains YH178 and YH179 scored more favourably during sensory analysis. Further, *T. delbrueckii* YH178 and YH179 were successfully used in mixed honey fermentations with *S. cerevisiae* WLP715 on a production scale. These data suggest that *T. delbrueckii* YH178 and YH179 have beneficial honey fermentation profile and are suitable for mead production by home-brewers and professional mead makers.

RESULTS AND DISCUSSION

Isolation of T. delbrueckii Strains YH178 and YH179

We initially set out to isolate ethanol-tolerant yeasts from multiple bee-related sources. Samples of honey, honeycomb, propolis, and fragments of a wooden beehive were surveyed for yeasts, but no microbes were recovered using our wild yeast enrichment protocol [7]. This failure was likely not due to the stringency (e.g., high ethanol and osmotic stress tolerance) of the method used because we previously recovered hundreds of wild ethanol-tolerant yeasts from a variety of environmental sources. However, as only small volumes of samples were surveyed from a single hive, a more extensive search may have yielded suitable yeasts. Alternatively, the lack of ethanol-tolerant yeasts may be attributed to the well-characterized hygienic behaviour of honey bees, which are known to keep their hives clean in order to avoid disease [20–23].

We next focused our bio-prospecting efforts on the bees themselves. We again failed to isolate ethanol-tolerant yeast from the outer surface of several bees via a quick rinse with yeast enrichment medium or prolonged incubation of a bee submerged in the same medium at 30°C with aeration. However, when a bee was submerged in yeast enrichment medium and its abdomen was ruptured, microbial growth was visible in the culture after 24 h. Plating for single colonies on Wallerstein Laboratories nutrient (WLN) agar revealed that two distinct strains of yeast were present, which differed in colour (Figure 1A). These strains were designated YH178 (darker green on WLN agar) and YH179 (pale green) as part of our larger yeast hunting efforts and were identified as *T. delbrueckii* isolates by rDNA sequencing (Figure 1B). For comparison, the *S. cerevisiae* strain WLP715 is shown, which attains an even lighter shade of green than YH179 on WLN agar (Figure 1A).

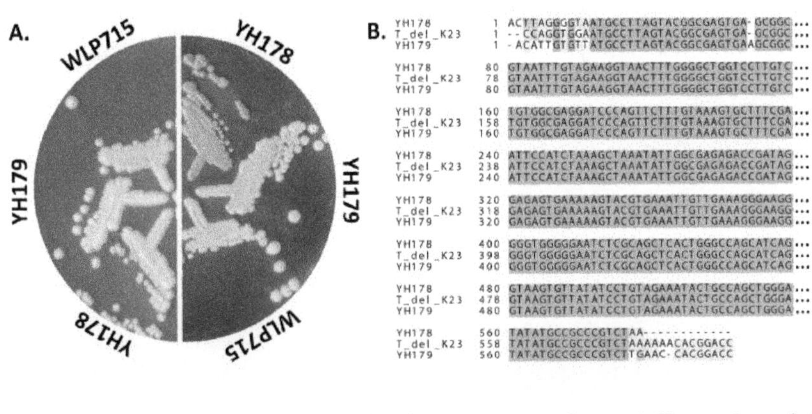

Figure 1. YH178 and YH179 are both strains of *T. delbrueckii*. (A) Colony morphology of YH178, YH179, and *S. cerevisiae* WLP715 on YPD (left) and WLN

(right) agar plates. The indicated strains were streaked for single colonies and grown at 30 °C for 2 days prior to capturing images of the plates with a flatbed scanner. (B) Sequence alignments revealed that YH178 and YH179 are isolates of *T. delbrueckii*. A portion of the 26S rDNA of YH178 and YH179 was PCR-amplified and sequenced, and the sequences are aligned to the closest related strain in the NCBI nucleotide database. T_del_K23 denotes the 26S rDNA gene of *T. delbrueckii* strain K23 (GenBank accession KP852445.1). The ellipsis at the end of each line represents the missing sequence that was truncated due to spatial constraints. The full alignment is shown in Figure S1. Green: fully conserved; yellow: conserved among two sequences; blue: similarity (pyrimidines); dashes: sequence gaps due to indels.

Although no colony size variation was noted on either YPD or WLN plates at the end of a 2-day incubation (Figure 1A), YH178 attained this colony size faster than YH179 when observed prior to the end point of growth (data not shown). This suggested that YH178 has a faster doubling time than YH179. However, plotting growth curves for these strains grown in YPD medium revealed that YH178 had a shorter lag phase (<4 h; Figure 2A) to exponential growth than YH179 (>6 h; Figure 2B). Otherwise, these strains displayed similar doubling times and final cell densities.

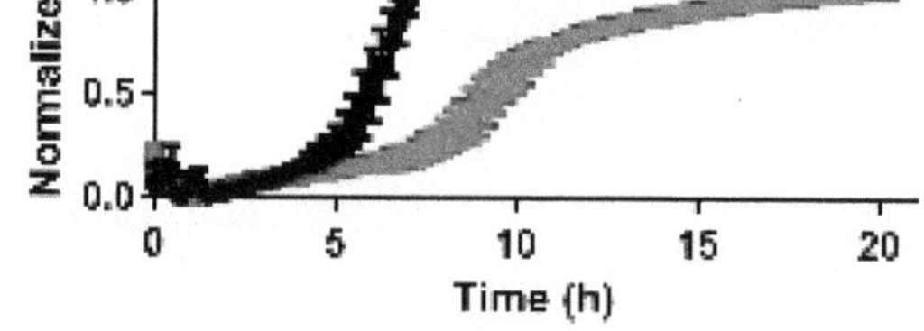

Figure 2. ***Cont.***

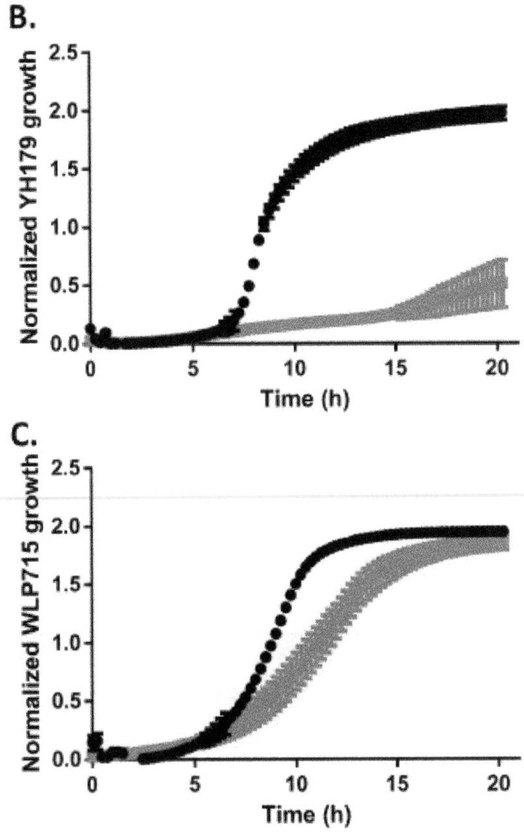

Figure 2. YH178 and YH179 grow poorly in the presence of maltose. YH178 (A), YH179 (B), and *S. cerevisiae* WLP715 (C) were grown to saturation overnight in YPD medium, back-diluted into YPD (black) or YPM (red) medium, and grown overnight in a 96-well plate. Growth was monitored by recording the OD660 of each culture and is plotted vs. time. The plotted points are the averages of three independent experiments, and the error bars are the standard deviations.

YH178 and YH179 Can Use Honey as a Carbon Source during Fermentation

To determine if YH178 and/or YH179 could ferment honey, we inoculated a diluted honey solution with an equivalent number of cells from the *T. delbrueckii* strains, singly and in combination. As a positive control for fermentation, we also used *S. cerevisiae* WLP715, which is marketed as an appropriate yeast for wine, mead, and cider fermentations [24]. The fermentations were monitored for 20 days, and the results are shown in Figure 3A. At approximately 12 h after inoculation, all cultures displayed signs of fermentation, and WLP715 quickly attenuated the honey solution to a final alcohol by volume (ABV) of 11.5% in 2 weeks (Table 1). The

$t1/2$ (time necessary for 50% attenuation) was <2 days. In contrast, YH178, YH179, and the combination thereof fermented the honey solution at a much more modest pace. The $t1/2$ for these fermentations was ~4–4.5 days, and while the final ABV for YH178 was identical to that of WLP715 (11.5%), the YH179 and YH178/YH179 combination fermentations only reached 11% and 11.2%, respectively (Table 1). Subsequent experiments did, however, reveal that additional incubation time of the YH179 and YH178/YH179 cultures led to the same terminal ABV of11.5% by 35 and 28 days, respectively.

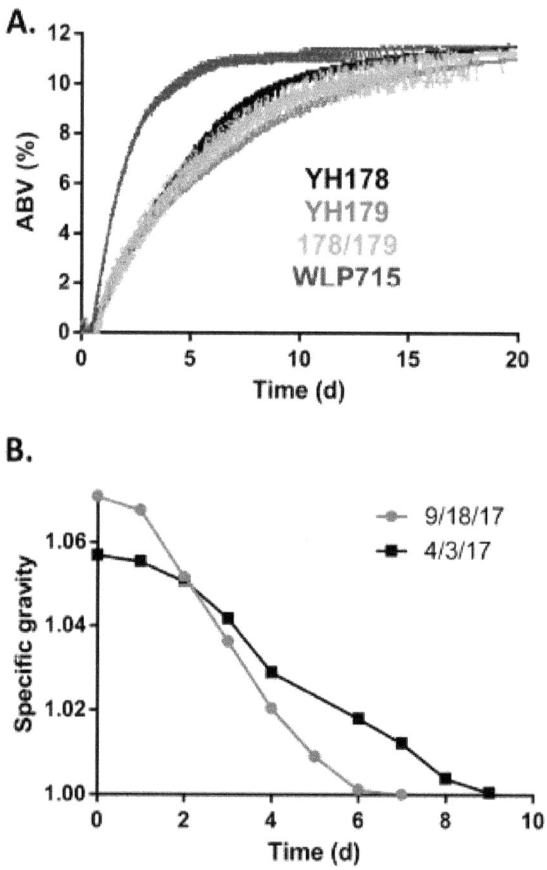

Figure 3. The *T. delbrueckii* isolates can ferment honey into mead. (A) The indicated strains were used to inoculate a honey solution, and fermentation of the honey was monitored with a digital hydrometer. The amount of alcohol produced is plotted vs. time. These results are representative of three independent fermentations for each strain or combination of strains (178/179 = a 1:1 mixture of YH178 and YH179 cultures). (B) Production-scale honey fermentations using a combination of YH178, YH179, and WLP715 yeasts. These fermentations were conducted in April (3 April 2017) and September (18 September 2017) of 2017 under similar conditions but with slightly different starting gravities of the honey solution.

Table 1. Experimental fermentations with *Saccharomyces cerevisiae*, *Torulaspora delbrueckii*, and *Hanseniaspora vineae* strains.

Final Gravity ABV 2 Final pH 3 Final Gravity ABV Final pH

Strain	Honey Fermentation 1			Beer Fermentation		
	Final Gravity	ABV 2	Final pH 3	Final Gravity	ABV	Final pH
S. cerevisiae WLP001	N.D.	N.D.	N.D.	1.011	4.33%	4.40
S. cerevisiae WLP715	0.992	11.5%	4.25	N.D.	N.D.	N.D.
T. delbrueckii YH178	0.992	11.5%	4.15	1.041	0.39%	5.40
T. delbrueckii YH179	0.996	11.0%	4.23	1.042	0.26%	5.40
YH178 + YH179	0.994	11.2%	4.20	N.D.	N.D.	N.D.
H. vineae YH72	1.018	8.2%	3.31	1.016	3.68%	3.26

1 The original gravity of the honey solution was 1.080, and that of the wort was 1.044. 2 Abbreviations: ABV: alcohol by volume; N.D.: not determined. 3 The initial pH of the honey solution was 4.62, and that of the wort was 5.43.

It is perhaps unsurprising that gut microbiota from the honey bee can metabolize honey, as this is a food source for *A. mellifera* itself. Honey is mainly composed of the hexose monosaccharides glucose and fructose, and most yeast species can utilize one or both as carbon sources for cellular metabolism [25]. Conversely, unadulterated honey contains no maltose [26], which is a glucose disaccharide and the primary sugar in beer wort [27]. We tested YH178 and YH179 for their ability to ferment wort and found that attenuation of the beer was extremely poor after two weeks of fermentation (Table 1). In contrast, the *S. cerevisiae* WLP001 ale yeast attenuated to expected levels. The basal level of fermentation displayed by YH178 and YH179 is likely attributable to the *T. delbrueckii* cells consuming the limited quantities of glucose, fructose, and/or sucrose found in typical worts [27]. Indeed, growth curves in YP medium supplemented with 2% maltose (YPM medium) rather than glucose (YPD medium) demonstrate that YH178 grows much more poorly in the YPM than YPD medium (Figure 2A). Similarly, YH179 displayed only minimal growth in YPM medium (Figure 2B). The WLP715 strain is able to utilize maltose though, showing only a modest decrease in growth rate in YPM medium compared to YPD medium and no reduction in final cell density (Figure 2C). While others have successfully isolated yeasts from wasps [28] that can ferment wort into beer, not all yeasts from hymenoptera gut microbiomes are capable of this feat. Thus, individuals interested in isolating yeasts for a particular fermentation milieu are cautioned to select for crucial phenotypes (e.g., maltose utilization) early in the bio-prospecting process.

YH178 and YH179 Produced Superior Meads Compared to WLP715

We next compared the sensory characteristics of the meads produced via the above fermentations. This analysis was performed by 10 panellists using a method developed to quantify eight organoleptic qualities of mead [29] on a 20-point scale. Table 2 shows the results of our sensory analysis of mead fermented with YH178, YH179, the combination of YH178 and YH179, and *S. cerevisiae* WLP715. Overall, the meads fermented with YH178 or YH179 were scored more highly than that produced by WLP715 (though not significantly preferred, $p \geq 0.052$). The WLP715 mead was notable for its very forward ethanol heat, which was more pleasantly masked by YH178 and YH179 despite having comparable ABVs. However, the sensory panel also detected obvious differences in the meads made with YH178 and YH179. For instance, the aroma of the YH178 mead was more reminiscent of honey than YH179 mead, which some panellists reported as "funky" or simply lacking in honey character. This again corroborates our findings that YH178 and YH179 are distinct strains of *T. delbrueckii*. The mead fermented with the combination of YH178 and YH179 scored highly on aroma but had a grassy flavour that some panellists did not enjoy. Regardless, all of these meads received a "standard" score (12–16 points) [29], indicating that YH178 and YH179 are suitable for home and commercial brewing of mead.

Table 2. Sensory evaluation of mead samples.

Yeast 1	Appear-ance	Aroma	Acidity	Balance	Body	Flavor	Finish	Overall Quality
YH178	2 ± 0 2	3.7 ± 0.9	1 ± 0.5	1 ± 0.5	0.9 ± 0.7	1.5 ± 0.3	1.3 ± 0.5 1.4 ± 0.2	12.8 ± 2.4
YH179	2 ± 0	3.7 ± 1.2	1.1 ± 0.7	1.7 ± 0.4	1.4 ± 0.5	1.8 ± 0.4	1.8 ± 0.4 1.8 ± 0.4	15.3 ± 2.4
178/179	2 ± 0	4.5 ± 1.2	0.8 ± 0.6	0.9 ± 0.2	0.8 ± 0.6	1 ± 0.5	1.1 ± 0.2 1.2 ± 0.2	12.3 ± 1.9
WLP715	2 ± 0	4.3 ± 0.8	0.8 ± 0.6	1.3 ± 0.4	0.6 ± 0.2	1.2 ± 0.6	1.1 ± 0.4 1.4 ± 0.3	12.7 ± 2.2
YH72	2 ± 0	5.2 ± 0.7	1.5 ± 0	2 ± 0	0.7 ± 0.2	2 ± 0	2 ± 0 2 ± 0	17.4 ± 0.8

[1] YH178 and YH179 are *Torulaspora delbrueckii* isolates, 178/179 is a mixture of both *T. delbrueckii* strains, WLP715 is a *Saccharomyces cerevisiae* strain, and YH72 is a lactic acid-producing isolate of *Hanseniaspora vineae* [8]. [2] The values presented are the averages ± standard deviations.

Knowing that acidity would be ranked, we also fermented mead with *Hanseniaspora vineae* strain YH72 (Table 2), which produces lactic acid during fermentation [8], as an additional control. As in beer fermentation, the final pH of the *H. vineae* YH72 mead dropped to pH = 3.31 due to the lactate created by this strain. Surprisingly, during our sensory analysis, the YH72 mead was ranked highest in all categories except one (Body) and achieved a quality score defining it as a "superior" (16–20 points) mead [29] ($p \leq 0.017$ vs. all other meads). This was in part due to its assertive peach aroma, pleasant acidic finish, and good balance. It should, however, be noted that the final gravity of this mead was only 1.018, yielding an ABV of 8.2%. Thus, comparing it to dryer meads across all sensory categories can be misleading. Indeed, previous large-scale sensory analysis of sweet and dry meads indicates that consumers prefer sweet meads [30].

YH178 and YH179 Were Used in the Creation of Honey Schnapps

Based on the performance of YH178 and YH179 in mead fermentation, we sought to determine how well these strains functioned in production-scale (>20 hL) fermentations. Two large, independent fermentations of diluted honey were conducted as described in the Materials and Methods. A mixed yeast culture containing approximately equal cell counts of YH178, YH179, and WLP715 was used in both cases. As shown in Figure 3B, this yeast blend was able to ferment the solution to a specific gravity of ~1.000 in <10 days during both trials. This is consistent with the kinetics of fermentation displayed by these strains individually in laboratory-scale fermentations (Figure 3A), indicating that YH178 and YH179 can be used for industrial honey fermentations without any additional "domestication" or adaptation, at least in the presence of WLP715. Future experiments with YH178 and YH179 alone are planned, as well as trials of serial re-pitching of these strains into fermentations to select for an increased fermentation rate. Indeed, the desirable rapidity with which WLP715 ferments honey is one of the reasons that it was added to the production-scale fermentations performed in this work. Regardless, the fermented honey solutions were distilled into a schnapps that was back-sweetened with honey and were well received by consumers [31].

MATERIALS AND METHODS

Strains and Culture Conditions

T. delbrueckii strains YH178 and YH179 were supplied by Wild Pitch Yeast, LLC (Bloomington, IN, USA), and *S. cerevisiae* strains WLP001 and WLP715 were purchased from White Labs (San Diego, CA, USA). All strains were routinely grown on yeast extract, peptone, and dextrose (YPD; 1% (*w*/*v*) yeast extract, 2% (*w*/*v*) peptone, and 2% (*w*/*v*) glucose) plates containing 2% (*w*/*v*) agar at 30°C and in YPD liquid culture at 30°C with aeration unless otherwise noted. WLN agar contained 4 g/L yeast extract, 5 g/L tryptone, 50 g/L glucose, 0.55 g/L KH_2PO_4, 0.425 g/L KCl, 0.125 g/L $CaCl_2$, 0.125 g/L $MgSO_4$, 2.5 mg/L $FeCl_3$, 2.5 mg/L $MnSO_4$, 22 mg/L bromocresol green, and 15 g/L agar. All strains were stored as 15% (*v*/*v*) glycerol stocks at -80°C. Media components were from Fisher Scientific (Pittsburgh, PA, USA) and DOT Scientific (Burnton, MI, USA). All other reagents were of the highest grade commercially available.

Yeast Isolation

Yeasts YH178 and YH179 were isolated from a honey bee sample as described [7]. Briefly, honey bees were collected into sterile glass jars and stored at 4°C before being processed in the laboratory. The bee containing YH178 and YH179 was submerged in 5 mL of YPD8E5 (1% yeast extract, 2% peptone, 8% glucose, and 5% (*v*/*v*) ethanol), crushed with a sterile pipet, and the culture was incubated at 30°C with aeration for 24 h. A small volume of the culture was then streaked onto a WLN agar plate and for 36–48 h at 30°C. To isolate pure strains, single colonies were picked and restreaked onto fresh WLN agar plates until a uniform colony morphology was obtained.

Strain Identification

YH178 and YH179 were identified as T. delbrueckii using the procedure described in [8]. Briefly, 100 μL of saturated overnight culture was mixed with an equal volume of lysis solution (0.2 M LiOAc and 1% SDS) and incubated in a 65°C water bath for ≥15 min to lyse the cells. The genomic (gDNA) was precipitated with 300 μL of 100% isopropanol and vortexing, and it was pelleted with cell debris for 5 min at maximum speed in a micro-centrifuge. The supernatant was removed, and the gDNA was resuspended in 50 μL TE buffer (10 mM Tris-HCl, pH 8, and 1 mM EDTA). The variable D1/D2 portion of the eukaryotic 26S rDNA was then amplified by PCR from the gDNA templates using oligosNL1 (GCATATCAATAAGCGGAGGAAAAG) and NL4 (GGTCCGTGTTTCAAGACGG) [32] and the following cycling conditions: 98°C for 5 min; 35 cycles of 98°C for 30 s, 55°C for 30 s, and 72°C for 30 s; and 72°C for 10 min. The amplified DNA was purified using a PCR Purification Kit (Thermo Scientific, Waltham, MA, USA) and sequenced by ACGT, Inc. (Wheeling, IL, USA) using primer NL1. The sequence was used to query the National Center for Biotechnology Information (NCBI) nucleotide database with the Basic Local Alignment Search Tool (BLAST; http://blast.ncbi.nlm.nih.gov/Blast.cgi? CMD=Web&PAGE_TYPE=BlastHome) to identify the most closely related species.

Growth Curves

The yeast strains were grown by inoculating 5 mL of YPD liquid medium with single colonies from YPD plates and incubation overnight at 30°C with aeration. The optical density at 660 nm (OD660) of each culture was determined using a Beckman Coulter DU730 UV/Vis Spectrophotometer. Then, the cells were diluted to an OD660 ≈ 0.1 in 200 μL of YPD medium or YPM medium (1% (*w*/*v*) yeast extract, 2% (*w*/*v*) peptone, and 2% (*w*/*v*) maltose) in round-bottom 96-well plates, overlaid with 50 μL of mineral oil to prevent evaporation and incubated at 30°C with shaking in a BioTek Synergy H1 plate reader. The OD660 of every well was measured and recorded every 15 min for ~20 h, and the normalized values (i.e., OD660 reads minus the initial OD660 value) were plotted vs. time to generate growth curves. All growth experiments were repeated three times, and the plotted points represent the average OD660 values (error bars represent the standard deviation).

Laboratory-Scale Honey Fermentation

Wild flower honey from Fort Wayne, IN was provided by Creek Ridge Honey, LLC and Cardinal Spirits. The honey was diluted approximately 1:5 with sterile deionized water and YP medium to a gravity of 1.080 and final concentrations of yeast extract and peptone of 0.25% and 0.5% (*w/v*), respectively. Fermentations of this honey solution were performed as described in [8]. Briefly, the yeast strains were grown to saturation (~200 × 10^6 cells/mL) in 50 mL of YPD liquid medium at 30°C with aeration and used to inoculate ~800 mL of honey solution (1.080 original gravity) in 1 L glass cylinders (30 cm tall, 7.5 cm inner diameter). Control fermentations were likewise set up in parallel using the same reagents but inoculating with 50 mL (~200 × 10^6 cells/mL) WLP715. These fermentations were incubated at an average temperature of 23.4 ± 0.3°C, and the gravity and alcohol by volume (ABV) were monitored in real time using BeerBug digital hydrometers (Sensor Share, Richmond, VA, USA) for 4 weeks.

Laboratory-Scale Wort Fermentation

Laboratory-scale wort fermentations were performed as described [8]. Briefly, blonde wort was prepared by mashing 65.9% Pilsner (2 Row) Bel and 26.9% white wheat malt at 65°C (149°F) for 75 min in the presence of 1 g/bbl $CaCO_3$ and 1.67 g/bbl $CaSO_4$. During the subsequent boil, 7.2% glucose and Saaz hops (to 25 international bittering units) were added. The original gravity (OG) of this wort was 1.044. The yeasts were grown to saturation (~200 × 10^6 cells/mL) overnight at 30°C with aeration in 5 mL YPD medium, and these starter cultures were used to inoculate approximately 400 mL of blonde ale wort in glass bottles capped with rubber stoppers and standard plastic airlocks. Control fermentations were set up identically but inoculated with 5 mL (~200 × 10^6 cells/mL) WLP001 culture. The fermentation cultures were incubated at room temperature for 2 weeks, and their final gravity was measured using a MISCO digital refractometer (Solon, OH, USA).

Mead Sensory Analysis

Sensory analysis was performed using the method developed by the University of California, Davis as described by [29]. Briefly, each mead was sampled in a blinded fashion by 10 or more individuals and immediately scored for appearance (2 points), aroma (6 points), and acidity, balance, body, flavour, finish, and overall quality (2 points each). Using these criteria, a score of 17–20 is superior, 13–16 is standard, 9–12 is below standard, and 1–8 is not acceptable.

Production-Scale Honey Fermentation and Distillation

Two large honey fermentations were performed at Cardinal Spirits, LLC (Bloomington, IN). The first was performed on 3 April 2017 with a solution of 378 L honey and 23.2 hL H_2O, which had an initial specific gravity of 1.0569 (14.02 Brix) and an initial pH of 4.4. Diammonium phophate (DAP,1.05 kg; BSG Wine, Napa, CA, USA) and Fermax (587 g; BSG HandCraft, Shakopee, MN, USA) were added to this solution as yeast nutrients, and then a 1:1:1 mixture of YH178, YH179, and WLP715 was added to inoculate the medium. The culture was maintained at a temperature of 28.5–30°C in a stainless steel fermenter, and gravity and pH were monitored daily. After 1 day of fermentation,1.05 kg DAP and 587 g Fermax were again added to the fermentation, and when the gravity decreased to <6 Brix, 587 g Fermax was added one last time. The fermentation proceeded for ~9 days, reaching a terminal gravity of 0.13 Brix (7.35% ABV) prior to distillation to approximately 150 proof.

The second large-scale fermentation was performed on 18 September 2017 with a solution of 549 L honey and 29.2 hL H_2O, which had an initial specific gravity of 1.0708 (17.24 Brix) and an initial pH of 5.04. The same nutrient addition schedule and yeast blend as above was used. The culture was maintained at a temperature of 28.5–31°C in a stainless steel fermenter, and gravity and pH were again monitored daily. The fermentation proceeded for ~7 days, reaching a terminal gravity of 0 Brix (9.14% ABV) prior to distillation to approximately 150 proof. In both this and the previous cases, the distillate was diluted to ~80 proof and back-sweetened with 0.45 kg honey/3.78 L distillate. After resting this liquid for 1 month on a fining agent to precipitate out proteins and pollen, the final product was a yellow-tinted clear liquid (visually reminiscent of white wine) and marketed as a 74.2 proof honey schnapps. Further details are available upon request.

Statistical Analysis

All statistical analyses were performed using GraphPad Prism 6 software. Differences between values were compared using Student's *t*-test, and *p*-values <0.05 were considered statistically significant.

SUMMARY

Yeasts are ubiquitous microbes found in virtually all environments. Many yeast species can ferment sugar into ethanol and CO_2, and humans have taken advantage of these characteristics to produce fermented beverages for thousands of years. As a naturally abundant source of fermentable sugar, honey has had a central role in such fermentations since Neolithic times. However, as beverage fermentation has become industrialized, the processes have been streamlined, including the narrow and almost exclusive usage of yeasts in the genus *Saccharomyces* for fermentation. We set out to identify wild honey- or honey-bee-related yeasts that can be used in honey fermentation. Here, we isolated two strains of *Torulaspora delbrueckii* from the gut of a locally collected honey bee. Both strains were able to ferment honey sugar into mead but failed to metabolize more than a modest amount of wort sugar in trial beer fermentations. Further, the meads fermented by the *T. delbrueckii* strains displayed better sensory characteristics than mead fermented by a champagne yeast. The combination of *T. delbrueckii* and champagne yeast strains was also able to rapidly ferment honey at an industrial scale. Thus, wild yeasts represent a largely untapped reservoir for the introduction of desirable sensory characteristics in fermented beverages such as mead.

REFERENCES

1. McGovern, P.E.; Mirzoian, A.; Hall, G.R. Ancient Egyptian herbal wines. *Proc. Natl. Acad. Sci. USA* **2009**, *106*, 7361–7366. [CrossRef] [PubMed]
2. McGovern, P.E.; Zhang, J.; Tang, J.; Zhang, Z.; Hall, G.R.; Moreau, R.A.; Nunez, A.; Butrym, E.D.; Richards, M.P.; Wang, C.S.; et al. Fermented beverages of pre- and proto-historic China. *Proc. Natl. Acad. Sci. USA* **2004**, *101*, 17593–17598. [CrossRef] [PubMed]
3. Crewe, L.; Hill, I. Finding Beer in the Archaeological Record: A Case Study from Kissonerga-Skalia on Bronze Age Cyprus. *Levant* **2012**, *44*, 205–237. [CrossRef]
4. Rogers, A. *Proof: The Science of Booze*; Mariner Books: Wilmington, DE, USA, 2015.
5. Iglesias, A.; Pascoal, A.; Choupina, A.B.; Carvalho, C.A.; Feas, X.; Estevinho, L.M. Developments in the fermentation process and quality improvement strategies for mead production. *Molecules* **2014**, *19*, 12577–12590. [CrossRef] [PubMed]
6. American Mead Makers Association. *What's the Buzz? 2017 Mead Industry Report*; American Mead Makers Association: Seattle, WA, USA, 2017.
7. Osburn, K.; Ahmad, N.N.; Bochman, M.L. Bio-prospecting, selection, and analysis of wild yeasts for ethanol fermentation. *Zymurgy* **2016**, *39*, 81–88.
8. Osburn, K.; Amaral, J.; Metcalf, S.R.; Nickens, D.M.; Rogers, C.M.; Sausen, C.; Caputo, R.; Miller, J.; Li, H.; Tennessen, J.M.; et al. Primary souring: A novel bacteria-free method for sour beer production. *Food Microbiol.* **2017**, *70*, 76–84. [CrossRef] [PubMed]
9. Rogers, C.M.; Veatch, D.; Covey, A.; Staton, C.; Bochman, M.L. Terminal acidic shock inhibits sour beer bottle conditioning by Saccharomyces cerevisiae. *Food Microbiol.* **2016**, *57*, 151–158. [CrossRef] [PubMed]

10. Independent We Stand. Craft Beer Goes Even More Local with the Farm-to-Glass Movement. 2017. Available online: https://www.independentwestand.org/craft-beer-and-the-farm-to-glass-movement/ (accessed on 28 February 2018).

11. Kallenberger, M. Local Sourcing from the Craft Beer Drinker's Perspective. *New Brew.* **2016**, *34*, 84–88.

12. Olaitan, P.B.; Adeleke, O.E.; Ola, I.O. Honey: A reservoir for microorganisms and an inhibitory agent for microbes. *Afr. Health Sci.* **2007**, *7*, 159–165. [CrossRef] [PubMed]

13. Silva, M.S.; Rabadzhiev, Y.; Eller, M.R.; Iliev, I.; Ivanova, I.; Santana, W.C. Microorganisms in Honey. In *Honey Analysis*; de Toledo, V.A.A., Ed.; InTech: Rijeka, Croatia, 2017.

14. Loira, I.; Vejarano, R.; Banuelos, M.A.; Morata, A.; Tesfye, W.; Uthurry, C.; Villa, A.; Cintora, I.; Suarez-Lepe, J.A. Influence of sequential fermentation with Torulaspora delbrueckii and Saccharomyces cerevisiae on wine quality. *LWT Food Sci. Technol.* **2014**, *59*, 915–922. [CrossRef]

15. Canonico, L.; Comitini, F.; Ciani, M. Torulaspora delbrueckii contribution in mixed brewing fermentations with different Saccharomyces cerevisiae strains. *Int. J. Food Microbiol.* **2017**, *259*, 7–13. [CrossRef] [PubMed]

16. Canonico, L.; Agarbati, A.; Comitini, F.; Ciani, M. Torulaspora delbrueckii in the brewing process: A new approach to enhance bioflavour and to reduce ethanol content. *Food Microbiol.* **2016**, *56*, 45–51. [CrossRef] [PubMed]

17. Michel, M.; Kopecka, J.; Meier-Dornberg, T.; Zarnkow, M.; Jacob, F.; Hutzler, M. Screening for new brewing yeasts in the non-Saccharomyces sector with Torulaspora delbrueckii as model. *Yeast* **2016**, *33*, 129–144. [CrossRef] [PubMed]

18. Pires, E.J.; Teixeira, J.A.; Branyik, T.; Vicente, A.A. Yeast: The soul of beer's aroma—A review of flavour-active esters and higher alcohols produced by the brewing yeast. *Appl. Microbiol. Biotechnol.* **2014**, *98*, 1937–1949. [CrossRef] [PubMed]

19. Etschmann, M.M.; Huth, I.; Walisko, R.; Schuster, J.; Krull, R.; Holtmann, D.; Wittmann, C.; Schrader, J. Improving 2-phenylethanol and 6-pentyl-α-pyrone production with fungi by microparticle-enhanced cultivation (MPEC). *Yeast* **2015**, *32*, 145–157. [CrossRef] [PubMed]

20. Al Toufailia, H.M.; Amiri, E.; Scandian, L.; Kryger, P.; Ratnieks, F.L.W. Towards integrated control of varroa: Effect of variation in hygienic behaviour among honey bee colonies on mite population increase and deformed wing virus incidence. *J. Apic. Res.* **2015**, *53*, 555–562. [CrossRef]

21. Bigio, G.; Al Toufailia, H.M.; Hughes, W.O.H.; Ratnieks, F.L.W. The effect of one generation of controlled mating on the expression of hygienic behaviour in honey bees. *J. Apic. Res.* **2014**, *53*, 563–568. [CrossRef]

22. Wilson-Rich, N.; Spivak, M.; Fefferman, N.H.; Starks, P.T. Genetic, individual, and group facilitation of disease resistance in insect societies. *Annu. Rev. Entomol.* **2009**, *54*, 405–423. [CrossRef] [PubMed]

23. Schoning, C.; Gisder, S.; Geiselhardt, S.; Kretschmann, I.; Bienefeld, K.; Hilker, M.; Genersch, E. Evidence for damage-dependent hygienic behaviour towards Varroa destructor-parasitised brood in the western honey bee, *Apis mellifera. J. Exp. Biol.* **2012**, *215*, 264–271. [CrossRef] [PubMed]

24. *WLP715 Champagne Yeast*; White Labs: San Diego, CA, USA, 2016. Available online: https://www.whitelabs. com/yeast-bank/wlp715-champagne-yeast (accessed on 7 February 2018).

25. Zimmermann, F.K.; Entian, K.-D. *Yeast Sugar Metabolism: Biochemistry, Genetics, Biotechnology, and Applications*; Technomic Publishing Company, Inc.: Lancaster, PA, USA, 1997.

26. Fujita, I. Determination of Maltose in Honey. *Int. J. Food Sci. Nutr. Diet.* **2012**, *1*, 1–2.

27. Briggs, D.E.; Hough, J.S.; Stevens, R.; Young, T.W. *Malting and Brewing Science*, 2nd ed.; St. Edmundsbury Press: Great Britain, UK, 1981.

28. Smith, L. Walk on the Wild Side With Beer Made From Wasp Yeast. In *The Plate*; National Geographic: Washington, DC, USA, 2015. Available online: http:// theplate. nationalgeographic.com/2015/09/29/walk-on-the-wild-side-with-beer-made-from-wasp-yeast/ (accessed on 28 February 2018).

29. Hernandez, C.Y.; Serrato, J.C.; Quicazan, M.C. Evaluation of Physicochemical and Sensory Aspects of Mead, Produced by Different Nitrogen Sources and Commercial Yeast. *Chem. Eng. Trans.* **2015**, *43*, 1–6. [CrossRef]

30. Gomes, T.; Dias, T.; Cadavez, V.; Verdial, J.; Sa Morais, J.; Ramalhosa, E.; Estevinho, L.M. Influence of Sweetness and Ethanol Content on Mead Acceptability. *Pol. J. Food Nutr. Sci.* **2015**, *65*, 137–142. [CrossRef]

31. Sagon, E. *Honey Schnapps: How We Made It*; Cardinal Spirits: Bloomington, IN, USA, 2017. Available online: http://cardinalspirits.com/thedrop/honey-schnapps-how-we-made-it (accessed on 2 March 2018).

32. Lee, Y.J.; Choi, Y.R.; Lee, S.Y.; Park, J.T.; Shim, J.H.; Park, K.H.; Kim, J.W. Screening wild yeast strains for alcohol fermentation from various fruits. *Mycobiology* **2011**, *39*, 33–39. [CrossRef] [PubMed]

CHAPTER-14

ANTIFUNGAL DRUG DISCOVERY: MICRO- AND NANOSCALE APPROACHES

INTRODUCTION: AN OVERVIEW

In the relentless pursuit of combating fungal infections, the field of antifungal drug discovery has evolved to harness the power of micro- and nanoscale approaches. The intricate and often resilient nature of fungal pathogens demands innovative strategies, and researchers are increasingly turning to micro- and nanotechnology to address the challenges posed by these infections. By employing techniques at the micro- and nanoscale, scientists can delve into the intricacies of fungal structures, interactions, and molecular processes with unprecedented precision. This paradigm shift in drug discovery not only facilitates a deeper understanding of fungal biology but also opens avenues for the development of more targeted and effective antifungal agents. This exploration into the realm of micro- and nanoscale approaches marks a pivotal moment in the ongoing quest to advance antifungal therapies and elevate the arsenal against fungal infections to new heights.

Fungal infections are an extremely important health problem. Fungi infect about 1.2 billion people every year, yet their contribution to the global burden of disease is largely unrecognised [1,2]. Over 600 different fungi have been reported to infect humans. Most are "relatively" minor infections, but millions contract diseases that kill at least as many people as tuberculosis or malaria. More than 300 million people are affected by serious fungal infections worldwide. Invasive fungal infections are responsible for about 1.5 million deaths per year. Fungi are present everywhere in our environment and are, usually, harmless for people with a normal immune system. Fungal infections can be topical and local, such as infections on the skin or in the vaginal tract. Systemic infections arise when the fungi enter and proliferate in the bloodstream. Systemic fungal infections affect people with an altered immune system due to medical interventions (such as cancer therapy, organ transplantation, and immune-modulatory medications), immunosuppressive diseases (such as Acquired Immune Deficiency Syndrome (AIDS)) [3], or malnutrition (under- and over nutrition) [4–6].

The choice of available antifungal drugs to treat invasive fungal infections is limited, since only three structural classes of compounds are available, i.e., polyenes, azoles, and echinocandins. Antifungal therapy has become progressively more effective since the 1990s, however, no new antifungal classes have been reported since 2006 [7]. Current antifungal drugs show some limitations:

Amphothericin B (a polyene antibiotic) displays a considerable toxicity and undesirable side effects [8,9], issues with pharmacokinetic properties (such as a short half-life of echinocandins) and activity spectrum, a small number of targets [10,11], and they can interact with other drugs, such as chemotherapy agents and immune-suppressants [12,13]. The last approved antifungal (i.e., anidulafungin) by the European Medicines Agency and the Food and Drug Administration (FDA) dates back to 2006 [14]. There is an urgent need for safer and more effective antifungal drugs. Multiple types of antifungal compounds are in clinical development and these new agents have been recently reviewed [15–17].

Over the last 20 years, several approaches to antifungal discovery have been explored. The traditional approach seeks, first, to identify active compounds from large compound libraries using a panel of fungal pathogens in standardised assays when possible. In the genetic, genomic, or bioinformatics approach, the objective is, initially, to identify broadly represented targets in fungal pathogens and non-pathogens [18]. In this "target-centric" genomic approach, up-front genetic, bioinformatics, and biochemical target prioritisation is performed and, subsequently, an in vitro-based screening of individual targets is achieved. However, an exceedingly high rate of failure has been observed when applying this approach [19]. Additionally, not all bioactive compounds act through a target-specific mechanism of action (MoA). Various compounds act non-selectively as alkylating agents, intercalators, detergents, etc. Therefore, an integrated approach has been recently proposed for antimicrobial lead discovery that is rooted in empirical whole-cell screening for small molecules with intrinsic bioactivity whose MoA may be determined using a variety of forward or reverse genomic platforms to identify and, subsequently, validate their target [20]. In this "compound-centric" strategy, it is proposed to use existing genetic methods to evaluate the therapeutic effect of inactivating the target (both in vitro and in vivo). These methods often use the principles of genetic interaction, relying on the idea of genetic modifiers (enhancers or suppressors) to generate target hypotheses. Direct biochemical methods or computational inference have also been used for target identification. In many cases, however, a combination of approaches may be required to fully characterise on-target and off-target effects, and understand the mechanisms of small-molecule actions [21].

Recently, micro- and nanoscale approaches have been introduced in the field of antifungal drug discovery. In this paper, microfluidic approaches for antifungal screening and antifungal susceptibility testing (AFST), nanosensor development for AFST, and antifungal nanoparticles are reviewed.

FUNGAL DISEASES AND ANTIFUNGAL DRUGS

The epidemiology of invasive fungal infection is evolving [22–25]. A growing population of immune-suppressed patients have been diagnosed with invasive fungal infections. The epidemiology of candida infections has shifted over the last decade [26]. Non-*albicans Candida* species, with a reduced susceptibility to antifungals, are becoming more and more responsible for invasive candidiasis [27]. Several species, such as *Candida glabrata* and *C. krusei*, which are less susceptible to azole antifungal therapy, are increasing in some settings. *C. auris* has emerged in hospitals as a global concern, since the strains demonstrated multidrug resistance. *Trichosporon* species infections are the second most common cause of fungaemia in patients with haematological malignant disease and they were the third most commonly isolated non-*candida* species from patients in the ARTEMIS Global Antifungal Surveillance Program [28]. These species show resistance to amphotericin and echinocandins. *Rhodotorula* species are emerging opportunistic yeasts that are responsible for catheter-related fungaemia and sepsis, and invasive infections, particularly in immune-suppressed or -compromised patients [24]. The incidence of infections with non-*neoformans* crytococci has increased over the past 40 years, especially in patients with advanced HIV infection or cancer who are undergoing transplant [29]. Since the 1990s, there have been a growing number of reports about *Saccharomyces cerevisiae* invasive infections and novel strains continue to be identified [24,30,31].

Aspergillus species have, historically, been one of the most common causative organisms associated with invasive mould diseases [32]. *Aspergillus fumigatus* species were responsible for most of infections.

Recently, the epidemiology has changed towards non-*A. fumigatus* species and other moulds, such as *Zygomycetes* [33–36], *Fusarium* [37,38], and *Scedosporium* [38,39] species. This is due to the occurrence of azole resistant *Aspergillus* species [40,41].

Antifungal development is challenging, since, apart from the fungal cell wall, fungi are metabolically similar to mammalian cells and, therefore, offer few pathogen-specific targets [42]. As explained above, only three molecular classes that target three different fungal metabolic pathways are commonly used in clinical practice to treat, essentially, systemic fungal infections: Polyenes, azoles, and echinocandins (Table 1) [43]. The fluoropyrimidine analog 5-fluorocystosine is also used, but only in combination with amphotericin B.

Table 1. Overview of the mostly used antifungal agents, mechanism of action (MoA), and spectrum of activity [43–50].

Class Compound	MoA	Spectrum of Activity	Comment
Polyenes			
Amphotericin B	Selective binding to ergosterol cause the formation of pores in the membrane.	Treatment of deep mycoses, candidiasis, cryptococcosis, histoplasmosis, blastomycosis, paracoccodioidomycosis, cocciodioidomycosis, aspergillosis, extracutaneous sporotrichichosis, and some cases of mucormycosis, hyalohyphomycosis, and phaeohyphomycosis, S. cerevisiae.	Fungicidal, broad spectrum, intravenous, little resistance observed, significant nephrotoxicity, indirectly affects action of many drugs.
Nystatin/Nyotran	Selective binding to ergosterol cause the formation of pores in the membrane.	Candidiasis	Nyotran is a liposomal formulationof nystatin with lowered toxicity
Natamycin	Binds to ergosterol in the plasma membrane, preventing ergosterol-dependent fusion of vacuoles, membrane fusion, and fission.	Keratinophilic fungi, corneal infections	
Azoles			
Fluconazole	Selective inhibition of fungal cytochrome P450-dependent lanosterol-14-α-demethylase	Candida immitis, C. neoformans, Paracoccidioides brasiliensis; lower activity against Aspergillus, Fusarium, Scedosporium, Penicillium species and other filamentous fungi.	Fluconazole resistant C. albicans and non-albicansstrains increasing.

Class Compound	MoA	Spectrum of Activity	Comment
Itraconazole	Selective inhibition of fungal cytochrome P450-dependent lanosterol-14-α-demethylase	Most Candida species, P. brasiliensis, H. capsulatum, Blastomyces dermatitidis, Aspergillus fumigatus, A. niger, Penicillium marneffei.	Better than fluconazole in the treatment of coccidioido-mycosis, not reaching the central nervous system; numerous drug interactions due to inhibition of CYP 3A4.
Voriconazole	Selective inhibition of fungal cytochrome P450-dependent lanosterol-14-α-demethylase	Candida species, including C. krusei, S. cerevisiae.	More potent than fluconazole; very rapid metabolism in children; numerous drug interactions due to inhibition of CYP 3A4.
Econazole	Selective inhibition of fungal cytochrome P450-dependent lanosterol-14-α-demethylase	Trichophyton rubrum, T. Mentagrophytes, Epidermophyton floccosum	Is an immidazole antifungal for the treatment of tinea pedis and crusis, pityriasis versicolor.
Tioconazole	Selective inhibition of fungal cytochrome P450-dependent lanosterol-14-α-demethylase.	C. albicans, Trichophyton sp., Epidermophyton sp.	Is an immidazole antifungal for topical treatment of superficial mycoses (ringworm, jock itch, athlete's foot, tinea versicolor.
Echinocandins			
Caspofungin	Fungal β-1,3-glucan synthase inhibitors	Candida species, Aspergillus species.	Fungicidal for Candida, fungistatic for Aspergillus; modest efficacy as first-line agent for invasive aspergillosis; intravenous formulation only; interacts with ciclosporin and rifampicin.
Anidulafungin	Fungal β-1,3-glucan synthase inhibitors.	Candida species, Aspergillus species	Fungicidal for Candida, fungistatic for Aspergillus; licensed for the treatment of invasive and esophageal candidiasis

Class Compound	MoA	Spectrum of Activity	Comment
Micafungin	Fungal β-1,3-glucan synthase inhibitors.	Candida species, Aspergillus species.	Fungicidal for Candida, fungistatic for Aspergillus; licensed invasive and esophageal candidiasis.
5-Fluoropyrimidine			
5-Fluorocystosine	Selective conversion of toxic intermediate (5-fluorouridine).	Cryptococcosis, candidiasis, chromoblastomycosis; high MIC for some strains of Aspergillus, Penicillium and several Zygomycetes, except for chromoblastomycosis.	5-Fluorocystosine is always used in combination with amphotericin B.

Recently, a lot of effort has also been made in analysing the antifungal action of natural compounds or natural bioactive compounds. The discovered natural antifungal derivatives or compounds derived from natural origins have been recently reviewed [17,51]. Some examples of these compounds include chitosan [52]; herbal compounds, such as thymol, carvacrol, eugenol, and menthol [53]; and extracts from plants, such as *Hypericum carinatum* [54], *Stenachaenium megapotamicum* [55], and *Acca sellowiana* [56].

Antifungal drug discovery methodologies rooted in omics-based approaches

The traditional approach seeks, first, to identify active compounds (generally from large libraries of synthetic small molecules or natural products) that inhibit the growth of the fungus. The most commonly used assay to identify antifungal leads is liquid growth inhibition assay in which microbial growth is measured by optical density (OD) of the culture. These assays have some limitations, such as, sometimes, a poor correlation between growth and OD for fungi that grow as filaments, and these assays are unable to distinguish between molecules that inhibit growth (fungistatic) from those that kill the organism (fungicidal), a feature that is very important for the treatment of some fungi (e.g., *Cryptococcus*) [57]. Recently, cell viability screening assays based on other readouts have been developed. The most widely adopted approach is the use of the dyes, Alamar Blue and tetrazolium salt (XTT), as reporters for metabolic activity. These dyes are converted to fluorescent molecules when metabolised by viable molecules and can be used for high-throughput multiwell screening [58,59]. The dye, resazurin, was used to develop an assay for high-throughput screening of *A. fumigatus* [60]. An Alamar-Blue-based high-throughput protocol was devised to identify molecules with fungicidal activity against *C. neoformans* [61]. A second type of viability assay that was recently developed is based on the detection of extracellular adenylate kinase (AK) as a reporter of cell lysis [62]. AK is a conserved enzyme that has been used as a cytotoxicity reporter in mammalian cell culture assays and as a reporter of brewing yeast autolysis in the beverage industry [63].

Another approach is referred to as genetic, genomic, or bioinformatics in which the objective is, initially, to identify broadly represented targets in fungal pathogens and non-pathogens. The operational code for this strategy consists of three basic rules for identifying potential targets:

(i) A bioinformatics-driven approach is used to identify pathogen-specific genes with the desired conservation and spectrum, and the absence of a human ortholog; (ii) genetic analyses are performed to confirmed that loss-of-function mutations result in a nonviable growth phenotype and/or a non-virulent phenotype under standard laboratory conditions and/or a non-virulent phenotype in a relevant animal model of infection; and (iii) there should be some evidence, derived from sequence, structure, or biochemical information that the target is druggable [64]. This operational code was used to guide genomic and target-based antimicrobial discovery. Forward and reverse chemical-genetic methods have been used to identify the target of drug-like bioactive compounds and MoA. Target-based approaches have been pursued using high-throughput screening for inhibitor of enzymes and/or proteins that perform essential functions in microbes. Molecular target-based screening has provided mediocre to poor results in the field of anti-infective drug discovery, suggesting that cell-based screening for antimicrobials remains the most effective strategy [19]. Gene knockout microorganisms and small molecules with well-defined mechanisms can each be used to alter the functions of putative targets, uncovering dependencies on activity. Reverse chemical screens that exploit genome-wide collections of mutants have been employed to map susceptibility phenotypes to specific genes by systematically screening antimicrobial agents against a defined (ideally) comprehensive mutant collection [20]. Mating of laboratory and wild yeast strains can reveal patterns of small-molecule sensitivity with specific loci [65]. In another method, molecularly barcoded libraries of open reading frames have been used to detect small-molecule-resistant clones that are then identified by microarray analysis [66].

A proteomic profiling approach, based on two-dimensional difference gel electrophoresis and mass spectrometric identification of the proteins, has been used to predict the target protein of small molecules of interest [67]. It was demonstrated that this proteomic profiling system could discriminate small molecules by their mechanism of action. Omics methods have also been used successfully to identify the MoA of compounds that were discovered using traditional screening approaches. A key illustrative case is the discovery of the orotomide F901318 compound [68]. A combination of genetic and biochemical approaches revealed the target of F901318 in *A. fumigatus*, which was initially discovered in a high-throughput screen of compound libraries. It was found that F901318 acts via the inhibition of the pyrimidine biosynthesis enzyme, dihydroorotate dehydrogenase, in a fungal-specific manner.

Although only around 1100 of the 6000 genes of *S. cerevisiae* are essential under nutrient-rich growth conditions [69], almost all genes become essential in specific genetic backgrounds in which another non-essential gene has been deleted or otherwise attenuated, an effect termed synthetic lethality [70]. Genome-scale surveys suggest that over 200,000 binary synthetic lethal gene combinations dominate the yeast genetic landscape [71]. The genetic buffering phenomenon is also manifested as a phalanx of differential chemical-genetic interactions in the presence of sublethal doses of bioactive compounds [72]. These observations highlight the inherent redundancy of genetic networks and frame the problem of interdicting network functions with single agent therapeutics [73]. This genetic network organisation suggests that judicious combinations of small molecule inhibitors of both essential and non-essential targets may elicit additive or synergistic effects on cell growth [74,75]. Compounds that enhance the activity of known agents in yeasts have been identified by small molecule library screens [76–79]. Direct tests of synergistic compounds have successfully yielded combinations that are active against pathogenic fungi, such as the combination of fluconazole with chemical inhibitors (such as 17-AAG and 17-DMAG) of Hsp90 [80], calcineurin (such as miconazole, ketoconazole), or adenosine diphosphate (ADP) ribosylation factor (ARF) (such as brefeldin A [81], the antibiotic polymyxin B [78]), and compounds selected from an off-patent drugs library [79]. Combinatorial antifungal therapies have many advantages, including a decrease in the rate of selection of resistant strains, a lower required dosage of individual drugs, a decrease in host toxicity, and enhanced antifungal activity [82]. Syncretic combinations of drugs with improved antifungal properties can be readily identified in both model fungal species and highly pathogenic clinical isolates [83]. Additionally, it was shown that synergistic combinations usually yield enhanced selectivity without adverse side effects [84].

These, recently developed, omics-based approaches require expensive, automated, robotic screening platforms when applied in high throughput. Additionally, these platforms are usually based on multiwell screening, which has some limitations. As discussed in the next section, further improvements can be achieved by integrating these methods in microfluidic screening platforms.

MICRO AND NANOSCALE APPROACHES

Microfluidic High-Throughput Antifungal Drug Discovery

Growing cells in 96-, 384-, or 1536-well plates has miniaturised cell assays for drug discovery. These experiments with multiwell plates are typically integrated in a robotic analysis platform. Major drawbacks of robotic platforms are the expense of the instrumentation, the cost of experimental consumables, the systems are closed (no flow through of reagents or cell culture medium), and still a relatively high consumption of reagents compared to recent developed microfluidic chips. Additionally, as cell collections are growing, there is a need to further miniaturise the assays to increase the parallelism of analyses.

In microfluidic lab-on-a-chips, fluids are manipulated at the micrometer length scale [85]. Reducing the scale evidently reduces the reagent consumption and, consequently, the cost to perform assays, which becomes significant for high-throughput drug screening. The physics at micro/nano scale are exploited in microfluidic chip designs for drug discovery. Other physical phenomena dominate at this small length scale compared to the macroscale [86]. Fluid flow in microchannels is characterized by a low Reynolds number (a dimensional criterion that determines the relative importance of inertial and viscous effects). The dominant role of viscous forces results in laminar flow behaviour and mass transport by diffusion between two adjacent fluids, resulting in the generation of stable concentration gradients [87]. Also, fast media and temperature changes can be obtained using the laminar flow [88]. Mass transport by diffusion becomes a fast mixing method when the length scale is reduced in small channels or wells. It has been demonstrated that, at these small length scales, time-resolved reactions have been conducted with millisecond resolution due to the extremely rapid diffusive mixing [89,90]. Surface tension plays an important role in the formation of small droplet emulsions in immiscible fluids in microfluidic channels [86]. Surface tension and viscous stresses destabilize the interface and create droplets when water is injected into a stream of oil at a T-junction or two perpendicular crossing microchannels [91]. Based on this principle, droplet-based microfluidic platforms have been developed to encapsulate and screen single cells [92].

Due to the ease of controlling the cellular environment, microfluidic technology has been used to perturb cellular physiology in screening arrays. Additionally, the cell number and density of a given area or volume can be controlled, which allows monitoring of a high spatial and temporal resolution and observation of the dynamic behaviour of many cells [93,94]. Parallellisation of experimental conditions and automatization in microfluidic chips has resulted in designs for high-throughput cell screening [95], such as living cell microarrays [96–101]. Microfluidic parallelization and single-cell monitoring allows the measurement and averaging of parameters on hundreds of individual cells, compared to measuring parameters of a whole cell population. New insights can be obtained by observing single cells, such as the monitoring of certain classes of proteins, using fluorescently tagged proteins [102,103] and give information about the cell-to-cell variation in a heterogeneous microbial population [104–106]. Automation of microfluidic cell culture systems also leads to standardized manipulation and monitoring, which allows the perfect timing of protocols to characterize dynamic processes at high temporal resolution to be performed [94].

Table 2. Examples of microfluidic platforms for high-throughput antifungal drug discovery.

High-Throughput Technology	Application	Microorganism	Characteristics	References
Cellular microarray	Antifungal biofilm screening	Candida albicans	Cells robotically printed, 768 (48 × 16 array) cultures of 50 nL biofilms in collagen.	[111,112]
Cellular microarray	Antifungal biofilm screening	C. albicans	Cells robotically printed, 1200 (60 × 20 array) cultures of 30 nL biofilms in alginate.	[113]
Cellular microarray	Antibiotic and antifungal biofilm screening	C. albicans, Staphylococcus aureus, Pseudomonas aeruginosa	Robotically printing of monoand polymicrobial biofilms, 576 (48 × 12 array) cultures of 30 nL biofilms in alginate	[114]
Droplet microfluidics	Antifungal drug screening	Phytophthora sojae	The plant pathogen spores and the drug wereencapsulated in liquid droplets. Phenotypic responses to the drug atdifferent concentrations were microscopically quantified.	[115]

Cellular microarray platforms have been developed for high-throughput antifungal drug discovery based on the screening of nanoliter biofilms that are created in hydrogels (Table 2). Robotic printing is used to fabricate the cellular arrays. The formation of biofilms complicates antifungal therapy, since the ability of fungal cells to form biofilms is an important reason for the emergence of severe resistance to most clinically available antifungal agents [107–110]. A better understanding of fungal biofilms provides new opportunities for the development of urgently needed novel antifungal agents and strategies.

Microfluidics for Antifungal Susceptibility Testing

Resistant microbial infections are becoming a major threat to public health, causing increasing mortality worldwide. The fast emergence of multi-resistant pathogens is caused by extensive, and sometimes unnecessary, use of antimicrobials and the lack of interest in developing new variants. The cost of antimicrobial resistance (AMR) is projected to increase significantly as some models predict a rise in global casualties from the present figure of one million to 10 million in 2050 [116]. A survey of antimicrobial susceptibility test (AST) methods available demonstrated a need for new approaches that would enable rapid, inexpensive, and sensitive tests that can quickly provide physicians with antibiotic profile [117]. A quick diagnostic method is necessary to prescribe a patient that is infected by a life-threatening microbe with an effective antimicrobial [118]. AST is widely applied to determine antimicrobial resistance profile of the microbial isolates and to help in the selection of an antimicrobial treatment option [117,119–121]. Currently, AST is usually not performed in the clinic, but in a clinical microbiology lab, which necessitates transportation of the patient samples [122]. Current susceptibility tests, that are based on cell culturing of the pathogen, can take several days (especially for slow growing microbes) and extends the time to make the correct diagnosis and decisions for appropriate and effective antimicrobial therapy. This leads to increased patient mortality and the use of broad-spectrum antimicrobials that promote resistance [120]. To survive this evolutionary war against microbial pathogens, we must pursue technologies that can rapidly perform AST to enable personalized therapies (narrow-spectrum antimicrobial administration) at the earliest possible treatment stage [122].

Standard antifungal susceptibility (AFST) methods rely on measuring fungal growth in the presence of antifungal over a few days. The broth microdilution (BMD) is the standard method for the evaluation of susceptibility to antifungal agents in *Candida* species [123–126]. Standardized micro-dilution-based procedures by the Clinical and Laboratory Standards Institute (CLSI) and the European Committee on Antibiotic Susceptibility Testing (EUCAST) are universally accepted for performing AFST [127]. However, these procedures are complex, time-consuming, and not intended for routine use.

Commercially available tests, such as Sensititre YeastOne (broth microdilution method), Etest (agar-based disk diffusion method), and the fully automated Vitek 2 (broth microdilution method) yeast susceptibility system, all easy-to-use modifications from CLSI/EUCAST reference methods, are widely used for testing antifungal susceptibility of relevant *Candida* and *Aspergillus* species [10]. New diagnostic approaches based on emerging technologies, such as flow cytometry, MALDI-TOF mass spectroscopy, and isothermal microcalorimetry, have been developed to expand, and potentially improve, the capability of the clinical microbiology laboratory to yield AFST results [126]. Flow cytometry is used to determine the effect of an antifungal compound by measuring changes in the viability of fungal cells that are fluorescently labeled [128,129]. A simple and rapid AFST assay based on MALDI-TOF was developed [130]. This approach facilitated the discrimination of the susceptible and resistant isolates of *C. albicans* after a 3-h incubation in the presence of "breakpoint" level drug concentrations of the echinocandin caspofungin (CSF). A microcalorimetry-based AFST assay has been developed for real-time susceptibility testing of *Aspergillus* spp. [131]. The method is based on measuring changes in growth-related heat production in the presence of the antifungal compound.

Semi-automated susceptibility systems (such as VITEK and MicroScan) decrease the turn-around-time and operator touch-time (culture and colony isolation are still required) compared to traditional culture-based methods, but can add significant cost to the tests [132]. The fastest test still requires 9 h and, therefore, these semi-automated systems do not provide information in time to influence initial treatment decisions [133,134].

Since microfluidics promises several advantages over existing macro-scale methods, several microfluidic platforms that can perform rapid antimicrobial susceptibility tests have been developed during the last years (Table 3) [132,135]. The recent development of microfluidic platforms was mostly focused on devices for antibiotics susceptibility testing and much less on antifungal susceptibility testing, although the same designs could, usually, be used for both (e.g., see [136,137]). Most systems rely on microscopic transmission or fluorescence observation of cells to quantify the effect of the antimicrobial on the cell growth or viability. The microorganisms are confined to a small volume (wells, channels, chambers, or droplets). In many systems, the cells are immobilized in a hydrogel, such as agarose or alginate (Table 3) (Figure 1).

Table 3. Examples of microfluidic platforms for antimicrobial susceptibility testing.

Microorganism	Measurement Principle	Description	Reference
Fungi			
Candida strains	Fluorescence-based distinction between living and dead cells	Cell Chip kit 1 used as cell sorter and the determination of fluorescence histograms of previously labelled cells.	[138]
Candida albicans	Immunosorbent ATP-bioluminescence assay	The microfluidic device employs a fiberglass membrane sandwiched between two polypropylene components, with capture antibodies immobilised on the membrane. Cells immobilised in alginate hydrogel.	[136]
Saccharomyces cerevisiae	Fluorescence staining and imaging after incubation.	Cell seeding and diffusive medium supply is provided by phase-guide technology, enabling operation of continuous culturing	[137]
Bacteria			
Escherichia coli	Optical imaging of single cell growth (number of cells).	Growth of cells in channels, with a large surface-to-volume ratio.	[139]
E. coli	Magnetic bead rotation, which is inversely proportional to bacterial mass	Droplet microfluidics where single cells are entrapped in liquid drops.	[140]
E. coli	Optical imaging of single cell growth (number of cells).	Individual cells grow in gas permeable (PDMS) microchannels, with dimensions comparable to a single cell.	[141]
E. coli	Fluorescence imaging of cells that express green fluorescent protein	Growth of cells in 12 sets of quadruplicate microfluidic chambers. Quantification of the effect of four antibiotics and their combinations.	[142]
E. coli	Reflectometric interference spectroscopy of pH-sensitive chitosan hydrogel measured the accumulation of metabolic products.	Growth of cells in microfluidic channels.	[143]

Microorganism	Measurement Principle	Description	Reference
E. coli	Fluorescence imaging of immunomicrobeads attached to the cells.	Growth in microfluidic chambers.	[144]
E. coli	Optical imaging.	The standard broth microdilution method was miniaturised in a microfluidic chip that generates an antibiotic concentration gradient and delivers antibiotic-containing culture media to eight 30-nL chambers for cell culture.	[145]
E. coli	Fluorescence imaging.	The microfluidic chip allows the carrying out of commonly executed antibiotic susceptibility assays in an array of nanoliter droplets.	[146]
E. coli	Optical phase-shift reflectometric interference spectroscopic measurements.	The use of biofunctionalised silicon micropillar arrays to provide both a preferable solid-liquid interface for bacteria networking and a simultaneous transducing element that monitors the response of bacteria when exposed to chosen antibiotics in real time.	[147]
E. coli, Nitrosomas europaea	Optical imaging of single cell growth (number of cells).	The cells grow in a layer of agarose upon which a gradient of the antibiotic is applied.	[149]
Enterobacter cloacae, E. coli, Klebsiella pneumoniae, P. aeruginosa, Acinetobacter baumannii	Optical imaging of bacterial replication.	A solid-phase microwell growth surface in a 384-well plate format was used, with inkjet printing-based application of both antimicrobials and bacteria at any desired concentrations.	[150]
Enterococcus faecalis, E. coli	Fluorescence staining and imaging after incubation.	Cell seeding and diffusive medium supply is provided by phase-guide technology, enabling operation of continuous culturing.	[137]
Mycobacterium tuberculosis	Optical imaging of single cell growth (number of cells).	Cells were immobilized in an agarose matrix, which was molded in a microfluidic chip.	[151]

Microorganism	Measurement Principle	Description	Reference
Pseudomonas aeruginosa	Fluorescence imaging of GFP-expressing cells.	The microfluidic chip generates a logarithmic concentration gradient through semidirect dilution in a zero-flow condition and cells grow in nanoliter reactors.	[135]
Staphylococcus aureus	Fluorescence imaging of viability indicator	Stochastic confinement of individual cells into liquid plugs (droplet microfluidics); distinction between sensitive and resistant bacteria	[152]
S. aureus	Optical imaging of single cell growth (number of cells).	Antibiotics diffuse into a microfluidic channel containing the cells.	[153]
S. aureus	Fluorescence imaging for dead cells (rates of killing).	Cells are covalently bound to the bottom of the channels and fluid flow shear stress activation of pathways that are targets of antibiotics	[154]
S. aureus	Optical imaging of single cell growth (number of cells).	Chip with 32 individual fluidic channels.	[155]
S. aureus, E. coli, K. pneumoniae, *P. aeroginosa*	Optical imaging of single cell growth (number of cells).	Antibiotics diffuse into a microfluidic agarose channels in 96-well format	[156]
S. aureus, S. epidermitis, *S. saprophyticus, E. coli,* *K. pneumonia, P. aeruginosa,* *P. mirabilis, Streptococcus pyrogenes,* *S. viridans*	Immunosorbent ATP-bioluminescence assay.	The microfluidic device employs a fiberglass membrane sandwiched between two polypropylene components, with capture antibodies immobilised on the membrane. Cells immobilised in alginate hydrogel.	[136]
Salmonella thyphimurium, E. coli, *S. aureus*	Optical imaging of single cell growth (number of cells).	Cells immobilised in agarose with a gradient of the antibiotic in the gel slab.	[157]

Microorganism	Measurement Principle	Description	Reference
P. aeruginosa	Optical imaging of single cell growth (number of cells).	Growth of cells in four parallel channels.	[158]

[1]Agilent Technologies.

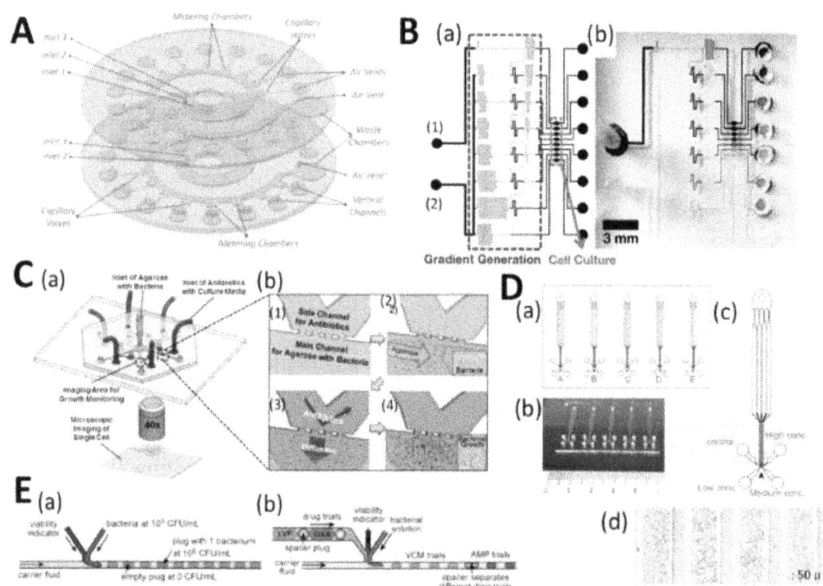

Figure 1. (A) An automated linear concentration gradient generator based on centrifugal microfluidics for antibiotic susceptibility testing (AST). The operation is based on the use of multi-layered microfluidics in which individual fluidic samples to be mixed together are stored and metered in their respective layers before, finally, being transferred to a mixing chamber. Reprinted with permission from [148]; (B) (a) The microfluidic microchip design: Black and red channels represent 149-μm and 4-μm tall features, respectively; (b) A photograph of the fabricated PDMS chip. Blue food dye solution and water were injected to upper (1) and lower (2) inlet ports. (a) Reprinted with permission from [145];Schematic diagram of the AST process for the microfluidic agarose channel (MAC) system. (a) The MAC chip was fabricated with PDMS on glass. An agarose–bacteria mixture solution was injected into the center of the chip, which flowed synchronously into the six main channels. Each interface between the agarose with the bacteria and antibiotic solutions was monitored microscopically to analyse bacterial cell growth. (b) (1) The empty channel before AST. (2) The bacteria were mixed with agarose and then injected into the main channels. (3) A sharp interface was generated due to the anchors and then liquid medium with different concentrations of antibiotic. (4) Time-lapse microscopic observation

of a bacterial cell. Reprinted with permission from [153]; (D) The structure of the AST microfluidic device for MIC determination of five drugs. (a) Design. (b) Actual image. (c) Precise structure of one set of fluids. (d) Microscopic image of *Pseudomonas aeruginosa* grown in the presence of piperacillin. Reprinted with permission from [158]; (E) Stochastic confinement of bacteria into plugs. Schematic drawing illustrates the increase in cell density, resulting from the stochastic confinement of an individual bacterium in a nanoliter-sized plug. (b) Screening of many antibiotics against the same bacterial sample. Antibiotics: Ampicillin (AMP), levofloxacin (LVF), vancomycin (VCN), and oxacillin (OXA). Reprinted with permission from [152].

Nanomotion Analysis for Evaluating Antifungal Suscessibility

Recently, very new sensitive sensor technologies based on microcantilevers have been developed [159–161]. Nanomechanical oscillators are being used for the detection of very small masses [162,163], for measuring buoyant mass, and determining the "instantaneous" growth rates of individual cells [164] for the quantitative time-resolved membrane protein (vancomycin)—ligand (mucopeptides) on cantilever arrays with 10 nM sensitivity and at clinically relevant concentrations in blood serum [165]. Cantilever nanosensors have been used to measure mass differences in the pico- to femtogram ranges in air [166] and liquid [167,168]. Many of the available systems are limited by the need to perform the measurements in air or in a humid environment [161].

Microcantilevers have been explored as nanosensors for living cell studies, since they offer many advantages, such as being highly sensitive, selective, label-free, performed in real time, and provide in situ detection capabilities [169]. Single cell detection and monitoring on the cantilever sensor has been reported for *S. cerevisiae* cells [164,170,171], *E. coli* [164,171,172], *Bacillus subtilis* [164,172], *Enterococcus faecalis* [171], HeLa cells [173], mouse lymphoblasts [164], and human lung carcinoma and mouse lymphocytic leukemia cells [171,174], and mouse and human T cells [171]. Cell growth detection has been demonstrated by monitoring resonance frequency changes of cantilevers as the mass increases from immobilized *S. cerevisiae* and fungal *A. niger* spores on the surface of the cantilevers in humid air [175]. *S. cerevisiae* cells were deposited onto the cantilever surface and its bending as a function of time and corresponded to the yeast growth behaviour [169]. Recently, serial microfluidic mass resonator arrays were used to measure single-cell growth of yeast, bacterial and mammalian cells in liquid and a higher throughput [171].

Recently, a nanomechanical detector, that can be used to assess the effects of chemicals on living organisms in a timeframe of minutes, was developed. In this technique, the living microorganisms are adsorbed to the surface of a nanomechanical sensor, i.e., an AFM cantilever, and its fluctuations are measured as a function of time (Figure 2). This approach requires only a few minutes without cell growth, since the decrease in cell activity in the presence of the antimicrobial is measured. Moreover, it is a label-free technique. These are major advantages compared to current methods. Antimicrobials can be introduced in the measuring chamber, allowing a rapid identification of the antimicrobial to which the microorganism is susceptible. The sensor position detection is similar to the one used in Atomic Force Microscopy (AFM): A laser beam is focused onto the sensor (cantilever) and ends its path onto a multi segment photodiode. Numerous proof-of-concept experiments have been performed involving several species of bacteria and fungi. By comparing this method with the traditional techniques in double-blind experiments, a success rate close to 90% could be achieved [176]. The variation of the nanovibration signal is directly proportional to the metabolic activity of the microbial cell. It allowed estimating with unprecedented speed (15 min per ampicillin dose), the minimum inhibitory (MIC) and bactericidal concentrations (MBC) for the effect of the antibiotic ampicillin on *E. coli* [177]. The high sensitivity of the nanosensor allowed differentiation between bacteriostatic (kanamycin) and bactericidal (ampicillin) effects when exposed to a resistant *E. coli.* The high sensitivity of the nanosensor made it possible to differentiate between bacteriostatic and bactericidal effects. Recently, this method was successfully applied to blood culture pellets to determine the antibiotic susceptibility against agents of bloodstream infection [178]. AFM-nanomotion detection (NMD) has also been applied for AFST, where the effect of a low (10 µg/mL) and high (40 µg/mL) caspofungin concentration on *C. albicans* was evaluated [179]. It was recently demonstrated that AFM-cantilever NMD is so sensitive that single-cell activities of *S. cerevisiae* cells could be detected [179], and even cell cycle progression could be observed [180].

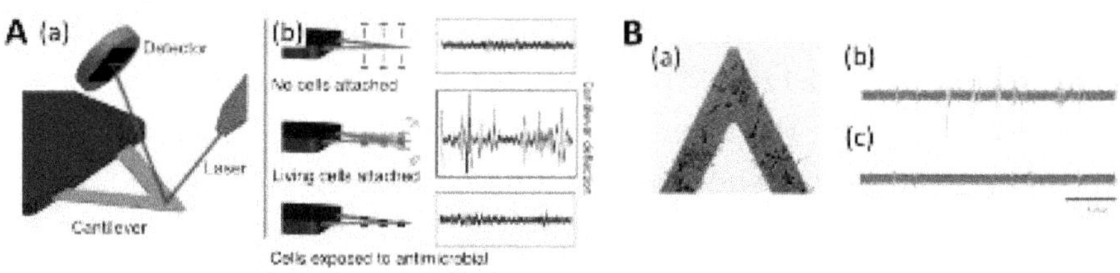

Figure 2. (A) Outline of experimental setup and description of experiments. (a) Schematics of nanomotion detector setup with a cantilever sensor; laser beam is focused on the surface of the sensor and reflection is used to monitor movements of the cantilever. (b) Representation of a typical nanomotion susceptibility test. When microorganisms are not attached to the sensor, fluctuations are driven only by thermal motion and are relatively low. After attachment of living cells, fluctuations are linked to their metabolic activity and are high. Finally, after exposure to an antimicrobial drug, the cells are nonviable and fluctuations return to low levels. Reprinted with permission from [178]. (B) (a) *C. albicans* deposited onto a cantilever. (b) Reprinted with permission from [161].

Antifungal Nanoparticles

Due to the development of antimicrobial resistance, pharmaceutical companies and researchers are searching for new antimicrobial agents. Nanoscale materials have emerged as novel antimicrobial agents, owing to their high surface area to volume ratio and their unique chemical and physical properties [181,182]. By definition, nanoparticles are structures that have dimensions in the 1–100 nm [183]. Silver nanoparticles (Ag-NPs) are one of the most commonly used nanomaterials in consumer and medical products because of their antimicrobial activity [184,185]. It is increasingly used in a variety of both medical and consumer products, resulting in an increase in human exposure [183,186]. A large number of in vitro studies indicate that Ag-NPs are also toxic to mammalian cells derived from the skin, liver, lung, brain, vascular system, reproductive organs, and the immune system. Although significant progress has been achieved on the elucidation of the antimicrobial mechanism of silver nanoparticles, the exact mechanism of action is still not completely known [187]. It has been demonstrated that non-cytotoxic doses of Ag-NPs could induce genes that are associated with cell cycle progression [188,189] and apoptosis in human hepatoma cells [189]. DNA damage by Ag-NPs in mammalian cells has also been reported [188,190]. The mechanisms for Ag-NP induced toxicity include the effects of this particle on cell membranes, mitochondria, and genetic material. It has been recently demonstrated that bacteria (*E. coli, P. aeruginosa*) can also develop resistance to silver nanoparticles after repeated exposure [191]. The biosynthesis of silver nanoparticles is now considered to be the most environmentally friendly and cost-effective method [185]. It can be achieved using a variety of organisms, such as by the fungi, *A. flavus* [185], *A. terreus* [192], *Arthroderma fulvum* [193], *Penicillium fellutanum* [194], and *P. expansum* [192].

Other nanostructures have been discovered that also show antifungal activity. It was demonstrated that cationic terephthalamide-bisurea molecules showed excellent microbial selectivity, with minimal host toxicity [195]. The terephthalamide-bisurea recognition motif facilitated spontaneous supramolecular self-assembly, with the formation of fibres in water. Antifungal activity against clinically isolated, drug-sensitive, and drug-resistant *Cryptococcus neoformans* strains was observed. These antifungal agents showed effectively dispersed *C. albicans* biofilms and excellent in vivo biocompatibility. Homogeneously dispersed copper nanoparticles (originating from n-Cu sepiolite fibres) in soda-lime glass powder showed high antibacterial properties against gram-positive (*Micrococcus luteus*) and gram-negative bacteria (*E. coli*), and antifungal (*Issatchenkia orientalis*) activity [196]. The observed high activity of the n-Cu glass powder was explained by the inhibitory synergistic effect of the Ca^{2+} lixiviated from the glass on the growth of the cells, since n-Cu sepiolite had no significant antifungal activity. Another approach of obtaining nanoparticles with antifungal activity is to immobilize, covalently, the antifungal amphotericin into nanomaterials, such as silicanano particles [197]. These antifungal nanoparticle conjugates demonstrated fungicidal activity against several strains of *Candida* sp., mainly by contact. In addition, they could be reused for up to five cycles without losing their activity. The results showed that the antifungal nanoparticle conjugates were more fungistatic and fungicidal than 10 nm colloidal silver.

SUMMARY

Fungal infections continue to appear as the population of people with an altered immune system increases. An impaired immune system arises due to medical intervention, immunosuppressive diseases, or malnutrition. The infections become life-threatening for systemic infections. The therapeutic options for invasive fungal infections are limited, since a very limited number of structural classes of drug compounds are available and some demonstrate significant limitations. The discovery of new antifungals is mostly achieved by the screening of natural or synthetic/semisynthetic chemical compounds [16,17]. Using the genomics approach, substantial progress has been made in antifungal drug development for a multitude of potential drug targets [198] and inhibitors [199,200]. Recent advances that could support and refine the antifungal pipeline were focused on elucidating fungal pathways, targets, and mechanisms of action that could lead to new antifungal therapies; antifungal compounds and immune strategies currently in development that could become new antifungal therapies; improved formulations of existing compounds; and the repurposing of drugs approved for other indications and could show potential antifungal activity [16]

Surely, there is a need for novel antifungal discovery approaches. The current antifungal tools that are available to tackle the invasive fungal epidemic occurring in clinics and hospitals have improved, but are still inadequate for use in all patient groups [16]. Recently, micro- and nanoscale approaches have been introduced in antifungal drug discovery. It has become increasingly clear in trends in antimicrobial drug discovery that microfluidic approaches will have an increasing role. The capability of manipulating fluids, flexibility on geometries and materials, manipulation of a large number of cells simultaneously and independently, and ease of integrating numerous analytical standard operations and large-scale integration makes microfluidic devices a versatile tool for antifungal drugs. One of the future challenges lies in the construction of extended cell microarrays, or single-cell droplet arrays, and integration in the microfluidic chip [100], since the recently developed bioinformatics approach for antifungal discovery could be further improved by integrating these methods in a microfluidic screening platform. Living cell arraying methods, based on closed microchambers arrays in microfluidic bioreactors, could increase throughput significantly compared to classical multiwell-plate cell assays, with significantly reduced amounts spent on expensive test reagents, cells, and chemical compounds, and without the need for expensive robotic multiwell-plate screening facilities.

Significant progress has been made in the development of microfluidic platforms for antimicrobial susceptibility testing (AST). Recent efforts are focused on the development of AFST methods that are independent of cell growth and can provide susceptibility results in a very short time frame. These methods rely on measuring the cell death in the presence of the antifungal. A promising technique is the AFM-based nanomotion detection, which can give an answer about the susceptibility in a few minutes. Further parallelization of this method is required to introduce it in the clinic.

Nanoscale materials have emerged as novel antimicrobial agents. One of the most extensively studied nanoparticles are silver nanoparticles and this is for their antifungal, but also antibacterial and antiviral, activities. In minute concentrations, it was found not to be toxic to humans and microorganisms [185]. Further research should focus on a better understanding the toxicity and determining the exact mechanism of the interaction between nanoparticles and cells.

Clinical needs for novel antifungal agents have increased due to the increase of people with a compromised immune system, the appearance of resistant fungi, and infections by unusual yeasts. The search for new molecular targets for antifungals has generated considerable research, especially using modern omics methods (genomics, genome-wide collections of mutants, and proteomics) and bioinformatics approaches. Recently, micro- and nanoscale approaches have been introduced in antifungal drug discovery. Microfluidic platforms have been developed, since they have a number of advantages compared to traditional multiwell-plate screening, such as low reagent consumption, the manipulation of a large number of cells simultaneously and independently, and ease of integrating numerous analytical standard operations and large-scale integration. Automated high-throughput antifungal drug screening is achievable by massive parallel processing. Various microfluidic antimicrobial susceptibility testing (AST) methods have been developed, since they can provide the result in a short time-frame, which is necessary for personalized medicine in the clinic. New nanosensors, based on detecting the nanomotions of cells, have been developed to further decrease the time to test antifungal susceptibility to a few minutes. Finally, nanoparticles (especially, silver nanoparticles) that demonstrated antifungal activity are reviewed.

REFERENCES

1. Vos, T.; Flaxman, A.D.; Naghavi, M.; Lozano, R.; Michaud, C.; Ezzati, M.; Shibuya, K.; Salomon, J.A.; Abdalla, S.; Aboyans, V.; et al. Years lived with disability (YLDs) for 1160 sequelae of 289 diseases and injuries 1990–2010: A systematic analysis for the Global Burden of Disease Study 2010. *Lancet* **2012**, *380*, 2163–2196. [CrossRef]
2. Brown, G.D.; Denning, D.W.; Levitz, S.M. Tackling human fungalinfections. *Science* **2012**, *336*, 647. [CrossRef] [PubMed]
3. Richardson, M.D. Changing patterns and trends in systemic fungal infections. *J. Antimicrob. Chemother.* **2005**,*56* (Suppl. 1), i5–i11. [CrossRef] [PubMed]
4. Huttunen, R.; Syrjänen, J. Obesity and the risk and outcome of infection. *Int. J. Obes.* **2013**, *37*, 333–340. [CrossRef] [PubMed]
5. Rytter, M.J.; Kolte, L.; Briend, A.; Friis, H.; Christensen, V.B. The immune system in children with malnutrition—A systematic review. *PLoS ONE* **2014**, *9*, e105017. [CrossRef] [PubMed]
6. Ibrahim, M.K.; Zambruni, M.; Melby, C.L.; Melby, P.C. Impact of childhood malnutrition on host defense and infection. *Clin. Microbiol. Rev.* **2017**, *30*, 919–971. [CrossRef] [PubMed]
7. Denning, D.W.; Bromley, M.J. Infectious Disease. How to bolster the antifungal pipeline. *Science* **2015**,*347*, 1414–1416. [CrossRef] [PubMed]
8. Lemke, A.; Kiderlen, A.F.; Kayser, O. Amphotericin B. *Appl. Microbiol. Biotechnol.* **2005**, *68*, 151–162. [CrossRef] [PubMed]
9. Bellmann, R. Pharmacodynamics and pharmacokinetics of antifungals for treatment of invasive aspergillosis.*Curr. Pharm. Des.* **2013**, *19*, 3629–3647. [CrossRef] [PubMed]
10. Pfaller, M.A. Antifungal drug resistance: Mechanisms, epidemiology, and consequences for treatment.*Am. J. Med.* **2012**, *125* (Suppl. 1), S3–S13. [CrossRef] [PubMed]
11. Pianalto, K.M.; Alspaugh, J.A. New Horizons in Antifungal Therapy. *J. Fungi.* **2016**, *2*, 26. [CrossRef] [PubMed]
12. Nivoix, Y.; Ubeaud-Sequier, G.; Engel, P.; Levêque, D.; Herbrecht, R. Drug-drug interactions of triazole antifungal agents in multimorbid patients and implications for patient care. *Curr. Drug. Metab.* **2009**, *10*, 395–409. [CrossRef] [PubMed]
13. Lewis, R.E. Current concepts in antifungal pharmacology. *Mayo Clin. Proc.* **2011**, *86*, 805–817. [CrossRef] [PubMed]

14. Campoy, S.; Adrio, J.L. Antifungals. *Mayo Clin. Proc.* **2017**, *133*, 86–96. [CrossRef] [PubMed]

15. McCarthy, M.W.; Walsh, T.J. Drugs currently under investigation for the treatment of invasive candidiasis.*Expert. Opin. Investig. Drugs* **2017**, *26*, 825–831. [CrossRef] [PubMed]

16. Perfect, J.R. The antifungal pipeline: A reality check. *Nat. Rev. Drug Discov.* **2017**, *16*, 603–616. [CrossRef] [PubMed]

17. Fuentefria, A.M.; Pippi, B.; Dalla Lana, D.F.; Donato, K.K.; de Andrade, S.F. Antifungals discovery: An insight into new strategies to combat antifungal resistance. *Lett. Appl. Microbiol.* **2018**, *66*, 2–13. [CrossRef] [PubMed]

18. Weig, M.; Brown, A.J. Genomics and the development of new diagnostics and anti-Candida drugs.*Trends Microbiol.* **2007**, *15*, 310–317. [CrossRef] [PubMed]

19. Payne, D.J.; Gwynn, M.N.; Holmes, D.J.; Pompliano, D.L. Drugs for bad bugs: Confronting the challenges of antibacterial discovery. *Nat. Rev. Drug Discov.* **2007**, *6*, 29–40. [CrossRef] [PubMed]

20. Roemer, T.; Boone, C. Systems-level antimicrobial drug and drug synergy discovery. *Nat. Chem. Biol.* **2013**,*9*, 222–231. [CrossRef] [PubMed]

21. Schenone, M.; Danc˘ík, V.; Wagner, B.K.; Clemons, P.A. Target identification and mechanism of action in chemical biology and drug discovery. *Nat. Chem. Biol.* **2013**, *9*, 232–240. [CrossRef] [PubMed]

22. Nucci, M.; Marr, K.A. Emerging fungal diseases. *Clin. Infect. Dis.* **2005**, *41*, 521–526. [CrossRef] [PubMed]

23. Castón-Osorio, J.J.; Rivero, A.; Torre-Cisneros, J. Epidemiology of invasive fungal infection. *Int. J. Antimicrob. Agents* **2008**, *32* (Suppl. 2), S103–S109. [CrossRef]

24. Miceli, M.H.; Díaz, J.A.; Lee, S.A. Emerging opportunistic yeast infections. *Lancet Infect. Dis.* **2011**, *11*, 142–151. [CrossRef]

25. Galimberti, R.; Torre, A.C.; Baztán, M.C.; Rodriguez-Chiappetta, F. Emerging systemic fungal infections.*Clin. Dermatol.* **2012**, *30*, 633–650. [CrossRef] [PubMed]

26. Enoch, D.A.; Yang, H.; Aliyu, S.H.; Micallef, C. The Changing Epidemiology of Invasive Fungal Infections. In *Human Fungal Pathogen Identification. Methods in Molecular Biology*; Lion, T., Ed.; Humana Press: New York, NY, USA, 2017; Volume 1508, pp. 17–65. ISBN 978-1-4939-6513-7.

27. Schwartz, I.S.; Patterson, T.F. The Emerging Threat of antifungal resistance in transplant infectious diseases.*Curr. Infect. Dis. Rep.* **2018**, *20*, 2. [CrossRef] [PubMed]

28. Pfaller, M.A.; Diekema, D.J.; Gibbs, D.L.; Newell, V.A.; Ellis, D.; Tullio, V.; Rodloff, A.; Fu, W.; Ling, T.A. Global Antifungal Surveillance Group. Results from the ARTEMIS DIS Global Antifungal Surveillance Study, 1997 to 2007: A 10.5-year analysis of susceptibilities of Candida Species to fluconazole and voriconazole as determined by CLSI standardized disk diffusion. *J. Clin. Microbiol.* **2010**, *48*, 1366–1377. [CrossRef] [PubMed]

29. Khawcharoenporn, T.; Apisarnthanarak, A.; Mundy, L.M. Non-neoformans cryptococcal infections: A systematic review. *Infection* **2007**, *35*, 51–58. [CrossRef] [PubMed]

30. Enache-Angoulvant, A.; Hennequin, C. Invasive *Saccharomyces* infection: A comprehensive review.*Clin. Infect. Dis.* **2005**, *41*, 1559–1568. [CrossRef] [PubMed]

31. Anoop, V.; Rotaru, S.; Shwed, P.S.; Tayabali, A.F.; Arvanitakis, G. Review of current methods for characterizing virulence and pathogenicity potential of industrial Saccharomyces cerevisiae strains towards humans. *FEMS Yeast Res.* **2015**, *15*, fov057-2. [CrossRef] [PubMed]

32. Lass-Flörl, C.; Cuenca-Estrella, M. Changes in the epidemiological landscape of invasive mould infections and disease. *J. Antimicrob. Chemother.* **2017**, *72* (Suppl. 1), i5–i11. [CrossRef] [PubMed]

33. Ribes, J.A.; Vanover-Sams, C.L.; Baker, D.J. Zygomycetes in human disease. *Clin. Microbiol. Rev.* **2000**,*13*, 236–301. [CrossRef] [PubMed]

34. Kontoyiannis, D.P.; Marr, K.A.; Park, B.J.; Alexander, B.D.; Anaissie, E.J.; Walsh, T.J.; Ito, J.; Andes, D.R.; Baddley, J.W.; Brown, J.M.; et al. Prospective surveillance for invasive fungal infections in hematopoietic stem cell transplant recipients, 2001–2006: Overview of the Transplant-Associated Infection Surveillance Network (TRANSNET) Database. *Clin. Infect. Dis.* **2010**, *50*, 1091–1100. [CrossRef] [PubMed]

35. Roilides, E.; Zaoutis, T.E.; Walsh, T.J. Invasive zygomycosis in neonates and children. *Clin. Microbiol. Infect.***2009**, *15* (Suppl. 5), 50–54. [CrossRef] [PubMed]

36. Lanternier, F.; Lortholary, O. Zygomycosis and diabetes mellitus. *Clin. Microbiol. Infect.* **2009**, *15* (Suppl. 5), 21–25. [CrossRef] [PubMed]

37. Nucci, M.; Anaissie, E. *Fusarium* infections in immunocompromised patients. *Clin. Microbiol. Rev.* **2007**,*20*, 695–704. [CrossRef] [PubMed]

38. Tortorano, A.M.; Richardson, M.; Roilides, E.; Diepeningen, A.V.; Caira, M.; Munoz, P.; Johnson, E.; Meletiadis, J.; Pana, Z.D.; Lackner, M.; et al. ESCMID & ECMM Joint Guidelines on Diagnosis and Management of Hyalohyphomycosis: *Fusarium* spp., *Scedosporium* spp., and others. *Clin. Microbiol. Infect.* **2014**, *20* (Suppl. 3), 27–46. [PubMed]

39. Guarro, J.; Kantarcioglu, A.S.; Horré, R.; Luis Rodriguez-Tudela, J.; Cuenca Estrella, M.; Berenguer, J.; Sybren De Hoog, G. *Scedosporium apiospermum*: Changing clinical spectrum of a therapy-refractory opportunist. *Med. Mycol.* **2006**, *44*, 295–327. [CrossRef] [PubMed]

40. Montesinos, I.; Dodemont, M.; Lagrou, K.; Jacobs, F.; Etienne, I.; Denis, O. New case of azole-resistant *Aspergillus fumigatus* due to TR46/Y121F/T289A mutation in Belgium. *J. Antimicrob. Chemother.* **2014**, *69*, 3439–3440. [CrossRef] [PubMed]

41. Buil, J.B.; Meis, J.F.; Melchers, W.J.; Verweij, P.E. Are the TR46/Y121F/T289A Mutations in Azole-Resistant Aspergillosis Patient Acquired or Environmental? *Antimicrob. Agents Chemother.* **2016**, *60*, 3259–3260. [CrossRef] [PubMed]

42. Ostrosky-Zeichner, L.; Casadevall, A.; Galgiani, J.N.; Odds, F.C.; Rex, J.H. An insight into the antifungal pipeline: Selected new molecules and beyond. *Nat. Rev. Drug Discov.* **2010**, *9*, 719–727. [CrossRef] [PubMed]

43. Vandeputte, P.; Ferrari, S.; Coste, A.T. Antifungal resistance and new strategies to control fungal infections.*Int. J. Microbiol.* **2012**, *2012*, 713687. [CrossRef] [PubMed]

44. Ballard, S.A.; Lodola, A.; Tarbit, M.H. A comparative study of 1-substituted imidazole and 1,2,4-triazole antifungal compounds as inhibitors of testosterone hydroxylations catalysed by mouse hepatic microsomal cytochromes P-450. *Biochem. Pharmacol.* **1988**, *37*, 4643–4651. [CrossRef]

45. DiDomenico, B. Novel antifungal drugs. *Curr. Opin. Microbiol.* **1999**, *2*, 509–515. [CrossRef]

46. Arévalo, M.P.; Carrillo-Muñoz, A.J.; Salgado, J.; Cardenes, D.; Brió, S.; Quindós, G.; Espinel-Ingroff, A. Antifungal activity of the echinocandin anidulafungin (VER002, LY-303366) against yeast pathogens: A comparative study with M27-A microdilution method. *J. Antimicrob. Chemother.* **2003**, *51*, 163–166. [CrossRef] [PubMed]

47. Mukherjee, P.K.; Sheehan, D.; Puzniak, L.; Schlamm, H.; Ghannoum, M.A. Echinocandins: Are they all the same? *J. Chemother.* **2011**, *23*, 319–325. [CrossRef] [PubMed]

48. Firooz, A.; Nafisi, S.; Maibach, H.I. Novel drug delivery strategies for improving econazole antifungal action.*Int. J. Pharm.* **2015**, *495*, 599–607. [CrossRef] [PubMed]

49. Denning, D.W.; Hope, W.W. Therapy for fungal diseases: Opportunities and priorities. *Trends Microbiol.* **2010**,*18*, 195–204. [CrossRef] [PubMed]

50. Terra, L.; Abreu, P.A.; Teixeira, V.L.; Paixão, I.C.; Pereira, R.; Leal, B.; Lourenço, A.L.; Rampelotto, P.H.; Castro, H.C. Mycoses and Antifungals: Reviewing the basis of a current problem that still is a biotechnological target for marine products. *Front. Mar. Sci.* **2014**, *1*, 12. [CrossRef]

51. Vengurlekar, S.; Sharma, R.; Trivedi, P. Efficacy of some natural compounds as antifungal agents.*Pharmacogn. Rev.* **2012**, *6*, 91–99. [CrossRef] [PubMed]

52. Lopez-Moya, F.; Colom-Valiente, M.F.; Martinez-Peinado, P.; Martinez-Lopez, J.E.; Puelles, E.; Sempere-Ortells, J.M.; Lopez-Llorca, L.V. Carbon and nitrogen limitation increase chitosan antifungal activity in Neurospora crassa and fungal human pathogens. *Fungal. Biol.* **2015**, *119*, 154–169. [CrossRef] [PubMed]

53. Abbaszadeh, S.; Sharifzadeh, A.; Shokri, H.; Khosravi, A.R.; Abbaszadeh, A. Antifungal efficacy of thymol, carvacrol, eugenol and menthol as alternative agents to control the growth of food-relevant fungi.*J. Mycol. Med.* **2014**, *24*, e51–e56. [CrossRef] [PubMed]

54. Barros, F.M.C.; Pippi, B.; Dresch, R.R.; Dauber, B.; Luciano, S.C.; Apel, M.A.; Fuentefria, A.M.; Von Poser, G.L. Antifungal and antichemotactic activities and quantification of phenolic compounds in lipophilic extracts of Hypericum spp. native to South Brazil. *Ind. Crops Prod.* **2013**, *44*, 294–299. [CrossRef]

55. Danielli, L.J.; Dos Reis, M.; Bianchini, M.; Camargo, G.S.; Bordignon, S.A.L.; Guerreiro, I.K.; Fuentefria, A.; Apel, M.A. Antidermatophytic activity of volatile oil and nanoemulsion of *Stenachaenium megapotamicum* (Spreng.) Baker. *Ind. Crops Prod.* **2013**, *50*, 23–28. [CrossRef]

56. Machado, G.R.; Pippi, B.; Dalla Lana, D.F.; Amaral, A.P.S.; Teixeira, M.L.; De Souza, K.C.B.; Fuentefria, A.M. Reversal of fluconazole resistance induced by a synergistic effect with *Acca sellowiana* in Candida glabrata strains. *Pharm. Biol.* **2016**, *54*, 2410–2419. [CrossRef] [PubMed]

57. Bicanic, T.; Muzoora, C.; Brouwer, A.E.; Meintjes, G.; Longley, N.; Taseera, K.; Rebe, K.; Loyse, A.; Jarvis, J.; Bekker, L.G.; et al. Independent association between rate of clearance of infection and clinical outcome of HIV-associated cryptococcal meningitis: Analysis of a combined cohort of 262 patients. *Clin. Infect. Dis.* **2009**, *49*, 702–709. [CrossRef] [PubMed]

58. Pierce, C.G.; Uppuluri, P.; Tristan, A.R.; Wormley, F.L., Jr.; Mowat, E.; Ramage, G.; Lopez-Ribot, J.L. A simple and reproducible 96-well plate-based method for the formation of fungal biofilms and its application to antifungal susceptibility testing. *Nat. Protoc.* **2008**, *3*, 1494–1500. [CrossRef] [PubMed]

59. LaFleur, M.D.; Lucumi, E.; Napper, A.D.; Diamond, S.L.; Lewis, K.N. Novel high-throughput screen against *Candida albicans* identifies antifungal potentiators and agents effective against biofilms.*J. Antimicrob. Chemother.* **2011**, *66*, 820–826. [CrossRef] [PubMed]

60. Monteiro, M.C.; de la Cruz, M.; Cantizani, J.; Moreno, C.; Tormo, J.R.; Mellado, E.; De Lucas, J.R.; Asensio, F.; Valiante, V.; Brakhage, A.A.; et al. A new approach to drug discovery: High-throughput screening of microbial natural extracts against *Aspergillus fumigatus* using resazurin. *J. Biomol. Screen* **2012**, *17*, 542–549. [CrossRef] [PubMed]

61. Rabjohns, J.L.A.; Park, Y.D.; Dehdashti, J.; Henderson, C.; Zelazny, A.; Metallo, S.J.; Zheng, W.; Williamson, P.R. A high-throughput screening assay for fungicidal compounds against *Cryptococcus neoformans*. *J. Biomol. Screen* **2014**, *19*, 270–277. [CrossRef] [PubMed]

62. DiDone, L.; Scrimale, T.; Baxter, B.K.; Krysan, D.J. A high-throughput assay of yeast cell lysis for drug discovery and genetic analysis. *Nat. Protoc.* **2010**, *5*, 1107–1114. [CrossRef] [PubMed]

63. Cameron-Clarke, A.; Hulse, G.A.; Clifton, L.; Cantrell, I.C. The Use of Adenylate Kinase Measurement to Determine Causes of Lysis in Lager Yeast. *J Am. Soc. Brew. Chem.* **2003**, *61*, 152–156. [CrossRef]

64. Roemer, T.; Krysan, D.J. Antifungal drug development: Challenges, unmet clinical needs, and new approaches. *Cold Spring Harb. Perspect. Med.* **2014**, *4*, A019703. [CrossRef] [PubMed]

65. Perlstein, E.O.; Ruderfer, D.M.; Roberts, D.C.; Schreiber, S.L.; Kruglyak, L. Genetic basis of individual differences in the response to small-molecule drugs in yeast. *Nat. Genet.* **2007**, *39*, 496–502. [CrossRef] [PubMed]

66. Pierce, S.E.; Fung, E.L.; Jaramillo, D.F.; Chu, A.M.; Davis, R.W.; Nislow, C.; Giaever, G. A unique and universal molecular barcode array. *Nat. Methods* **2006**, *3*, 601–603. [CrossRef] [PubMed]

67. Muroi, M.; Kazami, S.; Noda, K.; Kondo, H.; Takayama, H.; Kawatani, M.; Usui, T.; Osada, H. Application of proteomic profiling based on 2D-DIGE for classification of compounds according to the mechanism of action. *Chem. Biol.* **2010**, *17*, 460–470. [CrossRef] [PubMed]

68. Oliver, J.D.; Sibley, G.E.; Beckmann, N.; Dobb, K.S.; Slater, M.J.; McEntee, L.; du Pré, S.; Livermore, J.; Bromley, M.J.; Wiederhold, N.P.; et al. F901318 represents a novel class of antifungal drug that inhibits dihydroorotate dehydrogenase. *Proc. Natl. Acad. Sci. USA* **2016**, *133*, 12809–12814. [CrossRef] [PubMed]

69. Winzeler, E.A.; Shoemaker, D.D.; Astromoff, A.; Liang, H.; Anderson, K.; Andre, B.; Bangham, R.; Benito, R.; Boeke, J.D.; Bussey, H.; et al. Functional characterization of the S. cerevisiae genome by gene deletion and parallel analysis. *Science* **1999**, *285*, 901–906. [CrossRef] [PubMed]

70. Tong, A.H.; Evangelista, M.; Parsons, A.B.; Xu, H.; Bader, G.D.; Pagé, N.; Robinson, M.; Raghibizadeh, S.; Hogue, C.W.; Bussey, H.; et al. Systematic

genetic analysis with ordered arrays of yeast deletion mutants. *Science* **2001**, *294*, 2364–2368. [CrossRef] [PubMed]

71. Costanzo, M.; Baryshnikova, A.; Bellay, J.; Kim, Y.; Spear, E.D.; Sevier, C.S.; Ding, H.; Koh, J.L.; Toufighi, K.; Mostafavi, S.; et al. The genetic landscape of a cell. *Science* **2010**, *327*, 425–431. [CrossRef] [PubMed]
72. Hillenmeyer, M.E.; Fung, E.; Wildenhain, J.; Pierce, S.E.; Hoon, S.; Lee, W.; Proctor, M.; St Onge, R.P.; Tyers, M.; Koller, D.; et al. The chemical genomic portrait of yeast: Uncovering a phenotype for all genes. *Science* **2008**, *320*, 362–365. [CrossRef] [PubMed]
73. Hopkins, A.L. Network pharmacology: The next paradigm in drug discovery. *Nat. Chem. Biol.* **2008**,*4*, 682–690. [CrossRef] [PubMed]
74. Fitzgerald, J.B.; Schoeberl, B.; Nielsen, U.B.; Sorger, P.K. Systems biology and combination therapy in the quest for clinical efficacy. *Nat. Chem. Biol.* **2006**, *2*, 458–466. [CrossRef] [PubMed]
75. Lehár, J.; Stockwell, B.R.; Giaever, G.; Nislow, C. Combination chemical genetics. *Nat. Chem. Biol.* **2008**,*4*, 674–681. [CrossRef] [PubMed]
76. Borisy, A.A.; Elliott, P.J.; Hurst, N.W.; Lee, M.S.; Lehar, J.; Price, E.R.; Serbedzija, G.; Zimmermann, G.R.; Foley, M.A.; Stockwell, B.R.; et al. Systematic discovery of multicomponent therapeutics. *Proc. Natl. Acad. Sci. USA* **2003**, *100*, 7977–7982. [CrossRef] [PubMed]
77. Zhang, L.; Yan, K.; Zhang, Y.; Huang, R.; Bian, J.; Zheng, C.; Sun, H.; Chen, Z.; Sun, N.; An, R.; et al. High-throughput synergy screening identifies microbial metabolites as combination agents for the treatment of fungal infections. *Proc. Natl. Acad. Sci. USA* **2007**, *104*, 4606–4611. [CrossRef] [PubMed]
78. Zhai, B.; Zhou, H.; Yang, L.; Zhang, J.; Jung, K.; Giam, C.Z.; Xiang, X.; Lin, X. Polymyxin B, in combination with fluconazole, exerts a potent fungicidal effect. *J. Antimicrob. Chemother.* **2010**, *65*, 931–938. [CrossRef] [PubMed]
79. Spitzer, M.; Griffiths, E.; Blakely, K.M.; Wildenhain, J.; Ejim, L.; Rossi, L.; De Pascale, G.; Curak, J.; Brown, E.; Tyers, M.; et al. Cross-species discovery of syncretic drug combinations that potentiate the antifungal fluconazole. *Mol. Syst. Biol.* **2011**, *7*, 499. [CrossRef] [PubMed]
80. Cowen, L.E.; Singh, S.D.; Köhler, J.R.; Collins, C.; Zaas, A.K.; Schell, W.A.; Aziz, H.; Mylonakis, E.; Perfect, J.R.; Whitesell, L.; Lindquist, S. Harnessing Hsp90 function as a powerful, broadly effective therapeutic strategy for fungal infectious disease. *Proc. Natl. Acad. Sci. USA* **2009**, *106*, 2818–2823. [CrossRef] [PubMed]

81. Epp, E.; Vanier, G.; Harcus, D.; Lee, A.Y.; Jansen, G.; Hallett, M.; Sheppard, D.C.; Thomas, D.Y.; Munro, C.A.; Mullick, A.; et al. Reverse genetics in *Candida albicans* predicts ARF cycling is essential for drug resistance and virulence. *PLoS Pathog.* **2010**, *6*, e1000753. [CrossRef] [PubMed]

82. Sharom, J.R.; Bellows, D.S.; Tyers, M. From large networks to small molecules. *Curr. Opin. Chem. Biol.* **2004**,*8*, 81–90. [CrossRef] [PubMed]

83. Jansen, G.; Lee, A.Y.; Epp, E.; Fredette, A.; Surprenant, J.; Harcus, D.; Scott, M.; Tan, E.; Nishimura, T.; Whiteway, M.; et al. Chemogenomic profiling predicts antifungal synergies. *Mol. Syst. Biol.* **2009**, *5*, 338. [CrossRef] [PubMed]

84. Lehár, J.; Krueger, A.S.; Avery, W.; Heilbut, A.M.; Johansen, L.M.; Price, E.R.; Rickles, R.J.; Short, G.F., 3rd; Staunton, J.E.; Jin, X.; et al. Synergistic drug combinations tend to improve therapeutically relevant selectivity. *Nat. Biotechnol.* **2009**, *27*, 659–666. [CrossRef]

85. Whitesides, G.M. The origins and the future of microfluidics. *Nature* **2006**, *442*, 368–373. [CrossRef] [PubMed]

86. Squires, T.M.; Quake, S.R. Microfluidics: Fluid physics at the nanoliter scale. *Rev. Mod. Phys.* **2005**,*77*, 977–1026. [CrossRef]

87. Dertinger, S.K.W.; Chiu, D.T.; Jeon, N.L.; Whitesides, G.M. Generation of Gradients Having Complex Shapes Using Microfluidic Networks. *Anal. Chem.* **2001**, *73*, 1240–1246. [CrossRef]

88. Sackmann, E.K.; Fulton, A.L.; Beebe, D.J. The present and future role of microfluidics in biomedical research.*Nature* **2014**, *507*, 181–189. [CrossRef] [PubMed]

89. Knight, J.B.; Vishwanath, A.; Brody, J.P.; Austin, R.H. Hydrodynamic focusing on a silicon chip: Mixing nanoliters in microseconds. *Phys. Rev. Lett.* **1998**, *80*, 3863–3866. [CrossRef]

90. Pollack, L.; Tate, M.W.; Finnefrock, A.C.; Kalidas, C.; Trotter, S.; Darnton, N.C.; Lurio, L.; Austin, R.H.; Batt, C.A.; Gruner, S.M.; et al. Time resolved collapse of a folding protein observed with small angle X-ray scattering. *Phys. Rev. Lett.* **2001**, *86*, 4962–4965. [CrossRef] [PubMed]

91. Wen, N.; Zhao, Z.; Fan, B.; Chen, D.; Men, D.; Wang, J.; Chen, J. Development of Droplet Microfluidics Enabling High-Throughput Single-Cell Analysis. *Molecules* **2016**, *21*, 881. [CrossRef] [PubMed]

92. Clausell-Tormos, J.; Lieber, D.; Baret, J.C.; El-Harrak, A.; Miller, O.J.; Frenz, L.; Blouwolff, J.; Humphry, K.J.; Köster, S.; Duan, H.; et al. Droplet-based microfluidic platforms for the encapsulation and screening of Mammalian cells and multicellular organisms. *Chem. Biol.* **2008**, *15*, 427–437. [CrossRef] [PubMed]

93. Kim, D.; Wu, X.; Young, A.T.; Haynes, C.L. Microfluidics-based in vivo mimetic systems for the study of cellular biology. *Acc. Chem. Res.* **2014**, *47*, 1165–1173. [CrossRef] [PubMed]
94. Mehling, M.; Tay, S. Microfluidic cell culture. *Curr. Opin. Biotechnol.* **2014**, *25*, 95–102. [CrossRef] [PubMed]
95. Velve-Casquillas, G.; le Berre, M.; Piel, M.; Tran, P.T. Microfluidic tools for cell biological research. *Nano Today***2010**, *5*, 28–47. [CrossRef] [PubMed]
96. Castel, D.; Pitaval, A.; Debily, M.A.; Gidrol, X. Cell microarrays in drug discovery. *Drug Discov. Today* **2006**,*11*, 616–622. [CrossRef] [PubMed]
97. Chen, D.S.; Davis, M.M. Molecular and functional analysis using live cell microarrays. *Curr. Opin. Chem. Biol.***2006**, *10*, 28–34. [CrossRef] [PubMed]
98. Yarmush, M.L.; King, K.R. Living-cell microarrays. *Annu. Rev. Biomed. Eng.* **2009**, *11*, 235–257. [CrossRef] [PubMed]
99. Willaert, R.; Sahli, H. On-chip living-cell microarrays for network biology. In *Bioinformatics—Trends and Methodologies*; Mahdavi, M.A., Ed.; InTech-Open Access: Publisher Rejeka, Croatia, 2011; pp. 609–630. ISBN 978-953-307-282-1.
100. Willaert, R.G.; Goossens, K. Microfluidic bioreactors for cellular microarrays. *Fermentation* **2015**, *1*, 38–78. [CrossRef]
101. Jonczyk, R.; Kurth, T.; Lavrentieva, A.; Walter, J.G.; Scheper, T.; Stahl, F. Living cell microarrays: An overview of concepts. *Microarrays* **2016**, *5*, 11. [CrossRef] [PubMed]
102. Charvin, G.; Cross, F.R.; Siggia, E.D. Forced periodic expression of G1 cyclins phase-locks the budding yeast cell cycle. *Proc. Natl. Acad. Sci. USA* **2009**, *106*, 6632–6637. [CrossRef] [PubMed]
103. Bean, J.M.; Siggia, E.D.; Cross, F.R. Coherence and timing of cell cycle start examined at single-cell resolution.*Mol. Cell* **2006**, *21*, 3–14. [CrossRef] [PubMed]
104. Lidstrom, M.E.; Konopka, M.C. The role of physiological heterogeneity in microbial population behavior.*Nat. Chem. Biol.* **2010**, *6*, 705–712. [CrossRef] [PubMed]
105. Martins, B.M.; Locke, J.C. Microbial individuality: How single-cell heterogeneity enables population level strategies. *Curr. Opin. Microbiol.* **2015**, *24*, 104–112. [CrossRef] [PubMed]
106. Rosenthal, K.; Oehling, V.; Dusny, C.; Schmid, A. Beyond the bulk: Disclosing the life of single microbial cells. *FEMS Microbiol. Rev.* **2017**, *41*, 751–780. [CrossRef] [PubMed]
107. Pierce, C.G.; Srinivasan, A.; Uppuluri, P.; Ramasubramanian, A.K.; López-Ribot, J.L. Antifungal therapy with an emphasis on biofilms. *Curr. Opin. Pharmacol.* **2013**, *13*, 726–730. [CrossRef] [PubMed]

108. Borghi, E.; Borgo, F.; Morace, G. Fungal Biofilms: Update on Resistance. *Adv. Exp. Med. Biol.* **2016**, *931*, 37–47. [CrossRef] [PubMed]

109. Koo, H.; Allan, R.N.; Howlin, R.P.; Stoodley, P.; Hall-Stoodley, L. Targeting microbial biofilms: Current and prospective therapeutic strategies. *Nat. Rev. Microbiol.* **2017**, *15*, 740–755. [CrossRef] [PubMed]

110. Wu, S.; Wang, Y.; Liu, N.; Dong, G.; Sheng, C. Tackling Fungal Resistance by Biofilm Inhibitors. *J. Med. Chem.* **2017**, *60*, 2193–2211. [CrossRef] [PubMed]

111. Srinivasan, A.; Uppuluri, P.; Lopez-Ribot, J.; Ramasubramanian, A.K. Development of a high-throughput *Candida albicans* biofilm chip. *PLoS ONE* **2011**, *6*, e19036. [CrossRef] [PubMed]

112. Srinivasan, A.; Lopez-Ribot, J.L.; Ramasubramanian, A.K. *Candida albicans* biofilm chip (CaBChip) for high-throughput antifungal drug screening. *J. Vis. Exp.* **2012**, *18*, e3845. [CrossRef]

113. Srinivasan, A.; Leung, K.P.; Lopez-Ribot, J.L.; Ramasubramanian, A.K. High-throughput nano-biofilm microarray for antifungal drug discovery. *MBio* **2013**, *4*, E00331-13. [CrossRef] [PubMed]

114. Srinivasan, A.; Torres, N.S.; Leung, K.P.; Lopez-Ribot, J.L.; Ramasubramanian, A.K. nBioChip, a Lab-on-a-Chip Platform of Mono- and Polymicrobial Biofilms for High-Throughput Downstream Applications. *mSphere* **2017**, *2*. [CrossRef] [PubMed]

115. Yang, H.; Bharracharyya, M.K.; Dong, L. Plant Pathogen Spores Grow in Microfluidic Droplets: A High-Throughput Approach to Antifungal Drug Screening. In Proceedings of the IEEE Conference paper Transducers' 11, Beijing, China, 5–9 June 2011.

116. Michael, C.A.; Dominey-Howes, D.; Labbate, M. The antimicrobial resistance crisis: Causes, consequences, and management. *Front. Public Health* **2014**, *2*, 145. [CrossRef] [PubMed]

117. Jorgensen, J.H.; Ferraro, M.J. Antimicrobial susceptibility testing: A review of general principles and contemporary practices. *Clin. Infect. Dis.* **2009**, *49*, 1749–1755. [CrossRef] [PubMed]

118. Barenfanger, J.; Drake, C.; Kacich, G. Clinical and financial benefits of rapid bacterial identification and antimicrobial susceptibility testing. *J. Clin. Microbiol.* **1999**, *37*, 1415–1418. [PubMed]

119. Arikan, S. Current status of antifungal susceptibility testing methods. *Med. Mycol.* **2007**, *45*, 569–587. [CrossRef] [PubMed]

120. Daniels, R. Surviving the first hours in sepsis: Getting the basics right (an intensivist's perspective). *J. Antimicrob. Chemother.* **2011**, *66*, 11–23. [CrossRef] [PubMed]

121. Alcazar-Fuoli, L.; Mellado, E. Current status of antifungal resistance and its impact on clinical practice.*Br. J. Haematol.* **2014**, *166*, 471–484. [CrossRef] [PubMed]

122. Syal, K.; Mo, M.; Yu, H.; Iriya, R.; Jing, W.; Guodong, S.; Wang, S.; Grys, T.E.; Haydel, S.E.; Tao, N. Current and emerging techniques for antibiotic susceptibility tests. *Theranostics* **2017**, *7*, 1795–1805. [CrossRef] [PubMed]

123. Clinical and Laboratory Standards Institute. *Reference Method for Broth Dilution Antifungal Susceptibility Testing of Yeasts. Approved Standard M27-A3*, 3rd ed.; CLSI: Wayne, PA, USA, 2008.

124. Arendrup, M.C.; Cuenca-Estrella, M.; Lass-Florl, C.; Hope, W.; Eucast, A. EUCAST technical note on the EUCAST definitive document EDef 7.2: Method for the determination of broth dilution minimum inhibitory concentrations of antifungal agents for yeasts EDef 7.2 (EUCAST-AFST). *Clin. Microbiol. Infect.* **2012**, *18*, E246–E247. [CrossRef] [PubMed]

125. Pulido, M.R.; García-Quintanilla, M.; Martín-Peña, R.; Cisneros, J.M.; McConnell, M.J. Progress on the development of rapid methods for antimicrobial susceptibility testing. *J. Antimicrob. Chemother.* **2013**, *68*, 2710–2717. [CrossRef] [PubMed]

126. Posteraro, B.; Torelli, R.; De Carolis, E.; Posteraro, P.; Sanguinetti, M. Antifungal susceptibility testing: Current role from the clinical laboratory perspective. *Mediterr. J. Hematol. Infect. Dis.* **2014**, *6*, e2014030. [CrossRef] [PubMed]

127. Subcommittee on Antifungal Susceptibility Testing of the ESCMID European Committee for Antimicrobial Susceptibility Testing. EUCAST Technical Note on the method for the determination of broth dilution minimum inhibitory concentrations of antifungal agents for conidia-forming moulds. *Clin. Microbiol. Infect.* **2008**, *14*, 982–984.

128. Vale-Silva, L.A.; Buchta, V. Antifungal susceptibility testing by flow cytometry: Is it the future? *Mycoses* **2006**,*49*, 261–273. [CrossRef] [PubMed]

129. Vale-Silva, L.A.; Pinto, P.; Lopes, V.; Ramos, H.; Pinto, E. Comparison of the Etest and a rapid flow cytometry-based method with the reference CLSI broth microdilution protocol M27-A3 for the echinocandin susceptibility testing of Candida spp. *Eur. J. Clin. Microbiol. Infect. Dis.* **2012**, *31*, 941–946. [CrossRef] [PubMed]

130. Vella, A.; De Carolis, E.; Vaccaro, L.; Posteraro, P.; Perlin, D.S.; Kostrzewa, M.; Posteraro, B.; Sanguinetti, M. Rapid antifungal susceptibility testing by matrix-assisted laser desorption ionization time of flight mass spectrometry analysis. *J. Clin. Microbiol.* **2013**, *51*, 2964–2969. [CrossRef] [PubMed]

131. Furustrand Tafin, U.; Clauss, M.; Hauser, P.M.; Bille, J.; Meis, J.F.; Trampuz, A. Isothermal microcalorimetry: A novel method for real- time determination of antifungal susceptibility of *Aspergillus* species. *Clin. Microbiol. Infect.* **2012**, *18*, E241–E245. [CrossRef] [PubMed]

132. Campbell, J.; McBeth, C.; Kalashnikov, M.; Boardman, A.K.; Sharon, A.; Sauer-Budge, A.F. Microfluidic advances in phenotypic antibiotic susceptibility testing. *Biomed. Microdevices* **2016**, *18*, 103. [CrossRef] [PubMed]

133. Mittman, S.A.; Huard, R.C.; Della-Latta, P.; Whittier, S. Comparison of BD phoenix to vitek 2, microscan MICroSTREP, and Etest for antimicrobial susceptibility testing of Streptococcus pneumoniae. *J. Clin. Microbiol.* **2009**, *47*, 3557–3561. [CrossRef] [PubMed]

134. Chatzigeorgiou, K.S.; Sergentanis, T.N.; Tsiodras, S.; Hamodrakas, S.J.; Bagos, P.G. Phoenix 100 versus Vitek 2 in the identification of gram-positive and gram-negative bacteria: A comprehensive meta-analysis.*J. Clin. Microbiol.* **2011**, *49*, 3284–3291. [CrossRef] [PubMed]

135. Dai, J.; Suh, S.J.; Hamon, M.; Hong, J.W. Determination of antibiotic EC50 using a zero-flow microfluidic chip based growth phenotype assay. *Biotechnol. J.* **2015**, *10*, 1783–1791. [CrossRef] [PubMed]

136. Dong, T.; Zhao, X. Rapid identification and susceptibility testing of uropathogenic microbes via immunosorbent ATP-bioluminescence assay on a microfluidic simulator for antibiotic therapy. *Anal. Chem.* **2015**, *87*, 2410–2418. [CrossRef] [PubMed]

137. Puchberger-Enengl, D.; van den Driesche, S.; Krutzler, C.; Keplinger, F.; Vellekoop, M.J. Hydrogel-based microfluidic incubator for microorganism cultivation and analyses. *Biomicrofluidics* **2015**, *9*, 014127. [CrossRef] [PubMed]

138. Bouquet, O.; Kocsis, B.; Kilár, F.; Lóránd, T.; Kustos, I. Amphotericin B and fluconazole susceptibility of*Candida* species determined by cell-chip technology. *Mycoses* **2012**, *55*, e90–e96. [CrossRef] [PubMed]

139. Chen, C.H.; Lu, Y.; Sin, M.L.; Mach, K.E.; Zhang, D.D.; Gau, V.; Liao, J.C.; Wong, P.K. Antimicrobial susceptibility testing using high surface-to-volume ratio microchannels. *Anal. Chem.* **2010**, *82*, 1012–1019. [CrossRef] [PubMed]

140. Sinn, I.; Kinnunen, P.; Albertson, T.; McNaughton, B.H.; Newton, D.W.; Burns, M.A.; Kopelman, R. Asynchronous magnetic bead rotation (AMBR) biosensor in microfluidic droplets for rapid bacterial growth and susceptibility measurements. *Lab Chip* **2011**, *11*, 2604–2611. [CrossRef] [PubMed]

141. Lu, Y.; Gao, J.; Zhang, D.D.; Gau, V.; Liao, J.C.; Wong, P.K. Single cell antimicrobial susceptibility testing by confined microchannels and electrokinetic loading. *Anal. Chem.* **2013**, *85*, 3971–3976. [CrossRef] [PubMed]

142. Mohan, R.; Mukherjee, A.; Sevgen, S.E.; Sanpitakseree, C.; Lee, J.; Schroeder, C.M.; Kenis, P.J. A multiplexed microfluidic platform for rapid antibiotic susceptibility testing. *Biosens. Bioelectron.* **2013**, *49*, 118–125. [CrossRef] [PubMed]

143. Tang, Y.; Zhen, L.; Liu, J.; Wu, J. Rapid antibiotic susceptibility testing in a microfluidic pH sensor. *Anal. Chem.* **2013**, *85*, 2787–2794. [CrossRef] [PubMed]

144. He, J.; Mu, X.; Guo, Z.; Hao, H.; Zhang, C.; Zhao, Z.; Wang, Q. A novel microbead-based microfluidic device for rapid bacterial identification and antibiotic susceptibility testing. *Eur. J. Clin. Microbiol. Infect. Dis.* **2014**, *33*, 2223–2230. [CrossRef] [PubMed]

145. Kim, S.C.; Cestellos-Blanco, S.; Inoue, K.; Zare, R.N. Miniaturized Antimicrobial Susceptibility Test by Combining Concentration Gradient Generation and Rapid Cell Culturing. *Antibiotics* **2015**, *4*, 455–466. [CrossRef] [PubMed]

146. Derzsi, L.; Kaminski, T.S.; Garstecki, P. Antibiograms in five pipetting steps: Precise dilution assays in sub-microliter volumes with a conventional pipette. *Lab Chip* **2016**, *16*, 893–901. [CrossRef] [PubMed]

147. Leonard, H.; Halachmi, S.; Ben-Dov, N.; Nativ, O.; Segal, E. Unraveling Antimicrobial Susceptibility of Bacterial Networks on Micropillar Architectures Using Intrinsic Phase-Shift Spectroscopy. *ACS Nano* **2017**, *11*, 6167–6177. [CrossRef] [PubMed]

148. Tang, M.; Huang, X.; Chu, Q.; Ning, X.; Wang, Y.; Kong, S.K.; Zhang, X.; Wang, G.; Ho, H.P. A linear concentration gradient generator based on multi-layered centrifugal microfluidics and its application in antimicrobial susceptibility testing. *Lab Chip* **2018**, *17*. [CrossRef] [PubMed]

149. Li, B.; Qiu, Y.; Glidle, A.; McIlvenna, D.; Luo, Q.; Cooper, J.; Shi, H.C.; Yin, H. Gradient microfluidics enables rapid bacterial growth inhibition testing. *Anal. Chem.* **2014**, *86*, 3131–3137. [CrossRef] [PubMed]

150. Smith, K.P.; Richmond, D.L.; Brennan-Krohn, T.; Elliott, H.L.; Kirby, J.E. Development of MAST: A Microscopy-Based Antimicrobial Susceptibility Testing Platform. *SLAS Technol.* **2017**, *22*, 662–674. [CrossRef] [PubMed]

151. Choi, J.; Yoo, J.; Kim, K.J.; Kim, E.G.; Park, K.O.; Kim, H.; Kim, H.; Jung, H.; Kim, T.; Choi, M.; et al. Rapid drug susceptibility test of *Mycobacterium tuberculosis* using microscopic time-lapse imaging in an agarose matrix. *Appl. Microbiol. Biotechnol.* **2016**, *100*, 2355–2365. [CrossRef] [PubMed]

152. Boedicker, J.Q.; Li, L.; Kline, T.R.; Ismagilov, R.F. Detecting bacteria and determining their susceptibility to antibiotics by stochastic confinement in nanoliter droplets using plug-based microfluidics. *Lab Chip* **2008**, *8*, 1265–1272. [CrossRef] [PubMed]

153. Choi, J.; Jung, Y.G.; Kim, J.; Kim, S.; Jung, Y.; Na, H.; Kwon, S. Rapid antibiotic susceptibility testing by tracking single cell growth in a microfluidic agarose channel system. *Lab Chip* **2013**, *13*, 280–287. [CrossRef] [PubMed]

154. Kalashnikov, M.; Lee, J.C.; Campbell, J.; Sharon, A.; Sauer-Budge, A.F. A microfluidic platform for rapid, stress-induced antibiotic susceptibility testing of *Staphylococcus aureus*. *Lab Chip* **2012**, *12*, 4523–4532. [CrossRef] [PubMed]

155. Price, C.S.; Kon, S.E.; Metzger, S. Rapid antibiotic susceptibility phenotypic characterization of *Staphylococcus aureus* using automated microscopy of small numbers of cells. *J. Microbiol. Methods* **2014**, *98*, 50–58. [CrossRef] [PubMed]

156. Choi, J.; Yoo, J.; Lee, M.; Kim, E.G.; Lee, J.S.; Lee, S.; Joo, S.; Song, S.H.; Kim, E.C.; Lee, J.C.; et al. A rapid antimicrobial susceptibility test based on single-cell morphological analysis. *Sci. Transl. Med.* **2014**, *6*, 267ra174. [CrossRef] [PubMed]

157. Hou, Z.; An, Y.; Hjort, K.; Hjort, K.; Sandegren, L.; Wu, Z. Time lapse investigation of antibiotic susceptibility using a microfluidic linear gradient 3D culture device. *Lab Chip* **2014**, *14*, 3409–3418. [CrossRef] [PubMed]

158. Matsumoto, Y.; Sakakihara, S.; Grushnikov, A.; Kikuchi, K.; Noji, H.; Yamaguchi, A.; Iino, R.; Yagi, Y.; Nishino, K. A Microfluidic Channel Method for Rapid Drug-Susceptibility Testing of Pseudomonas aeruginosa. *PLoS ONE* **2016**, *11*, e0148797. [CrossRef] [PubMed]

159. Hansen, K.M.; Thundat, T. Microcantilever biosensors. *Methods* **2005**, *37*, 57–64. [CrossRef] [PubMed]

160. Fritz, J. Cantilever biosensors. *Analyst* **2008**, *133*, 855–863. [CrossRef] [PubMed]

161. Willaert, R.; Kasas, S.; Devreese, B.; Dietler, G. Yeast nanobiotechnology. *Fermentation* **2016**, *2*, 18. [CrossRef]

162. Braun, T.; Ghatkesar, M.K.; Backmann, N.; Grange, W.; Boulanger, P.; Letellier, L.; Lang, H.P.; Bietsch, A.; Gerber, C.; Hegner, M. Quantitative time-resolved measurement of membrane protein-ligand interactions using microcantilever array sensors. *Nat. Nanotechnol.* **2009**, *4*, 179–185. [CrossRef] [PubMed]

163. Godin, M.; Tabard-Cossa, V.; Miyahara, Y.; Monga, T.; Williams, P.J.; Beaulieu, L.Y.; Bruce Lennox, R.; Grutter, P. Cantilever-based sensing: The origin of surface stress and optimization strategies. *Nanotechnology* **2010**, *21*, 75501. [CrossRef] [PubMed]

164. Godin, M.; Delgado, F.F.; Son, S.; Grover, W.H.; Bryan, A.K.; Tzur, A.; Jorgensen, P.; Payer, K.; Grossman, A.D.; Kirschner, M.W.; et al. Using buoyant mass to measure the growth of single cells. *Nat. Methods* **2010**, *7*, 387–390. [CrossRef] [PubMed]

165. Ndieyira, J.W.; Watari, M.; Barrera, A.D.; Zhou, D.; Vögtli, M.; Batchelor, M.; Cooper, M.A.; Strunz, T.; Horton, M.A.; Abell, C.; et al. Nanomechanical detection of antibiotic-mucopeptide binding in a model for superbug drug resistance. *Nat. Nanotechnol.* **2008**, *3*, 691–696. [CrossRef] [PubMed]

166. Lang, H.P.; Baller, M.K.; Berger, R.; Gerber, C.; Gimzewski, J.K.; Battiston, F.M.; Fornaro, P.; Ramseyer, J.P.; Meyer, E.; Guntherodt, H.J. An artificial nose based a micromechanical cantilever array. *Anal. Chim. Acta* **1999**, *393*, 59–65. [CrossRef]

167. Braun, T.; Barwich, V.; Ghatkesar, M.K.; Bredekamp, A.H.; Gerber, C.; Hegner, M.; Lang, H.P. Micromechanical mass sensors for biomolecular detection in a physiological environment. *Phys. Rev. E Stat. Nonlin. Soft Matter Phys.* **2005**, *72*, 031907. [CrossRef] [PubMed]

168. Hosaka, S.; Chiyoma, T.; Ikeuchi, A.; Okano, H.; Sone, H.; Izumi, T. Possibility of a femtogram mass biosensor using a self-sensing cantilever. *Curr. Appl. Phys.* **2006**, *6*, 384–388. [CrossRef]

169. Liu, Y.; Schweizerb, L.M.; Wanga, W.; Reubena, R.L.; Schweizer, M.; Shu, W. Label-free and real-time monitoring of yeast cell growth by the bending of polymer microcantilever biosensors. *Sens. Actuator B-Chem.* **2013**, *178*, 621–626. [CrossRef]

170. Bryan, A.K.; Goranov, A.; Amon, A.; Manalis, S.R. Measurement of mass, density, and volume during the cell cycle of yeast. *Proc. Natl. Acad. Sci. USA* **2010**, *107*, 999–1004. [CrossRef] [PubMed]

171. Cermak, N.; Olcum, S.; Delgado, F.F.; Wasserman, S.C.; Payer, K.R.; Murakami, M.A.; Knudsen, S.M.; Kimmerling, R.J.; Stevens, M.M.; Kikuchi, Y.; et al. High-throughput measurement of single-cell growth rates using serial microfluidic mass sensor arrays. *Nat. Biotechnol.* **2016**, *34*, 1052–1059. [CrossRef] [PubMed]

172. Burg, T.P.; Godin, M.; Knudsen, S.M.; Shen, W.; Carlson, G.; Foster, J.S.; Babcock, K.; Manalis, S.R. Weighing of biomolecules, single cells and single nanoparticles in fluid. *Nature* **2007**, *446*, 1066–1069. [CrossRef] [PubMed]

173. Park, K.; Jang, J.; Irimia, D.; Sturgis, J.; Lee, J.; Robinson, J.P.; Toner, M.; Bashir, R. 'Living cantilever arrays' for characterization of mass of single live cells in fluids. *Lab Chip* **2008**, *8*, 1034–1041. [CrossRef] [PubMed]

174. Bryan, A.K.; Hecht, V.C.; Shen, W.; Payer, K.; Grover, W.H.; Manalis, S.R. Measuring single cell mass, volume, and density with dual suspended microchannel resonators. *Lab Chip* **2014**, *14*, 569–576. [CrossRef] [PubMed]

175. Nugaeva, N.; Gfeller, K.Y.; Backmann, N.; Lang, H.P.; Düggelin, M.; Hegner, M. Micromechanical cantilever array sensors for selective fungal immobilization

and fast growth detection. *Biosens. Bioelectron.* **2005**, *21*, 849–856. [CrossRef] [PubMed]

176. Stupar, P. Atomic Force Microscopy of Biological Systems: Quantitative Imaging and Nanomotion Detection. Ph.D. Thesis, EPFL Scientific Publications, Lausanne, Switzerland, 2018.

177. Longo, G.; Alonso-Sarduy, L.; Rio, L.M.; Bizzini, A.; Trampuz, A.; Notz, J.; Dietler, G.; Kasas, S. Rapid detection of bacterial resistance to antibiotics using AFM cantilevers as nanomechanical sensors. *Nat. Nanotechnol.* **2013**, *8*, 522–526. [CrossRef] [PubMed]

178. Stupar, P.; Opota, O.; Longo, G.; Prod'hom, G.; Dietler, G.; Greub, G.; Kasas, S. Nanomechanical sensor applied to blood culture pellets: A fast approach to determine the antibiotic susceptibility against agents of bloodstream infections. *Clin. Microbiol. Infect.* **2017**, *23*, 400–405. [CrossRef] [PubMed]

179. Stupar, P.; Yvanoff, C.; Chomicki, W.; Dietler, G.; Kasas, S.; Willaert, R. Exploring Nanoscale Motions of Yeast Cells. In Proceedings of the XIX Annual Linz Winter Workshop, Linz, Austria, 3–6 February 2017.

180. Vanden Boer, P.; Stupar, P.; Chomicki, W.; Dietler, G.; Kasas, S.; Willaert, R. Nanomotion Detection of Single Yeast Cell Growth. In Proceedings of the XX Annual Linz Winter Workshop, Linz, Austria, 2–5 February 2018.

181. Morones, J.R.; Elechiguerra, J.L.; Camacho, A.; Ramirez, J.T. The bactericidal effect of silver nanoparticles.*Nanotechnology* **2005**, *16*, 2346–2353. [CrossRef] [PubMed]

182. Kim, J.S.; Kuk, E.; Yu, K.N.; Kim, J.H.; Park, S.J.; Lee, H.J.; Kim, S.H.; Park, Y.K.; Park, Y.H.; Hwang, C.Y.; et al. Antimicrobial effects of silver nanoparticles. *Nanomed. Nanotechnol. Biol. Med.* **2007**, *3*, 95–101. [CrossRef] [PubMed]

183. Ahamed, M.; Alsalhi, M.S.; Siddiqui, M.K. Silver nanoparticle applications and human health.*Clin. Chim. Acta* **2010**, *411*, 1841–1848. [CrossRef] [PubMed]

184. Chen, X.; Schleusener, H.J. Nanosilver: A nanoproduct in medical application. *Toxicol. Lett.* **2008**, *176*, 1e12. [CrossRef] [PubMed]

185. Rai, M.; Yadav, A.; Gade, A. Silver nanoparticles as a new generation of antimicrobials. *Biotechnol. Adv.* **2009**,*27*, 76–83. [CrossRef] [PubMed]

186. De Jong, W.H.; Van Der Ven, L.T.; Sleijffers, A.; Park, M.V.; Jansen, E.H.; Van Loveren, H.; Vandebriel, R.J. Systemic and immunotoxicity of silver nanoparticles in an intravenous 28 days repeated dose toxicity study in rats. *Biomaterials* **2013**, *34*, 8333–8343. [CrossRef] [PubMed]

187. Durán, N.; Durán, M.; de Jesus, M.B.; Seabra, A.B.; Fávaro, W.J.; Nakazato, G. Silver nanoparticles: A new view on mechanistic aspects on antimicrobial activity. *Nanomedicine* **2016**, *12*, 789–799. [CrossRef] [PubMed]

188. Ahamed, M.; Karns, M.; Goodson, M.; Rowe, J.; Hussain, S.M.; Schlager, J.J.; Hong, Y. DNA damage response to different surface chemistry of silver nanoparticles in mammalian cells. *Toxicol. Appl. Pharmacol.* **2008**, *233*, 404–410. [CrossRef] [PubMed]

189. Kawata, K.; Osawa, M.; Okabe, S. In vitro toxicity of silver nanoparticles at noncytotoxic doses to HepG2 human hepatoma cells. *Environ. Sci. Technol.* **2009**, *43*, 6046–6051. [CrossRef] [PubMed]

190. AshaRani, P.V.; Low Kah Mun, G.; Hande, M.P.; Valiyaveettil, S. Cytotoxicity and genotoxicity of silver nanoparticles in human cells. *ACS Nano* **2009**, *3*, 279–290. [CrossRef] [PubMed]

191. Panácˇek, A.; Kvítek, L.; Smékalová, M.; Vecˇerˇová, R.; Kolárˇ, M.; Röderová, M.; Dycˇka, F.; Šebela, M.; Prucek, R.; Tomanec, O.; et al. Bacterial resistance to silver nanoparticles and how to overcome it. *Nat. Nanotechnol.* **2018**, *13*, 65–71. [CrossRef] [PubMed]

192. Ammar, H.A.; El-Desouky, T.A. Green synthesis of nanosilver particles by *Aspergillus terreus* HA1N and Penicillium expansum HA2N and its antifungal activity against mycotoxigenic fungi. *J. Appl. Microbiol.* **2016**, *121*, 89–100. [CrossRef] [PubMed]

193. Xue, B.; He, D.; Gao, S.; Wang, D.; Yokoyama, K.; Wang, L. Biosynthesis of silver nanoparticles by the fungus Arthroderma fulvum and its antifungal activity against genera of Candida, *Aspergillus* and *Fusarium*. *Int. J. Nanomedicine* **2016**, *11*, 1899–1906. [CrossRef] [PubMed]

194. Khan, N.T.; Jameel, N. Antifungal Activity of Silver Nanoparticles Produced from Fungus, *Penicillium fellutanumat* Different pH. *J. Microb. Biochem. Technol.* **2016**, *8*, 5. [CrossRef]

195. Fukushima, K.; Liu, S.; Wu, H.; Engler, A.C.; Coady, D.J.; Maune, H.; Pitera, J.; Nelson, A.; Wiradharma, N.; Venkataraman, S.; et al. Supramolecular high-aspect ratio assemblies with strong antifungal activity. *Nat. Commun.* **2013**, *4*, 2861. [CrossRef] [PubMed]

196. Esteban-Tejeda, L.; Malpartida, F.; Esteban-Cubillo, A.; Pecharromán, C.; Moya, J.S. Antibacterial and antifungal activity of a soda-lime glass containing copper nanoparticles. *Nanotechnology* **2009**, *20*, 505701. [CrossRef] [PubMed]

197. Paulo, C.S.; Vidal, M.; Ferreira, L.S. Antifungal nanoparticles and surfaces. *Biomacromolecules* **2010**,*11*, 2810–2817. [CrossRef] [PubMed]

198. Perfect, J.R. Fungal virulence genes as targets for antifungal chemotherapy. *Antimicrob. Agents Chemoth.* **1996**, *40*, 1577–1583.

199. Kitamura, A.; Someya, K.; Hata, M.; Nakajima, R.; Takemura, M. Discovery of a small-molecule inhibitor of β-1,6-glucan synthesis. *Antimicrob. Agents Chemoth.* **2009**, *53*, 670–677. [CrossRef] [PubMed]

200. Baxter, B.K.; DiDone, L.; Ogu, D.; Schor, S.; Krysan, D.J. Identification, in vitro activity and mode of action of phosphoinositide-dependent-1 kinase inhibitors as antifungal molecules. *ACS Chem. Biol.* **2011**, *6*, 502–510. [CrossRef] [PubMed]

INDEX

www.ingramcontent.com/pod-product-compliance
Ingram Content Group UK Ltd.
Pitfield, Milton Keynes, MK11 3LW, UK
UKHW050946290926
15470UKWH00005B/1677